Science
in Context

Science in Context

Readings in the Sociology of Science

Edited by

Barry Barnes and David Edge

The MIT Press

Cambridge, Massachusetts

First MIT Press paperback edition 1982

Copyright © 1982 The Open University Press

Library of Congress catalog card number: 82-081157

ISBN 0-262-52076-1

Printed and bound in Great Britain

Contents

Acknowledgements

1. From *The Scientific Community* by W.O. Hagstrom, 12—23. © 1965 by W.O. Hagstrom. By permission of Basic Books, Inc., Publishers, New York.

2. Reprinted from 'Cycles of credit' by B. Latour and S. Woolgar in *Laboratory Life*, 200—08. © 1979 Sage Publications, Beverly Hills and London, by permission of the publisher.

3. Reprinted from 'The TEA set: tacit knowledge and scientific networks' by H.M. Collins in *Science Studies*, 4, 165—86. © 1974 Sage Publications, Beverly Hills and London, by permission of the publisher.

4. Reprinted from 'The function of measurement in modern physical science' by T.S. Kuhn in *ISIS*, 52, 162—76. © 1961 *ISIS*, University of Pennsylvania, 215 South 34th Street/D6, Philadelphia, 19104, by permission of the publisher.

5. Reprinted from 'The seven sexes: a study in the sociology of a phenomemon, or the replication of experiments in physics' by H.M. Collins in *Sociology*, the journal of the British Sociological Association, 9, 205—24. © 1975 *Sociology*, by permission of the publisher.

6. Reprinted from *Knowledge and Social Imagery* by D. Bloor, 117—23. © 1976 Routledge & Kegan Paul Ltd., by permission of the publisher.

7. Reprinted from 'Interests, analogies and theory choice in high-energy physics' by A. Pickering. © 1981 A. Pickering, by permission of the author. (Research supported by the UK Social Science Research Council.)

8. Reprinted from 'The science-technology relationship as a historiographic problem' by O. Mayr in *Technology and Culture*, 17, 663—72. © 1976 The University of Chicago Press, by permission of the publisher.

9. Reprinted from 'The structures of publication in science and technology' by D.J. de S. Price in *Factors in the Transfer of Technology*, 91–104, W.H. Gruber and D.G. Marquis (eds), by permission of The MIT Press, Cambridge, Massachusetts. © 1969 The MIT Press.

10. Reprinted by permission from 'Relationship between science and technology' by M. Gibbons and C. Johnson from *Nature*, 227, 125–27. © 1970 Macmillan Journals Ltd.

11. Reprinted from 'The origins and structure of agricultural chemistry' by W. Krohn and W. Schäfer in *Perspectives on the Emergence of Scientific Disciplines*, 32–52, G. Lemaine, R. MacLeod, M. Mulkay and P. Weingart (eds). © 1976 Mouton Publishers, Division of Walter de Gruyter, Berlin and New York, by permission of the publisher.

12. Reprinted from 'Physics and psychics: science, symbolic action and social control in late Victorian England' by B. Wynne in *Natural Order: Historical Studies of Scientific Culture*, 167–86, B. Barnes and S. Shapin (eds). © 1979 Sage Publications, Beverly Hills and London, by permission of the publisher.

13. Reprinted from 'Cross-examination of chemists in narcotics and marijuana cases' by J.S. Oteri, M.G. Weinberg and M.S. Pinales in *Contemporary Drug Problems*, 2, 225–38. © 1973 Federal Legal Publications Inc., 157 Chambers St., New York, N.Y. 10007, by permission of the publishers.

14. Reprinted from 'Environments at risk' by M. Douglas in *Times Literary Supplement* (30 Oct. 1970), 1273–75. Reprinted as Chapter 15 in *Implicit Meanings*, 230–48. © 1970 M. Douglas, by permission of the author and Routledge & Kegan Paul Ltd.

15. Reprinted from *Controversy* by D. Nelkin, 14–18. © 1979 Sage Publications, Beverly Hills and London, by permission of the publishers.

16. Reprinted from 'The Science court on trial in Minnesota' by B. Casper and P. Wellstone in *The Hastings Center Report*, 8, 4 (August), 5–7. © 1978 Institute of Society, Ethics and the Life Sciences, 360 Broadway, Hastings-on-Hudson, N.Y. 10706. Reprinted with permission of The Hastings Center.

17. Reprinted from 'The fear of innovation' by D. Schon in *Uncertainty in Research, Management and New Product*

Development, R.M. Hainer, S. Kingsbury and D.B. Gleicher (eds) (Chapter 2), 70—78. (Originally published in *International Science and Technology*, Nov. 1966.) © 1966 Conover-Mast, by permission of the publisher.

18. Reprinted from 'Carcinogenic risk assessment in the United States and Great Britain: the case of Aldrin/Dieldrin' by B. Gillespie, D. Eva and R. Johnston in *Social Studies of Science*, 9, 265—301. © 1979 Sage Publications, Beverly Hills and London, by permission of the publisher.

Editors' note

We would like to extend our thanks to the staff at the Science
Studies Unit in Edinburgh, and those at the Open University Press,
whose efforts have made possible the rapid publication of this
collection of readings. We are grateful, too, to the many friends
and colleagues who generously proffered their help and advice
during its compilation.

The Readings have been assembled with as little alteration as
possible: each retains its original style of referencing and annota-
tion and is followed immediately by a list of cited sources, where
such a list was present in the original. The Bibliography at the end
of the book includes only those works cited in the various introductory sections, and the sources listed in the Bibliographical
Notes (which we refer to in the text by the abbreviation BN).

Finally, in all editorial material the convention is adopted
whereby 'he' stands for 'he or she', 'men' for 'men and women',
as is common practice.

Barry Barnes

David Edge

General introduction

There will always be controversy about the proper tasks of the social sciences, and hence about what an education in sociology should comprise. Nonetheless, few would deny that such an education must include a tour of the great social institutions: it must offer at least rudimentary accounts of our political, economic and military organizations, our forms of religion, our patterns of kinship, socialization and education, and, of course, our knowledge and science. Hence, the paucity of good teaching materials in the sociology of science has given rise to widespread concern, a concern broadened and intensified by recognition of the crucial social and cultural significance of modern science, and by spreading awareness of the value of the work produced in this field in the last decade.

In assembling the present volume we have done what we can to alleviate this unsatisfactory state of affairs. Needless to say, it is not enough: there is no way in which all the excellent work currently available can be adequately represented within a single volume. Nonetheless, our selection of readings, together with the Bibliography, should provide a tolerable indication of what is going on in the sociology of science, and, more importantly, of what kind of a social activity science is, and what its significance is. Moreover, we have laid particular stress, in the Bibliographical Notes (BN), upon sources which themselves cite or review important bodies of work, and which will accordingly guide the student further into those topics which excite particular interest.

All this, however, is far from representing the totality of our objectives. There are many ways of studying an institution, and innumerable reasons why it may invite study. In the case of science there is work on socialization, interaction and exchange, organization, hierarchy, cognition and cognitive change, interaction with other institutions, and so on through all the foci of interest characteristic of the social sciences. Hence, any attempt simply to survey the literature, or even to select the best examples

1

of all the diverse approaches within it, would have been bound to result in a superficial and aimless pot-pourri. We felt compelled to seek after more coherence and integration, particularly in our ordering and interpretation of material. How our selections are actually used remains a matter for the reader, but we ourselves concentrate upon their relationship to just a few of those fundamental themes which give the social sciences whatever integrity and cogency they possess.

The sociology of science has closely involved itself with many of these themes, so that there are a number of alternative ways in which its literature can be effectively ordered and interpreted. The initial development of the field was intimately bound up with the rise of sociological functionalism, to the extent that the analysis of science as an institution served as a crucial exemplification and vindication of the functionalist approach (cf. Merton, 1957, 1973). Even today in the United States, where functionalism has retained a considerable degree of credibility, it profoundly conditions what the sociology of science is taken to consist in, and how it is taught (Gaston, 1978; Gieryn, 1979). Similarly, in the last decade in Britain, an overriding concern with fundamental problems in the sociology of knowledge and culture has concentrated studies of science around just a few generally significant issues and hypotheses (cf. Barnes, 1974, 1977, 1982; Bloor, 1976; Mulkay, 1979; Collins, 1981a).

In this book, however, we adopt yet another, less restrictive, integrating strategy. We give prominence to the relationship between the sub-culture of science and the wider culture which surrounds it. Looked at from this perspective science is primarily a source of knowledge and competence: it is a repository of theories, findings, procedures and techniques which it makes generally available both directly, via expert intervention and consultation, and indirectly, via its interaction with technology and with specialized institutions in the economic and political structure. In addition, science operates as a source of cognitive authority: not only does it provide knowledge and competence, it is required also to evaluate the knowledge-claims and putative competences of those situated beyond its boundaries. Indeed, in modern societies, science is near to being *the* source of cognitive authority: anyone who would be widely believed and trusted as an interpreter of nature needs a licence from the scientific community.

Thus, to stress its interaction with its context not only highlights these aspects of science which most people find of overriding pragmatic interest, but also raises important basic issues concerning credibility, the distribution of authority in society, and

2

the nature of the interaction between different forms of culture. Moreover, this strategy is one of very few which does justice to the peculiar importance of the study of science. If one analyzes scientific roles, or communications, or hierarchies, or small-group interactions, one duly learns about these things—but only in the way that one learns about them when studying bureaucracies, or families, or gangs, or other professions. But in the study of credibility and cognitive authority science is the paradigm case. Scientists are our recognized men of knowledge: the touchstone of any hypothesis concerning the transmission of expertise and the basis of credibility must be how successfully it can be applied to them.

Given our stated objectives, we might be expected quickly to pass over material on the internal organization and esoteric culture of science. In fact, the first two sections of readings consist precisely of such material. But it is not included entirely for its own sake: it is also necessary in order to understand the interaction between science and the wider culture. The end is an understanding of science as a source of competence and of authority, but part of the means to that end must be an examination of the internal characteristics of science.

It is particularly important to obtain an understanding of some of the general characteristics of scientific research, and of the knowledge which it produces. But here it seems that we are immediately confronted by an insuperable difficulty. There are innumerable conflicting descriptions of scientific research and scientific knowledge. All the traditional conflicts of epistemology, between realism and instrumentalism, rationalism and empiricism, deductivism and inductivism, and many more, find expression in endless modifications and combinations, as competing accounts of science. And each such account implies a different foundation for the credibility and authority of science.

This difficulty, however, is illusory. Nearly all of these accounts of science are very heavily idealized, and represent the various utopias of our philosophers and epistemologists rather than what actually goes on in those places which we customarily call science laboratories. In contrast, the present need is for a general description which treats the beliefs and practices of scientists in a completely down-to-earth, matter-of-fact way, simply as a set of visible phenomena. Such descriptions, which spring from an empirical or historical orientation to science rather than from a philosophical orientation, are notably thin on the ground. The real difficulty we face is the daunting extent of our ignorance of the basic features of scientific activity and scientific inference, but

3

in truth this is no greater than that routinely faced by sociologists and anthropologists when they study other forms of culture.

Recent work in the sociology of science has relied very heavily upon one particular account of scientific culture, that initially offered by Thomas Kuhn in *The Structure of Scientific Revolutions* (1962) and developed and elaborated by him in a number of subsequent papers (Kuhn, 1977). Kuhn is a professional historian of science and his work is correspondingly detailed, complex and richly exemplified. It grows, however, from what is essentially a very simple picture — conveyed by his well-known description of *normal science*. This characterizes routine scientific investigation as puzzle-solving activity set in the context of a received tradition of research. Typically, the individual scientist is deeply committed to the received tradition, and thus to the specific modes of understanding and the specific research practices which it embodies. He uses the resources of the tradition, and in particular the accepted concrete problem-solutions or *paradigms* constructed by his scientific predecessors, as the means of solving specific narrowly-conceived problems or puzzles defined and imbued with significance by that same tradition. And where he is successful his solutions become themselves incorporated into the tradition and serve as taken-for-granted resources for use by later generations of scientists.

The themes, concepts and presuppositions of Kuhn's work have permeated the sociology of science at many levels, as will become clear later. Their importance in the field is an order of magnitude greater than that of any other external source. Since Kuhn's work is highly controversial and problematic, the wisdom of allowing it this seminal role can be questioned: critics have wondered whether its standing in the social sciences is not more a matter of its being sociological than of its being correct.

Nonetheless, there is much to be said in defence of the way that sociologists have responded to Kuhn's work. On the one hand, they have been appropriately circumspect in their treatment of its specific details, and have debated many of its controversial aspects. On the other hand, they have given it a seminal role since it embodies, as few alternatives do, many of what they take to be the fundamental characteristics of knowledge and culture. It is worth singling out two of these for explicit mention here, because of their pervasive significance in what follows.

First, nearly all sociologists would now agree that an account of knowledge must do full justice to its inherently theoretical and constructed character. No body of knowledge, nor any part of one, can capture, or at least can be known to capture, *the* basic pattern or structure inherent in some aspect of the natural world. Nature can be patterned in different ways: it will tolerate many different orderings without protest, as is shown by the great variety of orderings which have been and are successfully applied to it by experts, ancestors, aliens and deviants. No particular ordering is intrinsically preferable to all others, and accordingly none is self-sustaining. Specific orderings are constructed not revealed, invented rather than discovered, in sequences of activity which however attentive to experience and to formal consistency could nonetheless have been otherwise and could have had different results. Hence, in cleaving to its own specific body of knowledge a community commits itself to what is, in an entirely non-pejorative sense, a system of conventions. Even the lowest-level formulations of the verbal culture of natural science, its statements of matters of fact and its recorded findings, have a conventional character.

These themes do little more than rehearse current orthodoxy in the sociology, philosophy and history of science, and extensive argument in support of them can be found in all three fields (Hesse, 1980, provides a useful route to much relevant material). But although many accounts of science pay lip-service to them accordingly, scarcely any make wholehearted attempts to explore their consequences. In so far as knowledge has a conventional character there is a need to ascertain how the conventions are acquired, applied and sustained. Nobody has perceived this need more clearly than Kuhn. He describes how scientists acquire knowledge from their predecessors in the course of a dogmatic, textbook-based training, wherein the accepted authority of the teacher plays a major role. He describes precisely how they apply their received knowledge in the course of research, and the extent to which they take that knowledge for granted as they apply it. And he indicates how attacks upon orthodox doctrine, or idiosyncracies in its development, are systematically discouraged: he indicates, that is, the existence of mechanisms of social control in science. All this adds up to a comprehensive discussion of scientific fields as sub-cultures which has inestimable value for social scientists and is without obvious parallel elsewhere (see also Barnes, 1982).

Consider now the import of this for our primary concern, that of the relationship between science and the wider culture. If scientific knowledge, having a conventional character, is not self-sustaining within the scientific sub-culture, then clearly it cannot be self-sustaining within the wider society either. Scientific knowledge does not carry a revelation of its own correctness along with itself; rather its standing is inevitably bound up with such contingent factors as the degree of trust and authority possessed by its bearer, or by the institutions which sustain him and assert his competence and legitimacy. Like any other purveyor of knowledge-claims the scientist must face all the daunting ramifications of the problem of credibility, and how far and in what way he succeeds in overcoming them is invariably a complex empirical issue. In so far as he bears the verbally formulated doctrines of his field, the scientist must be considered formally as a transmitter of *lore*.

It is, however, a second general characteristic of knowledge that it always comprises more than what can be verbally formulated. When we study a culture empirically we are confronted with agents collectively engaged in practical activities and social interactions: that, at least, is how we describe it in the currently accepted observation language of the social sciences. To make sense of the activities and the interactions we speak of 'agents' knowledge'. This in itself raises difficulties and is perhaps less than justified, even if it is no more than agents themselves do in making sense of their own activities. But if as 'agents' knowledge' we allow only verbal statements — axioms, hypotheses and the like — then beyond any shadow of a doubt we handicap our understanding: we count as knowledge very much less than the culture knows.

Let us avoid the complex and esoteric issues which surround this point and simply note some of the more clearcut disadvantages of equating knowledge with a hypostatized set of verbal formulations. First, it takes no account of the linguistic and interpretative competences of the agents being studied and the extent to which the significance of verbal formulations may be a function of such competences; this gives rise to particularly acute difficulties when a single formulation is decoded or interpreted differently at different times, or in different contexts. Secondly, it largely ignores the practical competences of agents — the 'knowing-how' as opposed to the 'knowing-that' component of their knowledge. In science as elsewhere this is a major, indeed a predominant component, and it cannot be verbalized: there is always a difference between finding verbal instructions concerning a computation or a measurement intelligible, and actually knowing how to do the computation or the measurement. Finally, verbal formulations fail adequately to represent even the 'knowing-that' component of knowledge. Our physical environment is extraordinarily complex and rich in information. Even on the most favourable assumptions, only a few highly selected features of any situation find expression in terms of the few thousand words and signs which we possess: much more is apprehended and stored in the memory unverbalized.

Accordingly, men always know more than they can say; and often what they share in the way of unverbalized knowledge, the result of a shared environment and shared experiences, helps to account for the parallels in their actions and the effectiveness of their interactions. (See also Polanyi, 1958, and the references to 'tacit knowledge' in BN.)

Kuhn avoids many of the pitfalls of an over-verbal conception of science by characterizing the fundamental component of scientific culture as the *paradigm*. A paradigm is an accepted scientific achievement, a concrete example of actual scientific practice which includes laws, theories, applications and instrumentation within itself (cf. Kuhn, 1962: 10). For Kuhn, the history of a scientific field is an account of how its practitioners develop and elaborate their paradigms. To describe this is to describe actual scientific activity in a way which is sociologically interesting and informative.

It is also important to our larger purposes that we recognize the very limited extent to which a body of knowledge can be encapsulated in verbal formulations. This encourages us to look to people and their activities as the repository of knowledge. And it suggests that the transmission of culture is most completely and effectively brought about by the movement of people from place to place, rather than by the transmission of written or oral messages or by some abstractly conceived diffusion of ideas.

When we turn our attention directly to the processes whereby scientific knowledge and competence diffuse into the wider context, it is convenient to separate off those which involve technological innovation and economic exploitation, and which result in the first instance in the production of new artefacts and techniques. The greater part of this crucially important line of connection between science and its context will be omitted from consideration here, simply because it is already so satisfactorily covered by materials dealing with the economic and social effects of technological change, and indeed constitutes in itself the subject of a special branch of sociology. But the first segment of the line, the link between science and technology, is the subject of the third group of readings.

The main emphasis here is upon the science/technology relationship as an interaction between what are at least partially distinct forms of culture. The theoretical issues directly involved are not of the most fundamental kind, and much of the interest of the readings derives from the pragmatic importance of a proper understanding of the interaction, and from the persistence of seriously defective stereotypes of the interaction in many contexts.

Both everyday, practically-oriented social and political thinking and the speculations which represent the grandest levels of academic sociological theory rely heavily upon schematic models and stereotypes of the relationship between science and technology. But models which satisfactorily take into account the indications of current research remain less widely diffused than might be wished.

Both sections of the second half of the book are concerned with science as a direct source of knowledge and competence. The fourth section makes the important preliminary point that science cannot simply be assumed to be an independent, external agency, pumping expertise into the social order. Considered as an empirical entity, which is how sociologists must consider it, science is a sub-culture, or set of sub-cultures. In so far as it is bounded-off from the rest of society and isolated from its influence, it is through the activities and judgments of its practitioners. And the effect of the relevant activities and judgments is entirely an empirical matter, to be ascertained in every specific case by direct investigation and without preconceptions. Every point of contact between science and the wider society must be regarded as a point of potential interaction, with possible consequences for the esoteric culture of science as well as for the exoteric culture. One-way traffic from science to the wider society must be treated merely as a possible empirical finding in any particular case. And as the readings and references in this section clearly reveal, such a possibility is certainly not realized in *every* particular case.

The significance of these readings and references is enhanced, just as with those of the preceding section, by the extent to which they clash with aspects of existing models and stereotypes. It is often thought that there is a unique and specific kind of barrier which protects science, and particularly scientific judgment, against the intrusion of so-called 'external influences', and that the barrier fails only in highly exceptional and unusual circumstances which can properly be considered to be pathological. Opinions differ as to what precisely the barrier consists in: some speak of a rational commitment to scientific method, others of a strongly sanctioned normative order definitive of science, others again are as vague about its nature as they are insistent that it is there. For, needless to say, a prior faith in the superior cognitive authority of science, which is most readily justified if the barrier is indeed there, stimulates a degree of conviction in its presence and efficacy far surpassing anything which would be produced by habitual forms of inference from empirical evidence. Certainly, the evidence presently available suggests that 'external' influences upon scientific judgment are neither unusual nor necessarily pathological, and that the barrier which such influences have to penetrate is not fundamentally different from the boundaries surrounding other sub-cultures.

In the last section of the book the matter of the authority and credibility of scientific expertise is directly engaged. Some of the many contingent factors bearing upon credibility are identified, and their relevance is illustrated with specific cases. This conveys far more vividly than abstract argument the countless problems of justification which any putative expert faces if his pronouncements are actually called into question. And it helps to convey also the immense difficulty of identifying the actual basis of credibility in situations where such pronouncements are accepted.

In addition, we have attempted to go beyond actualities and to discuss what is possible, in ideal circumstances, by way of the justification of expertise. This is a crucial question. All current conceptions of rational decision and rational action ultimately rely upon the assumption that there is a single most rationally plausible appraisal of what appertains in any specific situation or set of circumstances. To deny the possibility of such appraisals is to deny that there is a best way of calculating which means are most appropriate to the attainment of specific ends or objectives. And without general acceptance that such calculations can be made, the legitimation of both specific political policies and ideal conceptions of social order becomes well-nigh impossible. Cognitive authority and political authority are inextricably intertwined: the recognition of technical expertise, of whatever kind, is fraught with political significance.

Needless to say, this has much to do with why strict hierarchies of cognitive authority are maintained in all societies, and why heavily idealized conceptions of science are insistently propagated in ours. For us, a rationalist stereotype of scientific expertise, which gives the doctrines, judgments and evaluations of scientists a privileged epistemological status, is highly expedient.

Consider, too, the consequences of turning instead to the only other developed stereotype of expertise we currently possess, that of the lawyer or advocate. The skills of the advocate are rhetorical: he is a maker-out of cases. His thought moves backward from a desired consequence to appropriate evidence and argument, which is then selected and presented accordingly. His duty is to rationalize. His devotion is given to whatever sequence of vested interests circumstances throw his way. His value, if such it is, only exists in a system wherein he is opposed by another such as himself, and where his discourse is assessed in its turn by another authority the basis of which is obscure and for the most part unexamined.

In today's increasingly differentiated societies, where political activity is very much the province of factions, vested interests and pressure groups, this alternative stereotype of expertise is indeed gaining ground. Every kind of competence is available to every kind of cause, to make its case and undermine that of its opponents. There is government expertise, industrialists' expertise and grassroots expertise. And these experts are recognized as advocates, and occasionally recognize themselves as such. Academic scientists are themselves sometimes perceived as advocates rather than as disinterested consultants, even as they overtly associate themselves with this or that institution or organization; and a few of them recognize that there is some justice in this, even if it is only by making a distinction between the 'disinterested' character of their academic work and the more 'involved' character of their other activities.

But if there is wide recognition that in our pluralist society the expert is tending to take the role of advocate, with the basis for decisions being contested between opposed bodies of expertise supporting opposed interests, there is a deafening silence concerning the matter of deciding *between* the conflicting claims of the experts involved. No doubt, like lawyers, our other advocate-experts hope and desire that such decisions are made 'rationally'. But it is no random eccentricity in our lawyers that they retain a profound and intense indifference to all that occurs in the jury-room.

To conceive of expert knowledge solely in terms of advocacy is to leave untouched central problems concerning its credibility. Nor can these problems be completely resolved by the (perfectly plausible) suggestion that one rationalization is preferred to another to the extent that it favours the plans of dominant powers and interests. Such a preference can only be the product of judgments which are themselves informed by accredited knowledge. The problem of what knowledge-claims people find credible is at least as deep as that of what actions and policies they find expedient. However tempting such a step may be in particular instances or case studies, the general problem of credibility cannot be completely reduced to a matter of what beliefs are immediately expedient, or immediately relevant to vested interests. Indeed the perceived degree of disinterest of a knowledge source, the extent to which it is decoupled from the vested interests surrounding the issue upon which it pronounces, remains of major importance as a determinant of the credibility of its pronouncements. Were academic natural scientists not perceived as generally disinterested in this sense, the current high general level of credibility they enjoy would be greatly diminished. Simply to be perceived in terms of a stereotype of advocacy is weakening to credibility.

As far as a sociological understanding is concerned, expertise cannot be satisfactorily conceptualized simply in terms of the stereotype of advocacy. No doubt this is something for which to be thankful. The dismal vision produced when this stereotype and its consequences are thoroughly explored is so profoundly disturbing that the intensity and passion with which the rationalist alternative is defended needs no further explanation.

Unfortunately, however, our chosen readings and the gist of our subsequent remarks offer no support to the rationalist stereotype either: rationalists will take our conclusions, with every reason, as profoundly pessimistic ones. Expertise may be more than a matter of simple advocacy; but it is nonetheless always the outcome of conventional procedures and conventional modes of understanding. Even as science itself has a conventional character, so too must scientific expertise have such a character. There is no way in which such expertise can be guaranteed by reference to reason *rather than* habitual inference, nature *rather than* culture. Hence there is no way in which the reception of such expertise can be understood without appeal to authority, interest and analogous contingent factors. And the associated normative questions of how expert knowledge is best assessed, and how experts themselves are best evaluated and kept under a modicum of control, raise such intractable and viciously circular problems as to strangle speech.

This raises a further point, which forces us to end on a yet more pessimistic note. If the credibility of scientific experts is in some respects problematically based, nonetheless, on the whole it stands high. Those who study that expertise themselves possess very much less credibility. And rightly so. Sociological studies of expertise frequently raise more problems than they attempt to solve, and sometimes exhibit such transparent partiality in their arguments and conclusions as to forfeit all respect (cf. Pfetsch, 1979). So long as a significant proportion of social scientists unashamedly tailor their arguments to desired consequences, and even argue that knowledge generally should develop as an auxiliary to some specific political orientation, so long ought the credibility of the social sciences to remain low (even if we ourselves are thereby involved in some little local difficulty). Nonetheless, even if the social sciences were to pull themselves together and eventually acquire the kind of general credibility which other experts enjoy, that credibility would rest upon the same contingencies as have already been described. Their studies would still offer no obvious solutions to the normative problems involved in the evaluation of expertise. Far from social scientists then being in a position to offer a privileged evaluative perspective upon the claims of this or that body of experts, they would have joined the ranks.

PART ONE

The organization of academic science: communication and control

If, in the modern world, science is primarily a source of knowledge and competence, on which other institutions increasingly rely, how is that knowledge refined and validated, and that competence legitimated?

It has always been a basic tenet of the sociology of science that there is an intimate connection between the growth of knowledge and the social organization of the scientific community. Robert Merton stressed the point in his pioneering work, and the important school of functionalist sociologists which followed him tended to take it for granted. This led them to explore the interactions between scientists themselves for evidence of the social processes which maintain order, confer rewards and recognition, and allow for the exercise of control over the content and quality of certified scientific knowledge. Further, the Mertonian tradition emphasized the role of *academic* science as the fullest embodiment of scientific ideals and practices, and the ultimate source of cognitive authority.

That emphasis is still widely accepted. Those scientists engaged in 'pure' or 'basic' research are routinely distinguished from researchers whose work is carried out in an 'applied' context. 'Pure' scientists are taken to be a community of peers, whose members constitute their own audience. They determine their own research goals; they themselves recognize competence and reward achievement; they deny the legitimacy of outside involvement in their judgments, and systematically discourage attempts at such 'interference'. Historically, cadres of such 'natural philosophers' have tended to be located in universities, and to share traditional academic values. They support learned societies and journals, which maintain academic interchange and scholarly ('disciplinary') standards: here the credibility of any knowledge claim is formally certified.

In contrast, scientists who work in an 'applied' context are generally held to experience different social constraints. Their

13

goals will, at least in broad terms, be set by their employers, who will judge the success of their research, and reward it, by their own practical and economic criteria. In such circumstances, the control exercised by the 'academic' community, and the rewards it offers, are of secondary significance. Credibility and recognition both tend to be a function of immediate, concrete results.

Since the Second World War, public support for science has increased dramatically. 'Pure' scientists are now strikingly out-numbered by their 'applied' brethren. And even their basic research has itself often so expanded in scale, scope and expense as to earn the title of 'Big Science' (Price, 1963; Greenberg, 1967; Sklair, 1973: Part One). There are at least three aspects of this new context which attract interest (and frequently concern). The first is related to the *size* of the modern scientific community. As special-ists and esoteric skills proliferate, *differentiation* increases. Large teams are often needed to carry out simple experiments (Weinberg, 1970; Swatez, 1970). Journals, and other less formal media, multiply, raising problems of collegial scrutiny and mutual compre-hension — let alone the practical problem of 'keeping up with the literature'. The second aspect concerns the *capital cost* of research techniques. Senior researchers must commit themselves to long-term research programmes and spend an increasing proportion of their time operating in effect as business administrators and entre-preneurs: while junior researchers become mere employees, 'proletarianized', ripe for unionization. And the third emphasises the extent to which the whole research community is permeated by *applied* concerns. An increasing proportion of research funds is earmarked for specific 'missions' by government, the military and industry (Sklair, 1973: Part One; Pavitt and Worboys, 1978; Norman, 1979). Even the 'purest' scientists can find themselves accountable to a constituency wider than their peers. The tra-ditional base of academic science is thus less clearcut than it was, and its autonomy has been eroded (see Krohn, 1971; Blume, 1974).

Considerations such as these have led to the notion that we are witnessing 'the industrialization of science':

> This means, in the first place, the dominance of capital-intensive research, and its social consequences in the concentration of power in a small section of the community. It also involves the interpenetra-tion of science and industry, with the loss of boundaries which enabled different styles of work, with their appropriate codes of behaviour and ideals, to coexist. Further, it implies a large size, both in particular units and in the aggregate, with the consequent loss of networks of informal, personal contacts binding a community. Finally, it brings into science

14

the instability and sense of rapid but uncontrolled change, characteristic of the world of industry and trade in our civilization. (Ravetz, 1971: 31)

Ravetz is not alone in regretting these developments. Part of this regret is moral: important values, ideals and standards seem to be compromized and degraded. But part is practical: the processes of 'quality control', which monitor the knowledge and competence on which political decisions increasingly depend, appear to be disrupted or threatened.

The Mertonian model of the academic science community is obviously an 'ideal type'. Even as such, many have questioned its relevance. Nonetheless, it is an ideal type of real value which, by focussing attention on processes of communication, recognition and reward among relatively stable associations of active researchers, has exposed to sociological inspection the assessment of knowledge-claims. It has generated a sequence of detailed studies without which our knowledge of the social institution of science, and of its mechanisms of differentiation and growth, would be much the poorer (see Bibliographical Notes [BN]). (A similar focus has guided much research outside the strictly Mertonian tradition, especially case studies of specialty formation [BN].)

The major achievement of the Mertonian tradition has been to explore how the academic science community operates an institutionalized system which is *simultaneously* a communication system *and* a reward system *and* what might be called a system for 'distributing property'. This system rests on peer judgment. The contribution of a scientist is assessed by an 'audience' of colleagues, who are potentially in a position to make use of it. If they judge it to be original and significant, allow it to be published, and capitalize on it in their own work (citing the contribution meanwhile in the references to their papers), then the scientist receives the award of *recognition*. At its most mundane, this can bring promotion and better conditions of work; it also enhances reputation. At its highest level, recognition is formalized in such honorific awards as Nobel Prizes, Fellowships of the Royal Society, and the like: there is some evidence that scientists who achieve such awards tend to form an élite (Groeneveld, Koller and Mullins, 1975; Mulkay, 1976b); not only do greater resources flow to them to further their work and consolidate their position, but they also gain access to those government and foundation committees which decide the fate of research grant proposals, map the future of whole disciplines, and offer specialist advice (Greenberg, 1967). The reward, communication and allocation systems in science are therefore intimately bound together: and their operation tends to produce *stratification* (Cole and Cole, 1973).

Warren Hagstrom has analyzed this process as a form of 'gift-giving' — a primitive system of barter, in which information is 'traded' for the recognition which allegedly motivates individual scientists. In selecting his statement of this position, we appreciate its somewhat atypical place within the broad Mertonian tradition. On the one hand, in giving an account of how an exchange system of this kind can generate a commitment to values, Hagstrom is underpinning a central Mertonian concern (see below): but, on the other hand, his model tends towards a psychological theory of motivation. The mainstream of functionalism, however, recognizes that what is involved here is primarily *institutional,* not psychological. The primary object is not to hypothesize what the 'real' motives of scientists are; rather, it is to characterize an ideal institutional form which *channels* the various, even idiosyncratic, motives of individual scientists, and whose operation is compatible with a wide range of such motives. We could say that recognition is the 'currency' which channels a diversity of individual motives in science, just as the £ sterling (or US $) is the currency which channels individual motives in a car factory. Both currencies are the *routes* to a range of specific rewards which satisfy the specific wants of individuals. As Merton wrote in 1957, in discussing the demonstrable eagerness with which scientists claim priority for important discoveries:

> It is not only the institution of science, of course, that instils and reinforces the concern with recognition: in some degree, all institutions do . . . The point is, rather, that with its emphasis on originality, the institution of science greatly reinforces this concern and indirectly leads scientists to vigorous self-assertion of their priority. (Merton, 1973: 294)

It is clear that Merton has always appreciated this crucial point, even if those whom he has influenced have sometimes lapsed from grace.

Latour and Woolgar (1979: Chapter 5) offer another analysis at this institutional level. They criticize the models offered by both Hagstrom and Bourdieu (1975) and, in their own search for a characterization of this institutionalized system, offer a variant of the pervasive 'market' metaphor. They see scientists as continually engaged in converting 'capital' from one form to another, as 'moves' become necessary in the scientific field. 'Credibility' is the 'credit' which allows scientists to convert capital: the basic aim of a scientist's strategy is therefore to maximize *credibility.*

Recognition of a recent achievement increases credibility, which can then be 'cashed' in the form of a research grant, converted into equipment (or 'inscription devices') and recruited colleagues, then into data, argumentation, fresh publications, more credibility . . . and so on. Credibility is not itself the reward: it is the credit *which makes possible a variety of moves* (and hence of rewards, of motives and needs satisfied). We reproduce a key section of their chapter here.

These theories all employ an overtly economic metaphor (Law, 1980): 'recognition' or 'credibility' is seen as a kind of 'money' that can be 'earned', 'credit' that can be 'accumulated', 'capital' that can be 'invested'. And a common feature of all these various 'economic' models is that they take scientific interactions to be mediated by a *special* form of 'currency'. Scientists may only directly acquire the special currency of the scientific community — 'recognition' or 'credibility' — and must not orient their activity specifically toward the accumulation of financial reward. This restriction is made intelligible as an institutional arrangement which protects the autonomy of the scientific community of peers, and helps to insulate it from external direction: users of a single currency constitute a *closed* social system. Knorr-Cetina (1982) has recently challenged the whole range of such models on precisely this point: citing recent ethnographic studies in scientific laboratories, she emphasizes the potential importance of the interaction of scientists *with non-scientists* in the social control and development of scientific specialties. And, indeed, more attention is now being paid to scientists in applied contexts (BN, and Part 4: see also Mulkay, 1976; Johnston and Robbins, 1977; Friedkin, 1978); for now 'it is important that sociological analysis of the production and use of scientific knowledge takes full account of the "incorporation" and "industrialization" of contemporary science, and avoids the implicit assumption of a closed social system' (Robbins and Johnston, 1976: 352).

The achievement of Merton's school in exploring the reward/ communication system in science is widely recognized. Much more controversial is the Mertonian account of the institutionalized norms of science (see BN). Merton took the behaviour of scientists to be guided and ordered by a set of institutionalized 'norms', conformity with which maximizes the chances of the community efficiently attaining its goal of the production of 'certified knowledge'. Merton's four original norms were 'universalism', 'communality', 'disinterestedness' and 'organized scepticism': others (such as 'originality', 'humility', 'rationality' and 'individualism') have been added since. But such 'moral' norms will not be effective unless conformity to them can be rewarded, and deviance punished: and, as Mulkay points out:

> [T]here is no clear evidence that the award of recognition is institutionally linked to conformity to the values of universalism, disinterestedness, etc. Rather, recognition is furnished in response to the provision of information which is judged to be valuable in the light of currently accepted cognitive and technical standards. (Mulkay, 1977: 132)

Mulkay suggests that Merton's norms may be treated as part of a *repertoire* of moral rhetoric upon which scientists can draw in order to characterize their own behaviour as 'proper' (and others' as 'improper') — a set of resources which they can actively manipulate in situation where justification of their actions and practices is required. (For a clear statement of this view, see Mulkay, 1976a; cf. also Mitroff, 1974a.)

One result of the operation of this communication/reward system which is specially important for our later purposes is to erect *boundaries* — between 'science' and 'non-science'; between scientists in different disciplines and specialties; and between accredited professionals and lay 'outsiders'. Any scientist who can earn recognition or credibility within the system can be seen to be 'playing the game', and has established a claim to be treated as an *expert* in the particular field in question. This status is consolidated, and the boundaries reinforced, by command of esoteric *language* and skills: but the status is by no means unproblematically defined or attained. The matter is of practical consequence, since many disputes, both within and outside science, raise such questions as whether or not claims are 'scientific' (see Collins and Pinch, 1979); or who is to 'count' as a 'relevant expert'; or whether or not an expert witness has 'exceeded the limits of his area of competence'. Later we shall suggest that such judgments cannot ever be decisively and unproblematically justified, by objective or rational means: they are endemically *contingent*.

As well as the boundary between science and non-science, those between the different 'fields' or 'areas' of science are of great sociological interest. The scientific community is not a monolithic, homogeneous institution: it is subdivided into a complex of smaller units (Hagstrom, 1965: Chapter IV). Scientific disciplines are broad federations of research areas, which have in common only a loose allegiance to widely-defined subject matters, characteristic concepts or techniques: members tend to share few, if any, specific research interests. Since learned societies and university departments tend to be defined on disciplinary criteria, disciplines have an overall political significance, and disciplinary élites can exert some influence on scientific strategy. But collegial contact is, usually attained through smaller groupings. Specialties consist of researchers sharing more narrowly-defined concerns, but, even here, differentiation can obviate the need for professional interchange: many specialties are subdivided into topic areas, by subject and technique, that have little in common. Sociologists have therefore increasingly tended to locate the central processes of communication and control within science in *research networks* — relatively small groups of scientists who, because of their active engagement with similar research tasks, are in dialogue. As Mulkay puts it:

> These networks are of fundamental importance in science because research findings can become certified knowledge only after they have been communicated to, and recognised as valid by, those most competent to judge them: that is, in the first instance, the members of such a network. (Mulkay, 1977: 132)

Such associations go by several names: 'invisible colleges' (Price, 1963; Crane, 1972); 'problem areas' and 'social circles' (Crane, 1969); 'networks' and 'clusters' (Mullins, 1972); 'coherent social groups' (Griffith and Mullins, 1972); when they are at the heart of scientific controversies, and hence exposed fully to the uncertainties of scientific negotiation over knowledge-claims, Collins (1981b) has called them 'core-sets'; Edge and Mulkay (1976: 127, 436) talk of 'transient networks'. Several studies have been made of research groups at single sites (for a summary, see Geison, 1981); and citation and co-citation analyses have been used to 'map' sets of interacting scientists (BN). All this research has attempted to locate and study those social networks within which credibility is earned, expertise secured, and the directions of research strategy determined.

However, these social arrangements are fluid: networks come and go as research develops, and topics and techniques emerge and disappear; they overlap, with many scientists engaged, at any one time, in more than one; and their membership changes over time (Mullins, 1968, 1972; Crane, 1969, 1972; Mulkay, Gilbert and Woolgar, 1975). Worse: participants themselves differ in their perceptions of the boundaries and membership of such groupings; and it is difficult, or perhaps impossible, for the outside observer to find any objective criteria with which to define them unambiguously (Woolgar, 1976b). If it is impossible, then there is an *essential* imprecision in the definition of expertise. But even if it is not, that definition still has to face the uncertainties of collegial negotiations on which credibility rests. We will later discuss the nature of such negotiations: the full difficulty of defining expertise will then become clearer. (Of course, all this process is invisible to

19

the lay 'outsider': a 'scientist' may be held, by the public, to possess a *very general* expertise and authority; and the credibility of any expert view may, in any case, be more a function of its acceptability than its accreditation [see Part 5].)

From the available network studies we have chosen one by Harry Collins because it illustrates with unique clarity two particularly important points about scientific communication. The first is that the effect of competition, which is a necessary concomitant of a reward system of the kind we have discussed, may be to limit communication (Hagstrom, 1965: Chapter II).

> One source of difficulty is that the relationships among research scientists involve a delicate balance between cooperation in the pursuit of shared problems and competition for the recognition associated with the solution of these problems. (Mulkay, 1977: 133)

Where 'communality' seems to demand complete openness, competitive pressures lead scientists to adopt 'a reasonable degree of secrecy'.

Secondly, and more importantly, Collins' study emphasizes the extent to which scientific knowledge is *tacit* knowledge, embodied in human agents (Burns, 1969), and not amenable to the 'paper transfer' which underlies the formal academic reward and communication system. Any analysis of the role of science as a source of knowledge and competence in modern society must do justice to this point. One corollary is the importance of the *mobility* of trained personnel in scientific innovation. The phenomenon is now well established: 'mobile' physicists, for instance, were involved in the emergence of specialties in both biology (Mullins, 1972; Law, 1973) and astronomy (Edge and Mulkay, 1976; Edge, 1977); expertise in new techniques was crucial in these cases. Ben-David (1960) has analyzed innovations in medicine, and Ben-David and Collins (1966) the emergence of psychology as an academic discipline, in terms of cross fertilization by mobility, and 'role-hybridization'; Mulkay and Turner (1971) have argued that such innovative mobility often arises from the overproduction of trained personnel; and Mulkay (1974) has illustrated the ways in which such mobility can produce innovation by metaphorical transfer. These considerations arise again in the interaction of science with technology and technical innovation, and we will return to them in Part 3. Meanwhile, Burns expresses their significance:

> If it is agents rather than agencies which are the vectors, then we are dealing with a world of scientific and technological activity, which presents itself in another light than that of alienable, transferable, marketable ideas. Reality now is where the action is. (Burns, 1969: 20)

1

Warren O. Hagstrom

Gift giving as an organizing principle in science

Manuscripts submitted to scientific periodicals are often called 'contributions.' and they are, in fact, gifts. Authors do not usually receive royalties or other payments, and their institutions may even be required to aid in the financial support of the periodical.[1] On the other hand, manuscripts for which the scientific authors do receive financial payments, such as textbooks and populariz- ations, are, if not despised, certainly held in much lower esteem than articles containing original research results.

Gift-giving by scientists is thus similar to one of the most common modes of allocating resources to science, for this often takes the form of gifts from wealthy individuals or organizations. This has been true from the time of Cosimo de Medici to today, the time of the Rockefeller and Ford foundations. The gift status of moneys spent by industrial firms and governments on research is ambiguous; usually money seems to be spent with specific goals in mind, but the vast sums spent on space programs, particle accel- erators, radiotelescopes, and so forth often seem like a potlatch by the community of nations. Neil Smelser has suggested that the gift mode of exchange is typical not only of science but of all institu- tions concerned with the maintenance and transmission of common values, such as the family, religion, and communities.[2]

In general, the acceptance of a gift by an individual or a com- munity implies a recognition of the status of the donor and the existence of certain kinds of reciprocal rights.[3] These reciprocal rights may be to a return gift of the same kind and value, as in many primitive economic systems, or to certain appropriate sentiments of gratitude and deference. In science, the acceptance by scientific journals of contributed manuscripts establishes the donor's status as a scientist—indeed, status as a scientist can be

21

achieved *only* by such gift-giving—and it assures him of prestige within the scientific community.

The organization of science consists of an exchange of social recognition for information. But, as in all gift-giving, the expectation of return gifts (of recognition) cannot be publicly acknowledged as the motive for making the gift. A gift is supposed to be given, not in the expectation of a return, but as an expression of the sentiment of the donor toward the recipient. Thus, in the kula expeditions of the Melanesians:

> The ceremony of transfer is done with solemnity. The object given is disdained or suspect; it is not accepted until it is thrown on the ground. The donor affects an exaggerated modesty. Solemnly bearing his gift, accompanied by the blowing of a conch-shell, he apologizes for bringing only his leavings and throws the objects at his partner's feet. . . . Pains are taken to show one's freedom and autonomy as well as one's magnanimity, yet all the time one is actuated by the mechanisms of obligation which are resident in the gifts themselves.[4]

Gift-giving is capable of cynical manipulation; if this is publicly expressed, however, the exchange of gifts ceases, perhaps to be succeeded by contractual exchange. Consequently, scientists usually deny that they are strongly motivated by a desire for recognition, or that this desire influences their research decisions. A biochemist gave a typical reply when asked whether scientists compete for recognition:

> Most scientists have sincere interests in the advancement of science, more than in their own recognition. I don't think honor is the greatest ambition of professionals generally; rather, they want to solve a problem. Professionals are a relatively idealistic group. You might find extremists of all sorts, but by and large their real interest is in their work and in advancing knowledge. They are motivated by curiosity. These people think that rewards will take care of themselves; they are fatalistic in that respect. . . There is, after all, a kind of achievement in just getting to be a professor [i.e., in attaining the status of a scholar-scientist in the larger community], which is likely to be satisfying.

Some of my informants allowed themselves to be pressed into admitting that recognition was a source of gratification. For example, a theoretical physicist responded, when told that another scientist was not disturbed at all when he was anticipated and thereby prevented from publishing:

> I think I would admit to not having such pure interests. I must admit to a desire for recognition. I suppose it doesn't make much difference whether one wants to glorify himself or have others glorify him.

It is only the exceptional scientist, however, who sees the desire for recognition as a prime motivating force for himself and his colleagues. A mathematician, for example, said:

> A field in mathematics may become popular if the more popular mathematicians, the big shots, become interested in it. Then it grows rapidly. Junior mathematicians want recognition from big shots and, consequently, work in areas prized by them.

This man was something of a social isolate, capable of taking a detached view of the system.[5]

Nevertheless, the public disavowal of the expectation of recognition in return for scientific contributions should no more be taken to mean that the expectation is absent than the magnanimous front of the kula trader can be taken to mean that he does not expect a return gift. In both instances, this is made clearest when the expected response is not forthcoming. In primitive societies, failure to present return gifts often means warfare.[6] In science, the failure to recognize discovery may give rise, if not to warfare, at least to strong antagonisms and, at times, to intense controversy. A historical summary and analysis of priority controversies has been given by Robert K. Merton,[7] who pointed out that the failure to recognize previous work threatens the system of incentives in science. The pattern is not infrequent today. Of my seventy-nine informants, at least nine admitted to having been involved in questions of disputed priority either as the culprit or the victim. (The question was not asked in all seventy-nine interviews.) For example, an eminent theoretical physicist testified as follows:

> [This priority dispute] happened through an unfortunate habit of mine of not publishing things, delaying a year or two, talking about them, but not publishing. Therefore it has not always been clear to me whether something was known as my work or not. Under those conditions it is rather easy to get in priority disputes. It is much more satisfactory if one simply publishes what one does as he goes along. But a feeling of perfectionism frequently interferes with that. You don't want to publish something that's wrong. So you do the work, think about it for a couple of years, talk about it to everyone, but don't publish it. By that time somebody else has thought about it, and of course it's impossible to tell whether there was any influence from your work or not. It's a situation that's not very good. . . . *Did this result in some hostility against you?* Yes. One or two people—there haven't been many—were quite acrimonious. *They saw you about it?* Yes. . . . It was a completely personal matter and unfortunate.[8]

Another man, a little-known experimental physicist, was the victim in such a situation. A departmental colleague told about it first, when he was asked about the consequences of failure to recognize work:

> It causes a fatalistic attitude. . . . A professor on our faculty concerned himself with the X effect long before it became popular in the early nineteen fifties. [He did his work in the late nineteen thirties.] His work was referred to *once* by a review from a large laboratory, and in the review his name was omitted—'a professor at Y university' was the way they expressed it. This was deplorable. No action was ever taken on that. The man was just disappointed.

The experimental physicist himself described the same sequence of events, omitting the specific failure to recognize his accomplishments. Something similar to this had happened earlier in his career, when a grant he had requested was rejected, and shortly afterward someone else had become famous for doing essentially what he had proposed to do. This scientist was the most secretive man interviewed and was relatively isolated from his colleagues. In this case the individual affronted took no action; he may have reacted with hostility, but this was not communicated to the offenders. In other cases, as noted, affronted parties will communicate directly with the offenders. This may lead to a public recognition of priority—a 'Notice of Priority,' as it is called in the mathematical periodicals—or to an expression of recognition in succeeding papers by the offender or his collaborators. Thus, a mathematical statistician said that he was once anticipated in 'a serious matter':

> . . . I hastened to recognize the other. In 1934 . . . I overlooked a paper by a Russian, published twelve years before in an Italian journal. It was a very messy paper. In 1950 somebody from Iowa wrote me and informed me of these results, so I wrote a note of 'Recognition of Priority.'

The desire to obtain recognition induces scientists to publish their results. 'Writing up results' is considered to be one of the less pleasant aspects of research—it is not intrinsically gratifying in the way that other stages of a research project are. Some respondents were asked about the source of the greatest gratifications in research work. Generally the response was that most gratification came when the problem was essentially solved—when one became confident that an experiment would be successful, or when the outlines of a mathematical proof became clear—although details might remain to be cleaned up. For example, a theoretical physicist said: '[I get most pleasure] when the problem has been solved in principle but when some hard work remains to be done—when you have enough security to know you're not wasting your time but while there is some challenge left in the problem.'

An experimental physicist said one receives most gratification: '. . . when you find the effect you're looking for—everything else is anticlimax. Also in seeing some new and unexpected effect—in seeing new phonomena before others do.'

Research is in many ways a kind of game, a puzzle-solving operation in which the solution of the puzzle is its own reward.[9] 'Everything else'—including the communication of results—'is anticlimax.' Writing up results, 'cleaning up loose ends,' may be an irksome chore. A mathematical statistician said he was often pleased to discover that the result he had obtained was already in the literature:

> Being anticipated doesn't bother me. Maybe it does once in a while, when I have something really nice. But if someone else has published the result it means I don't have to write it out, and that is gratifying. The real pleasure in the work comes in working the problem out. Publishing is necessary for money and is nice in a way. I try to publish what I find amusing and what I hope others will find amusing as well. . . . Actually, I shouldn't give the impression that one publishes only in order to survive. People who are in really secure positions publish anyway. For example, [a very eminent man in the department] publishes about five papers annually, and he wouldn't have to publish at all any more.

The desire to obtain social recognition induces the scientist to conform to scientific norms by contributing his discoveries to the larger community. Thomas Sprat, writing near the dawn of modern science, perceived the importance of this: 'If neither *Chance,* nor *Friendship,* nor *Treachery* of Servants, have brought such Things out; yet we see *Ostentation* alone to be every day powerful enough to do it. This Desire of Glory, and to be counted Authors, prevails on all. . . .'[10]

Not only does the desire for recognition induce the scientist to communicate his results; it also influences his selection of problems and methods. He will tend to select problems the solution of which will result in greater recognition, and he will tend to select methods that will make his work acceptable to his colleagues.

The range of acceptable methods varies. In mathematics, for example, the standards of rigor have changed steadily. E. T. Bell has pointed this out in noting that eighteenth-century mathematicians were lucky, by later standards, not to have made more mistakes than they did:

> How did the master analysts of the 18th century—the Bernoullies, Euler, Lagrange, Laplace—contrive to get consistently right results in by far the greater part of their work in both pure and applied mathematics? What these great mathematicians mistook for valid reasoning at the very beginning of the calculus is now universally regarded as unsound.[11]

In this field, and most others, the change of standards is one of progress; the later standards can be shown to be technically superior. However, the definition of appropriate standards is not a technical matter only. For example, one informant was relatively famous for a method he had devised for making a certain kind of biochemical analysis. This method depended on distinctive nutritional requirements of certain bacteria. He noted that his method, while clearly superior to its alternatives for some purposes, and while very widely used, tended to be neglected by chemists:

> [The method] fell into some disrepute because we're using organisms, and the chemists said 'Huh. Nobody can do the quantitative work with an organism.' . . . There are other methods which chemists adopt. Chemists are a peculiar breed. They feel it is slightly debasing to use organisms. They just couldn't do that. . . . Most chemists use [other methods] for psychological reasons if nothing else. But I would never agree that other methods are generally better, except for specific purposes.

That is, social recognition in biochemical work done by chemists induces them to use techniques defined as distinctively chemical.[12]

Similarly, in mathematics the style of a proof, its 'elegance,' is often considered as important to its merit as the truth of the theorem proved. While there are technical reasons for this, there are distinctively social ones as well. A mathematician described a mixed case:

> Let me tell you a story about a famous problem in topology called the Poincaré conjecture. Various proofs were talked about, maybe they were published, but errors were found in all of them. Then, three years ago a Japanese mathematician published a hundred-page proof of it. A hundred-page proof is very unusual. Nobody has gone through it yet and found an error—this would be very difficult—but nobody believes in it. The man didn't have any new methods, and people didn't think he could do it with what he had.

Conformity with methodological standards is necessary if social recognition is to be given for contributions.

Similarly, the goals of science as they are specified in particular disciplines at particular times cover a restricted range, and the process of the reward of social recognition tends to produce individual conformity to the differentiated goals. As John R. Baker has put it:

> The scientist is able to construct a sort of scale of scientific values and to decide that one thing or theory is relatively trivial and another relatively important, quite apart from any question of practical applications. There is, as Poincaré has well said, *'une hierarchie des faits.'* Most scientists will agree that certain discoveries or propositions are more important because more widely significant than others, though

around any particular level on the scale of values there may be disagreement.[13]

Erwin Schrödinger, the eminent theoretical physicist, has indicated that this has been true in his experience as well:

> ...it might be said that scientists all the world over are fairly well agreed as to what further investigations in their respective branches of study would be appreciated or not.... The argument applies to the research workers all the world over, but only of *one* branch of science and of *one* epoch. These men practically form a unit. It is a relatively small community, though widely scattered, and modern methods of communication have knit it into one. The members read the same periodicals. They exchange ideas with one another. And the result is that there is a fairly definite agreement as to what opinions are sound on this point or that.... In this respect international science is like international sport.... Just as it would be useless for some athlete in the world of sport to puzzle his brain in order to initiate something new—for he would have little or no hope of being able to 'put it over,' as the saying is—so too it would, generally speaking, be a vain endeavor on the part of some scientist to strain his imaginative vision toward initiating a line of research hitherto not thought of.[14]

Sanctions to enforce conformity in this respect, as well as with regard to appropriate techniques, are of two general kinds. First, works that deviate too far from the norm will be refused publication in scientific journals. A mathematician interviewed had a paper rejected for this reason:

> A couple of us did some work that another person thought was merely trivial. And there was a little cat fighting going on here.... This was rejected by one journal, we felt for poor reasons; we rewrote it and sent it in again, and it was again rejected, we felt because of a sloppy refereeing job. I have been tempted and may yet write a complaining letter to the editor, whom I know, not asking that the paper be reconsidered but that the referee's work be evaluated, because the paper was not well refereed.[15]

Such an exercise of sanctions makes it impossible for the great mass of scientists to evaluate for themselves the importance and validity of the information presented. Delegating considerable power to a few authorities obviously infringes on the norms of independence in science. For this reason, editors and referees tend to be tolerant, basing their decisions on estimates that others will find manuscripts interesting even if they themselves do not. This and the fact that there are many journals, some of them unrefereed, means that the sanction is of little importance to most scientists most of the time. A more important sanction is the social recognition published work receives; this sanction is exercised by the community at large and applies to all published research.

Another type of sanction is not primarily important in science, although it is often alleged to be. This consists of extrinsic rewards, primarily position and money.[16] It is alleged that scientists publish, select problems, and select methods in order to maximize these rewards. University policies that base advancement and salary on quantity of publication sometimes seem to imply that this is true, that scientists' research contributions are not freely given gifts at all but are, instead, services in return for salary. While it is important for extrinsic rewards to be more or less consistent with recognition, the ideal seems to be that they should follow recognition, and this seems to be the general practice. In any case, an explanation of scientific behavior in terms of extrinsic rewards is weakened by the fact that many scientists in élite positions, whose extrinsic rewards will be unaffected by their behavior, continue to be highly productive and to conform to scientific goals and norms. Furthermore, scientists usually feel that it is degrading and improper to submit manuscripts for publication primarily to gain position without really caring if the work is read by others.

But why should gift-giving be important in science when it is essentially obsolete as a form of exchange in most other areas of modern life, especially the most distinctly 'civilized' areas? Gift-giving, because it tends to create particularistic obligations, usually reduces the rationality of economic action. Rationality is maximized when 'costs' of alternative courses of action can be assessed, and such costs are usually established in free-market exchanges or in the plans of central directing agencies. When participants are paid a money wage or salary for their efforts, and when this effectively controls their behavior, the system is more flexible than when controls derive from traditional or gift obligations.[17] Why, then, does this frequently inefficient and irrational form of control persist in science? To be sure, it also tends to persist in other professions. Professionals are expected to be motivated by a desire to serve others.[18] For example, physicians do receive a fee for service, yet they are expected to have a 'sliding scale' and serve the indigent at reduced fees or for no fees at all. The larger community recognizes two types of public dependence on professions: professional services are regarded as essential, concerned with values that should be realized regardless of a client's ability to pay; and nonprofessionals are unable to evaluate professional services, which makes them vulnerable to exploitation by unqualified persons. The rationale for the norm of service is usually the former type of dependence. In science, for example, the fact that a community has no one willing and able to pay for an important item of useless knowledge is not supposed to interfere with its ability to acquire the knowledge. But the idea of the gift and the norm of service is also related to the dependence of the public that follows from its inability to evaluate services.

The rationality of professional services is not the same as the rationality of the market. In contractual exchanges, when services are rewarded on a direct financial or barter basis, the client abdicates, to a considerable degree, his *moral control* over the producer. In return, the client is freed from personal ties with the producer, and he is able to choose rationally between alternative sources of supply. In the professions, and especially in science, the abdication of moral control would disrupt the system. The producer of professional services must be strongly committed to higher values. He must be responsible for his products, and it is fitting that he not be alienated from them. The scientist, for example, must be concerned with maintaining and correcting existing theories in his field, and his work should be oriented to this end. The exchange of gifts for recognition tends to maintain such orientations. On the one hand, the recipient of the gifts finds it difficult to refuse them (they are 'free'), and, on the other, the donor is held responsible for adhering to central norms and values. The maxim, *caveat emptor,* is inapplicable.[19] Furthermore, the donor is not alienated from his gift, but retains a lasting interest in it. It is, in a sense, his property.[20] One indication of this is the frequent practice of eponymy, the affixing of the name of the scientist to all or part of what he has found.[21]

Emphasis on gifts and services occurs frequently in social life, and we can get at the root of this generality by focusing on certain paradoxical elements implicit in the argument presented thus far. It has been argued that scientists are oriented to receiving recognition from colleagues and that this orientation influences their research decisions. Yet evidence that scientists themselves deny this has also been presented. There is a normative component to this denial, one that appears more clearly in analyzing scientific fashions. It is felt that, if a scientist's decisions are influenced by the probability of being recognized, he will tend to deviate from certain central scientific norms—he will fail to be original and critical. Thus, while it is true that scientists are motivated by a desire to obtain social recognition, and while it is true that only work on certain types of problems and with certain techniques will receive recognition at any particular time, it is also true that, if a scientist were to admit being influenced in his choices of problems and techniques by the probability of being recognized, he would be considered deviant. That is, if scientists conform to norms about problems and techniques as a result of this specific form of social control, they are thereby deviants.

This apparent paradox, that people deviate in the very act of conforming, is common whenever people are expected to be strongly committed to values. In general, *whenever strong commitments to values are expected, the rational calculation of punishments and rewards is regarded as an improper basis for making decisions.* Citizens who refrain from treason merely because it is against the law are, by that fact, of questionable loyalty; parents who refrain from incest merely because of fear of community reaction are, by that fact, unfit for parenthood; and scientists who select problems merely because they feel that in dealing with them they will receive greater recognition from colleagues are, by that fact, not 'good' scientists. In all such cases the sanctions are of no obvious value: they evidently do not work for the deviants, and none of those who conform admit to being influenced by them. But this does not mean that the sanctions are of no importance; it does mean that more than overt conformity to norms is demanded, that inner conformity is regarded as equally, or more, important.

Thus, the gift exchange (or the norm of service), as opposed to barter or contractual exchange, is particularly well suited to social systems in which great reliance is placed on the ability of well-socialized persons to operate independently of formal controls. The prolonged and intensive socialization scientists experience is reinforced and complemented by their practice of the exchange of information for recognition. The socialization experience produces scientists who are strongly committed to the values of science and who need the esteem and approval of their peers. The reward of recognition for information reinforces this commitment but also makes it flexible. Recognition is given for kinds of contributions the scientific community finds valuable, and different kinds of contributions will be found valuable at different times.

The scientist's denial of recognition as an important incentive has other consequences related to those already mentioned. When peers exchange gifts, the denial of the expectation of reciprocity in kind implies an expectation of gratitude, a highly diffuse response.[22] This kind of gift exchange occurs among scientists, although the more important form of scientific contribution is directed to the larger scientific community.[23] In this case the denial of the pursuit of recognition serves to emphasize the universality of scientific standards: it is not a particular group of colleagues at a particular time that should be addressed, but all possible colleagues at all possible periods. These sentiments were expressed with his typical fervor by Johannes Kepler:

I have robbed the golden vessels of the Egyptians to make out of them a tabernacle for my God, far from the frontiers of Egypt. If you forgive me, I shall rejoice. If you are angry, I shall bear it. Behold, I have cast the dice, and I am writing a book either for my contemporaries, or for posterity. It is all the same to me. It may wait a hundred years for a reader, since God has also waited six thousand years for a witness.[24]

While this orientation is consistent with the scientists's need for autonomy—being dependent on the favors of particular others is terrifying—it also contains a strong element of the tragic. Scientists learn to *expect* injustice, the inequitable allocation of rewards. Occasionally one of them makes this explicit. Max Weber addressed students on 'Science as a Vocation' in the following way:

I know of hardly any career on earth where chance plays such a role. . . . If the young scholar asks for my advice with regard to habilitation, the responsibility of encouraging him can hardly be borne . . . one must ask every . . . man: Do you in all conscience believe that you can stand seeing mediocrity after mediocrity, year after year, climb beyond you, without becoming embittered and without coming to grief? Naturally, one always receives the answer: 'Of course, I live only for my calling.' Yet, I have found that only a few men could endure this situation without coming to grief.[25]

More common than such an explicit statement is the myth of the hero who is recognized only after his death. This myth is important in science, as in art, because it strengthens universal standards against tendencies to become dependent on particular communities. Thus, mathematics has such heroes as Galois, who wrote the major part of his great mathematical opus the night before he was killed in a duel at the age of twenty-one; Abel, who died of tuberculosis as his greatness was coming to be recognized; and Cantor, who died mad, believing his ideas were spurned by others.[26] The stories of Copernicus, receiving his revolutionary book the day he died, and Mendel, rediscovered years after his death, are well known in the larger society, where they perhaps serve a function similar to that performed in science, albeit more general.[27]

The distinctive functions of the system by which gifts of information are exchanged for recognition can be seen within science. Textbook-writing and the preparation of popularizations are expected to be neither original in the scientific sense nor critical of existing theories. Since texts draw on what is already known, teachers who adopt them are usually competent to judge the validity of the material presented. Consequently, writing of this sort is regarded as a technical skill and not a highly responsible task, and it has little effect on a scientist's reputation.[28] A dean in

31

a leading university said: 'We classify text writing under a man's teaching, and it may help him [in appointment and promotion]. However, if he has written nothing at all but texts, they will have a null value or even negative value.'

Because authors of texts and popularizations need not be highly committed to the values of the scientific community, their activities have little bearing on their standing within it. As a result, incentives for such tasks must be of a more general nature. Royalties are such a generalized reward; unlike recognition, cash can be used outside the community of pure science. The exchange of information for recognition, on the other hand, binds donors and recipients in a community of values.

Thus, gift-giving in exchange for recognition is an appropriate method of social control in science, and it is apparently relatively effective.[29]

NOTES AND REFERENCES

1.   When this is so, editorial decisions to publish are kept independent of the possibility of payment. Thus, in 1962, only 78 per cent of the pages published in the *Journal of Mathematical Physics* were paid for by the authors' institutions. See Henry A. Barton, 'The Publication Charge Plan in Physics Journals,' *Physics Today*, 16, 6 (June 1963), 45–57.

2.   Neil J. Smelser, 'A Comparative View of Exchange Systems,' *Economic Development and Cultural Change*, 7 (1959), 173–182.

3.   Cf. Alvin W. Gouldner, 'The Norm of Reciprocity,' *American Sociological Review*, 25 (1960), 161–178; and Marcel Mauss, *The Gift: Forms and Functions of Exchange in Primitive Societies* (Glencoe, Ill.: Free Press, 1954), pp. 40 f., 73, *et passim*.

4.   Mauss, *op. cit.*, p. 21, reporting the work of Malinowski.

5.   In a series of interviews with twenty eminent American biologists, Anne Roe was given the same impression about the suppression of the wish for recognition: '. . . the concentration is on the work primarily as an end in itself, not for economic or social ends, or even for professional advancement and recognition, although they are not indifferent to these.' Roe, 'A Psychological Study of Eminent Biologists,' *Psychological Monographs*, 65 (1951), p. 65. Bernice T. Eiduson makes a similar report based on her study of forty scientists in her *Scientists: Their Psychological World* (New York: Basic Books, 1962), pp. 162 and 178f. See also Charles Darwin, *The Autobiography of Charles Darwin, 1809–1882*, Nora Barlow, ed. (London: Collins, 1958), p. 141, for another scientist's disavowal of desire for recognition.

6.   Mauss, *op. cit.*, p. 3 *et passim*.

7.   Merton, 'Priorities in Scientific Discovery,' *American Sociological Review*, 22 (1957), 635–659.

8.   This may be a common failing of eminent men. Leopold Infeld wrote, regarding Einstein; 'Before we published our paper I suggested to Einstein that I should look up the literature to quote scientists who had worked on this subject before. Laughing loudly, he said: "Oh yes. Do it by all means. Already I have sinned too often in this respect" ' *Quest: The Evolution of a Scientist* (New York: Doubleday Doran, 1941), p. 277.

9. See T. S. Kuhn, *The Structure of Scientific Revolutions* (Chicago: University of Chicago Press, 1962), pp. 35—42. The layman can get some ideas of the gratification involved by reading such novels as C. P. Snow's *The Search* or Sinclair Lewis' *Arrowsmith*.
10. Sprat, *The History of the Royal Society of London* (London, 1673), pp. 74f. See also Karl Mannheim on the importance of the desire for recognition in science and other cultural pursuits: *Essays on the Sociology of Knowledge* (London: Routledge and Kegan Paul, 1952), ch. VI, especially pp. 239, 242—243, 272.
11. Eric T. Bell, *The Development of Mathematics* (2nd ed.; New York: McGraw-Hill, 1945), p. 153.
12. Compare the reception by chemists of chromatographic techniques, discovered by Tswett, a Russian botanist, in 1906: '. . . the chromatographic method got off to a bad start . . . for a lowly botanist to assault thus the whole chemical profession was unthinkable! . . . the chromatographic method fell largely into disrepute, and . . . Tswett spent the later part of his life in misery and poverty.' The importance of the technique was only recognized in the late nineteen twenties. James E. Meinhard, 'Chromatography: a Perspective,' *Science*, 110 (1949), 387—392.
13. *Science and the Planned State* (London: George Allen and Unwin, 1945), p. 33.
14. Schrödinger, in *Science and the Human Temperament*, James Murphy, trans. (London: George Allen and Unwin, 1935), pp. 76—80.
15. Cf. the comment of a referee of a biochemical journal: 'If the editor reversed what I said he would have to look for a new referee awfully fast, because I don't take that job lightly. And I've turned down many papers, too.'
16. Simon Marcson shows how organizations 'structure recognition by means of appropriate symbols, including titles, size of office, accessibility, financial rewards, and so on.' *The Scientist in Industry* (New York: Harper, 1960), p. 73. However, the difference between recognition and other rewards should be kept clear, for otherwise it is difficult to analyze social control and to specify the source of control. In this work 'recognition' means only the written and verbal behavior and the 'expressive gestures' of scientists that indicate their approval and esteem of a colleague because of his research accomplishments.
17. Put another way, flexibility and rationality are maximized when workers are alienated from their products. Cf. Talcott Parsons, '*Voting* and the Equilibrium of the American Political System,' in Eugene Burdick and Arthur J. Brodbeck, eds., *American Voting Behavior* (Glencoe, Ill.: Free Press, 1959), p. 89. See also Max Weber's stress on 'formally free' and actually alienated labor as a defining characteristic of capitalism: *General Economic History* (Glencoe, Ill.: Free Press, 1927), p. 277.
18. Talcott Parsons, *Essays in Sociological Theory* (Glencoe, Ill.: Free Press, 1954), ch. II.
19. This does not mean that scientists are not supposed to be skeptical; that they should be skeptical about their own work as well as that of their colleagues is one of the more important institutionalized norms of science. Cf. Robert K. Merton, *Social Theory and Social Structure* (Glencoe, Ill.: Free Press, 1949), pp. 315 f. It does mean that unlike the consumer in the free market, the 'consumer' of scientific products can hold the producer morally responsible for 'defective products.'
20. Cf. Merton, 'Priorities in Scientific Discovery,' *op. cit.*, pp. 640 f., and *Social Theory and Social Structure, op. cit.*, pp. 312 f. Mauss pointed out that members of some 'archaic societies' felt gifts somehow remained part of the donor and that this belief was reinforced by further existential beliefs, e.g., the gift itself possessed the power to harm the recipient if it was not reciprocated. Mauss, *op. cit.*, pp. 41 ff. *et passim.*
21. Cf. Merton, 'Priorities in Scientific Discovery,' *op. cit.*, pp. 642—644. See also Edwin G. Boring, 'Eponym as Placebo,' in his *History, Psychology, and Science: Selected*

Papers (New York: Wiley, 1963), pp. 5–28, where it is suggested that eponymy has psychological benefits even for scientists who cannot hope to be remembered this way themselves.

22.   As Georg Simmel says, gratitude 'establishes the bond of interaction, of the reciprocity of service and return service, even where they are not guaranteed by external coercion.' *The Sociology of Georg Simmel,* Kurt Wolff, ed. and trans. (Glencoe, Ill.: Free Press, 1950), p. 387. He goes on to note that persons make great efforts to avoid receiving gifts in order not to make such commitments to others. Something like this may occur in science.

23.   In other words, the scientific contribution is closely approximated by the sacrificial model of the gift. Cf. Émile Durkheim, *The Elementary Forms of Religious Life,* J. W. Swain, trans. (London: George Allen and Unwin, 1915), pp. 342 f.: 'The sacrifice is partially a communion; but it is also, and no less essentially, a gift and an act of renouncement.'

24.   Quoted in Arthur Koestler, *The Sleepwalkers* (London: Hutchinson, 1959), pp. 393 f.

25.   Hans H. Gerth and C. Wright Mills, trans. and eds. *From Max Weber: Essays in Sociology* (New York: Oxford University Press, 1946), pp. 132, 134. Weber was partly concerned with the particular aspects of science in German universities, but the entire essay shows that he was also concerned with the more universal aspects of science as a profession.

26.   Cf. Eric T. Bell, *Men of Mathematics* (New York: Simon and Schuster, 1937), for these and others.

27.   The story of Job is the classic version of the hero who serves without reward.

28.   Serious, unorthodox errors in a text might harm an author's reputation, although certain kinds of orthodox errors may be permitted as 'pedagogical license.' In any case, although an author's reputation may help sales of a text, it is possible for a person who has no reputation as a research scientist to write a text that will be widely adopted. One of my informants was such an author.

29.   Recognition by colleagues is probably a more effective means of social control among scientists, whose products are necessarily public knowledge, than it is among some other professionals. The limitations of colleague control in medicine are pointed out in Eliot Freidson and Buford Rhea, 'Processes of Control in a Company of Equals,' *Social Problems,* 11 (1963), 119–131.

2

Bruno Latour and Steve Woolgar

The cycle of credibility

Figure 1 illustrates the cycle of credibility. The notion of credibility makes possible the conversion between money, data, prestige, credentials, problem areas, argument, papers, and so on. Whereas many studies of science focus on one or other small section of this

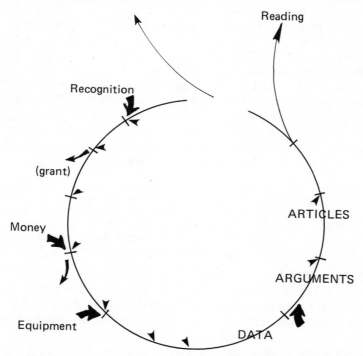

Reading

Recognition

(grant)

Money

Equipment

ARTICLES

ARGUMENTS

DATA

Figure 1   This figure represents the conversion between one type of capital and another which is necessary for a scientist to make a move in the scientific field. The diagram shows that the complete circle is the object of the present analysis, rather than any one particular section. As with monetary capital, the size and speed of conversion is the major criterion by which the efficiency of an operation is established. It should be noted that terms corresponding to different approaches (for example, economic and epistemological), are united in the phases of a single cycle.

circle, our argument is that each facet is but one part of an endless cycle of investment and conversion. If, for example, we portray scientists as motivated by a search for reward, only a small minority of the observed activity can be explained. If instead we suppose that scientists are engaged in a quest for credibility, we are better able to make sense both of their different interests and of the process by which one kind of credit is transformed into another.[1]

In order to understand the full force of the difference between reward and credibility, it is necessary to distinguish between the process by which reward is bestowed and the process by which credibility is assessed. Both reward and credibility originate essentially from peers' comments about other scientists. Thus, even the award of a Nobel Prize depends on various submissions, recommendations, and assessments by working scientists. But what form do these evaluative comments take in the laboratory itself? Two features are readily apparent. Firstly, evaluative comments made by scientists make no distinction between scientists as people and their scientific claims. Secondly, the main thrust of these comments turn on an assessment of the credibility which can be invested in an individual's claim. The possibility of bestowing reward is a marginal consideration. A striking illustration of this is provided by the following example: C and Parine were in the bioassay room when C asked Glenn to synthesize a peptide which another colleague, T, had claimed to be more active than endorphin. When a syringe of the peptide had been prepared, C made ready to inject a rat on the surgery table:

> I bet you the peptide is going to do nothing . . . this is the confidence I have in my friend T. [C squeezed the syringe and enjoined the rat]: O.K., Charles T., tell us. [A few minutes passed.] See, nothing happened . . . if anything the rat is even stiffer [sigh]. Ah, my friend T . . . I went to his laboratory in New York and saw his records . . . which lead to publication . . . it made me feel uncomfortable (V, 53).

This incident underscores the common conflation of colleague and his substance: the credibility of the proposal and of the proposer are identical. If the substance had the desired effect on the rat, T's credibility would have increased. If, on the other hand, he had had more confidence in T, C would have been surprised by the result. This is made particularly clear in the following:

> Last week, my prestige was very low, X said I was not reliable, that my results were poor and that he was not impressed. . . . Yesterday, I showed him my results . . . good lord, now he is very nice, he says, he was very impressed and that I will get a lot of credit for that (XI, 85).

For a working scientist, the most vital question is not 'Did I repay my debt in the form of recognition because of the good paper he wrote?' but 'Is he reliable enough to be believed? Can I trust him/ his claim? Is he going to provide me with hard facts?' Scientists are thus interested in one another not because they are forced by a special system of norms to acknowledge others' achievements, but because each needs the other in order to increase his own production of credible information.

Our discussion of the demand for credible information contrasts with two influential models of the exchange system in science proposed by Hagstrom (1965) and Bourdieu (1975b). Both models have obviously been influenced by economics. Hagstrom's model employs the economics of preindustrial societies and portrays the relation between two scientists as that of gift exchange. According to Hagstrom, however, the expectation of exchange is never made explicit:

> [T]he public disavowal of the expectation of recognition in return for scientific contributions should no more be taken to mean that the expectation is absent than the magnanimous front of the kula trader can be taken to mean that he does not expect a return gift (Hagstrom, 1965: 14).

Explicit reference to the expectation of exchange occurred in many of the cases we observed. There was no suggestion that our scientists had to maintain a fiction that they were not expecting any return gift. Consequently, the basic argument that scientists are gift givers does not seem warranted. Indeed, we can pose the same question which Hagstrom himself asked:

> But why should gift giving be important in science when it is essentially obsolete as a form of exchange in most other areas of modern life, especially the most distinctly 'civilized' areas? (Hagstrom, 1965: 19)

Hagstrom provides no reasons for the survival of this antiquated tradition in the scientific community other than the fact that the same phenomenon is evident in other professional spheres. In all such professional spheres, Hagstrom argues,

> the gift exchange (or the norm of service), as opposed to barter or contractual exchange, is particularly well suited to social systems in which great reliance is placed on the ability of well-socialized persons to operate independently of formal controls (Hagstrom, 1965: 21).

For Hagstrom, then, the archaic system of gift exchange is functionally corequisite to the maintenance of social norms. In other words, the archaic system of potlatch is seen as a way of supporting the central system of norms. Even scientists' publication strategies are manifestations of conformity to norms through participation in gift exchange:

> The desire to obtain social recognition induces the scientist to conform to scientific norms by contributing his discoveries to the larger community (Hagstrom, 1965: 16).

Scientific activity is governed by norms, and the enforcement of these norms entails the existence of a special system of gift giving. But this system is never made mention of by participants. Indeed, if scientists deny they are expecting a return gift, this can be taken as proof of the success of their training and their rigorous conformity to the norms. We have here an explanation of an exchange system in terms of norms which is both empirically undemonstrable and which the author himself considers an inexplicable and paradoxical archaism.

Why should Hagstrom use a primitive exchange analogy to explain relationships between scientists? We had the distinct impression that the constant investment and transformation of credibility taking place in the laboratory mirrored economic operations typical of modern capitalism. Hagstrom was struck by the apparent absence of transfers of money. But this feature should not lead to the formulation of a model designed to preserve the existence of norms. Do scientists read each other out of deference to norms? Does one individual read a paper so as to force its author to read his work in return? Hagstrom's exchange system has the aura of a rather contrived fairy tale: scientists read papers as a matter of courtesy, and similarly thank their authors out of politeness. Let us look at one more example of scientific exchange in order to show that this view is needlessly complicated.

One of the main problems in studying diabetes was the difficulty of discriminating between the effects of insulin and glucagon in a diabetic patient's glucose level. In other words, attempts to study the effects of insulin were foiled because of the 'noise' generated by glucagon, the effects of which it was impossible to suppress. In 1974, however, a new substance called somatostatin was isolated (in a completely unrelated field), which was found to inhibit the secretion of both growth hormones and glucagon (Brazeau and Guillemin, 1974). Somatostatin was immediately imported into the field of diabetic study and used to decrease the effect of glucagon.

> The discovery of GH releasing inhibiting hormone, Somatostatin, might open the way to an objective evaluation of the role of glucagon in diabetes. It will soon be possible to follow diabetic patients with completely suppressed glucagon secretion.

This passage, written by a clinician, indicates the potential importance of glucagon. If at this point, someone had told the clinicians that he knew the structure of a glucagon suppressor substance, he would have been seized violently by the lapels. Why? Because the clinician would have felt overcome by a desire to reward this individual for his contribution? Or because he felt a debt of honour to the individual's achievement? *No.* The clinician's violent reaction would stem from his ability, once armed with the new information, to rush to his bench or his ward and set up a protocol in which one of the causes of noise in his inscription device can be controlled. The clinician would not be obliged to disburse credit to the bearer of information, nor even to cite his paper. The utility of the information for the generation of fresh information is crucial, whereas the subsequent bestowal of recognition is only a secondary concern to the scientist.

Bourdieu's model of scientific exchange compares scientists' behaviour with that of modern businessmen rather than precapitalist dealers and traders. The absence of money in scientific exchange does not cut any ice with him because of his experience in studying exchange systems in fields other than science. For Bourdieu (1975b), economic exchange can include the accumulation and investment of resources other than money. By using the idea of symbolic capital Bourdieu describes the investment strategies in fields such as education or art in terms of modern capitalism. Even business strategies are analyzed from the point of view of accumulation of symbolic (rather than just monetary) capital. By contrast with Hagstrom, Bourdieu (1975b) does not attempt to explain scientists' behaviour in terms of norms. Norms, the socialisation processes, deviance, and reward are the consequences of social activity rather than its causes. Similarly, Bourdieu takes the position that science can be studied without forging ad hoc explanations and in terms of other more usual rules of economics. For Bourdieu, then, the cause of social activity is the set of strategies adopted by investors wanting to maximise their symbolic profit.

> The scientific field is the locus of a competitive struggle, in which the specific issue at stake is the monopoly of scientific authority, defined inseparably as technical capacity and social power (Bourdieu, 1975b: 19).

Investors' strategies are likened to any other businessman's strategy. However, it is not made clear why scientists should be interested in one another's production. Bourdieu simply asserts:

> [T]he transmutation of the anarchic antagonism of particular interests into a scientific dialectic becomes more and more complete as the interest that each producer of symbolic goods has in producing products

that, as Fred Reif puts it 'are not only interesting to himself but also important to others' . . . comes up against competitors more capable of applying the same means (Bourdieu, 1975b: 33).

This tautological explanation of interest is worsened by the absence of any reference to the content of the science produced. In particular, there is no analysis of the way in which technical capacity is linked to social power. This absence might not be a problem in the study of 'haute couture' (Bourdieu, 1975a), but it is absurd in science.

Neither Bourdieu nor Hagstrom helps us understand why scientists have any interest in reading each other. Their use of economic models, derived respectively from capitalist and precapitalist economies, fails to consider *demand*. This failure corresponds to their failure to deal with the contents of the science. As Callon (1975) has argued, economic models can be applied only if this accounts for the content of science. Hagstrom and Bourdieu provide useful explanations of the repartition of credit as a sharing process but they contribute little to an understanding of the *production of value*.

Let us suppose that scientists are investors of credibility. The result is the creation of a *market*. Information now has value because, as we saw above, it allows other investigators to produce information which facilitates the return of invested capital. There is a *demand* from investors for information which may increase the power of their own inscription devices, and there is a *supply* of information from other investors. The forces of supply and demand create the *value* of the commodity, which fluctuates constantly depending on supply, demand, the number of investigators, and the equipment of the producers. Taking into account the fluctuation of this market, scientists invest their credibility where it is likely to be most rewarding. Their assessment of these fluctuations both explains scientists' reference to 'interesting problems,' 'rewarding subjects,' 'good methods,' and 'reliable colleagues' and explains why scientists constantly move between problem areas, entering into new collaborative projects, grasping and dropping hypotheses as the circumstances demand, shifting between one method and another and submitting everything to the goal of extending the credibility cycle.[2]

It would be mistaken to take the central feature of our market model as the simple exchange of goods for currency. Indeed, at the preliminary stage of fact production, the straightforward exchange of information for reward is hindered by the fact that the scientist and the claim are *not* distinguished. What then is the equivalent of buying in our economic model of scientific activity?

40

Our scientists only rarely assessed the success of their operations in terms of formal credit. For example, they had little idea of the extent to which their work was cited. They were not normally concerned about the distribution of awards, and they were only marginally interested in questions of credit and priority.[3] Indeed, our scientists had a much more subtle way of accounting success than simply measuring returns in currency. The success of each investment was evaluated in terms of the extent to which it facilitated the rapid conversion of credibility and the scientist's progression through the cycle. For example, a successful investment might mean that people phone him, his abstracts are accepted, others show interest in his work, he is believed more easily and listened to with greater attention, he is offered better positions, his assays work well, data flow more reliably and form a more credible picture. The objective of market activity is to extend and *speed up the credibility cycle as a whole.* Those unfamiliar with daily scientific activity will find this portrayal of scientific activity strange unless they realise that only rarely is information itself 'bought.' Rather, the object of 'purchase' is the scientist's ability to produce some sort of information in the future. The relationship between scientists is more like that between small corporations than that between a grocer and his customer. Corporations measure their success by looking at the growth of their operations and the intensity of the circulation of capital.[4]

Before using this model to interpret the behaviour of our laboratory scientists, it is important to stress its complete independence of any argument concerning motivations. Explanations using the notion of reward required us to suppose that scientists routinely hide their real motivations when they fail to reveal an explicit interest in credit and recognition. By contrast, our credibility model can accommodate a variety of types of motivations. It is not necessary, therefore, to doubt the motivations expressed in informants' accounts. Scientists are thus free to report interest in solving difficult problems, in getting tenure, in wanting to alleviate the miseries of humanity, in manipulating scientific instruments, or even in the pursuit of true knowledge. Differences in the expression of motivation are matters of psychological make-up, ideological climate, group pressure, fashion, and so on.[5] Since the credibility cycle is one single circle through which one form of credit can be converted into another, it makes no difference whether scientists variously insist on the primacy of credible data, credentials, or funding as their prime motivating influence. No matter which section of the cycle they choose to emphasize or consider as the objective of investment, they will necessarily have to go through all the other sections as well.

NOTES

1.   One main advantage of the notion of cycle, is that it frees us from the necessity of specifying the ultimate motivation behind the social activity which is observed. More precisely, one might suggest that it is the formation of an endless cycle which is responsible for the extraordinary success of science. Marx's (1867: Ch. 4) comments on the sudden conversion from use value to exchange value, could well apply to the scientific production of facts. The reason so many statements are produced is that each is without use value, but has an exchange value which enables conversion and accelerates the reproduction of the credibility cycle. This view also has implications for the so-called relations between science and industry (Latour, 1976).

2.   This is typical of the double standard of some analysts of science. When a businessman gives up and sells a bankrupt company, this is taken as an obvious manifestation of greed and interested motives. However, when a scientist gives up a dying area or a discredited hypothesis (which means that no one is going to 'buy' the argument any more), this is considered as an indication of conformity to the ethos of scientific disinterestedness.

3.   As noted earlier, the laboratory chosen for our study was characterized by an almost pathological concern for credit. It became clear, however, that the *'point d'honeur'* of credit receipt was not itself at stake. Because of the modification of the field, each participant adopted different strategies: the struggle concerned, not credit but space, research programmes, and equipment. As long as they agreed on these points, there was little quarrel about who received credit. When they differed on these points, the tangible focus of conflict was a bitter argument about credit sharing.

4.   This comparison is viable in so far as the notion of economics is not restricted to the circulation of money. It should be extended instead to all activities permeated by the existence of a valueless capital, the sole purpose of which is accumulation and expansion. This differs from the efforts of the Chicago School to portray activities in economic terms even where no capital is involved. The link between the scientific production of facts and modern capitalist economics is probably much deeper than a mere relation.

5.   A related problem is the extent to which the scientists' activities we portray are conscious and explicit strategies. This is a problem we cannot resolve in abstract because each scientist is also engaged in a debate to make logical, explicit, or necessary his career's choices. We do not wish to say that scientists are 'really' interested although they do not avow it, or that they are 'really' determined by the field although they think they have some freedom and merit in having chosen this or that way. We leave questions like the notion of motivations entirely open to psychologists and historians. Some scientists try to show that it was their conscious decision to choose this subject, while simultaneously arguing that a colleague could not do otherwise because the time was ripe. On another occasion the same informant may try to persuade you that he was not conscious at all and that it was some kind of artistic intuition, only to inform you a few days later that the whole thing was quite logical and that he did not have much choice. This consideration is important because we certainly do not wish to propose a model of behaviour in which individuals make calculations in order to maximize their profits. This would be Benthamian economics. The question of the calculation of resources, of maximization, and of the presence of the individual are so constantly moving that we cannot take them as our points of departure.

REFERENCES

BOURDIEU, P. (1975a). 'Le coutourier et sa griffe,' *Actes de la Récherche en Sciences Sociales,* 1, 1.

BOURDIEU, P. (1975b). 'The Specificity of the Scientific Field and the Social Conditions of the Progress of Reason', *Social Science Information,* 14, 6: 19—47.

BRAZEAU, P. and GUILLEMIN, R. (1974). 'Somatostatin: Newcomer from the Hypothalamus', *New England Journal of Medicine,* 290 : 963—64.

CALLON, M. (1975). 'L'opération de traduction comme relation symbolique', *in* P. Roqueplo (ed.), *Incidence des rapports sociaux sur le développement scientifique et technique.* Paris : CNRS.

HAGSTROM, W.O. (1965). *The Scientific Community.* New York : Basic Books.

LATOUR, B. (1976). 'Including Citations Counting in the Systems of Actions of Scientific Papers'. Unpublished paper delivered to first meeting of the Society for Social Studies of Science, Cornell University, Ithaca, NY, November.

MARX, K. (1867). *The Capital,* Volume 1. New York : Random House (1977).

3

H.M. Collins

Tacit knowledge
and scientific networks

INTRODUCTION: METHODOLOGICAL AND THEORETICAL ARGUMENT

Thomas Kuhn's concept of 'paradigm'[1] has attracted a lot of attention from sociologists and historians of science. In particular, some recent work has involved the search for groups of scientists which are taken to be the social analogue of this idea. I will argue here that the boundaries of those 'social circles' of scientists which *have* been found are not likely to correspond with the boundaries of groups sharing a paradigm unless the term be construed in a restricted sense. This is because the research methods used in most cases are unsuitable for the investigation of *cognitive* specificity and discontinuity. But it is precisely because paradigm groups are seen as conceptually homogeneous and bounded that the idea has excited interest in the sociology of science as a branch of the sociology of knowledge.

An impressive number of studies[2] have been published which try to show that workers in scientific specialties are organized in a 'social circle'.[3] Such groups have been identified with Price's notion of 'Invisible College',[4] and have also been seen as the social location of distinctive sets of 'technical and cognitive norms'.[5] The claim that a set of actors exists as a social 'group' is, of course, partly a matter of definition. A 'social circle' is distinguished by the greater density of relations between its members than between members and non-members, and this depends on the definition of the relations which are held to be significant. In turn, the discovery of such patterns of relations depends in part on the research methods used to locate them. In the last resort, empirical research can only discover the existence of operationalizations of a relation, not the relation itself. In the case of the recent work by Crane and others, these operationalizations have been based on ideas of the nature of scientific information and influence which are philosoph-

44

ically undeveloped and likely to be sociologically uninteresting. This seems to be the result of a conflation of 'information science' and the 'sociology of science'[6] —a conflation which is well illustrated by a sentence on the first page of Crane's recent book. She writes:

> The growth of scientific knowledge like that of most natural phenomena takes the form of the logistic curve.[7]

But this growth pattern has been shown, at best, only for what Price has called 'any reasonable definition of science',[8] not for 'scientific knowledge'.[9] The term 'Invisible College' itself was used by Price for hypothesized groups of scientists whose interest in meeting together would be to overcome the problems of the 'Information Explosion'. Within this context, research techniques such as the use of interval scale bibliometric indices and questionnaires are quite appropriate, for 'Information Science' treats information as though it can be contained in discrete visible packages of roughly equal value. However, in spite of the operational attractiveness of these techniques, the sociologist has to be concerned, at least at first, with the actor's interpretation of these items, before he can treat them as sociologically relevant. It follows that the 'groups' defined by relations based upon the information scientist's research techniques (that is, groups constructed around relations based upon *their own* perceptions of information flow) may not be groups which have any particular interest for the sociologist concerned with the boundaries of networks whose members share a common conceptual map.

Crane and others have missed the point that learning to become part of, or helping in the conceptual development of, a particular paradigm group, is 'doing' something, in the same sense that absorbing the conceptual structure that makes, say, logical inference 'natural' is learning 'to do' something.[10] What is more, there is no reason to suppose that it is possible to formulate the knowledge required to do the activity in question, any more than it is possible to formulate the knowledge required in order to infer logically. Achilles had this problem in his conversation with the tortoise.[11] Because cognitive influences are often intangible it is unlikely that the associations between scientists discovered through the correlation of questionnaire responses or in bibliographic interconnections will reflect them. This has been appreciated in studies of the transfer of technology, where it is more clear that knowledge consists of the ability to do something; thus Burns has written that technological knowledge is the property of people rather than documents.[12] I am suggesting (with Ravetz)[13] that this element is present in all knowledge, however pure, but is perhaps less noticeable elsewhere.

I will repeat the above point, for it is the most important element of the theoretical argument of the paper.

All types of knowledge, however pure, consist, in part, of tacit rules which may be impossible to formulate in principle. For instance, the ability to solve an algebraic equation includes such normally non-articulated knowledge as that the symbol 'x' usually means the same whether it is written in ball point, chalk, or print, or if spoken, irrespective of the day of the week, or the temperature of the air. But in another sense, x stands for anything at all and may only mean the same—exactly (e.g. 2.75 g; 27 inches, etc.) —on co-incidental and unimportant occasions. Again, sometimes a capital X or an italicized *x* may have a distinctive meaning. X in the equation x = 5y is the *same* as x in the equation 5y = x, but is not the same as x in x = 5z, unless y = z: but, on the other hand, 'x' is being used in the same way in all the equations. This list of tacit rules as it is extended becomes more confusing, and comes to resemble a list of all the examples of the uses of x which have ever been made. But such a list cannot serve at all as a guide for the use of x in the future.[13] Learning algebra consists of more than the memorization of sets of formal rules; it involves also knowing how to *do* things (e.g. use 'x' correctly; use logical inference) which may have been learned long before. This means that while it is quite sensible to say 'Mr Jones taught me to solve equations in the third week of January 1947' (he taught me: 'change side change sign'; 'get all the x's on one side'; 'add all the x's up'; 'divide throughout by the number in front of the x'; etc.), it is not sensible to say, 'Mr Jones taught me to use symbols' during a particular week. Now, an important difference between members of different paradigm groups (as I am using the idea) lies in the contents of their tacit understandings of the things that they may legitimately do with a symbol or a word or a piece of apparatus. Because the process of learning, or building up tacit understandings, is not like learning items of information, but is more like learning a language, or a skill, it must be investigated differently. To ask respondents directly for the sources of their tacit knowledge, or to assume that sociograms based on responses to questions about articulated knowledge necessarily picture the diffusion or development of tacit knowledge, is to confuse concept formation with information exchange.

There is a second type of criticism to be made of sociometric studies of scientific groups: this concerns the appropriateness of questionnaire research, even where articulated knowledge is concerned. Crane, reporting her survey of agricultural innovation diffusion researches, writes that some respondents seemed to have difficulty (or were unwilling) to distinguish the influences upon them of other scientists. Gaston quotes a theoretical physicist:

> There are many people who are not very much aware of who they get their ideas from, and six months after they hear the idea they forget who suggested it.

These reservations are serious, but might be less so were it the case that only the most immediately obvious sources of influence and information were significant. Crane writes:

> The use of a questionnaire to elicit some of this information probably has the advantage of obtaining the most important influences, rather than a complete list of major and minor influences.

But there is serious doubt that the most important contributors to the ideas of a scientist are necessarily the most obvious contributors. In fact, a recent article seems to imply that the opposite might be the case.

Mark S. Granovetter[15] argues that weak sociometric links are likely to be more important than strong ones for the transmission of influences over long (sociometric) distances and between groups which are not densely connected. If this is true, those sociometric ties which are furthest from the forefront of a respondent's mind could, for some innovatory networks, be the most important. Granovetter argues, from a premise drawn from the theory of structural balance, that where two groups are connected by only one link, or where there is one link between the two which is part of a much shorter sociometric path than any alternative, that link will be weak. It follows (and he has some empirical evidence to back this up) that the tracing out of sequences of weak links from a node in a social network will define a far larger area of the network than will the tracing out of strong links. Sequences of strong links will soon return to the original node and thus define local groups only[16] Granovetter suggests that this may explain why marginal individuals can be particularly effective diffusers of ideas. It also seems to follow that where a postulated mechanism of innovation is the transfer of ideas from one scientific field (social group) to another, weak sociometric links may form an essential part of the diffusion path. (In commonsensical terms, if ideas brought into one field from another are genuinely new, there can have been little contact, and therefore no strong links, between the fields previously.) It seems probable that Granovetter's ideas will be important in future sociometric studies of the diffusion of innovations. In particular, they throw further serious doubt on the validity of the questionnaire response as a direct indicator of the flow of real scientific innovatory influence.

In order not to be unfairly critical of the authors in question, it must be pointed out that they do show some sensitivity to these issues. Crane concerns herself with student-teacher relationships, as well as with sociometric choice and published collaboration relations, and of course the former relation may be the seat of the transmission and development of much tacit understanding. She writes:

> From each of these types of ties among scientists is obtained a somewhat different picture of the extent to which members of a research area are linked to one another. Nevertheless, if one uses the various indicators of linkage separately and then in combination, one is provided with a fairly complete picture of the amount of relatedness that exists.

Unfortunately, we are never shown the networks which are defined by these different relations, nor are expected differences systematically discussed. Without a clear understanding of the social relations defining the network, it is difficult to see what is meant by the phrase '. . . a fairly complete picture of the amount of relatedness that exists.' Relatedness cannot exist unqualified, any more than can equality. Earlier, Crane writes:

> The use of citation linkages between scientific papers is an approximate rather than an exact measure of intellectual debts. . . . Sociometric choices can also be criticized as unreliable indicators of relationships between scientists since it is obvious that the scientists may not recall all such contacts and may be biassed toward reporting contacts with more prestigious individuals and ignoring those with less prestigious individuals. In the absence of other equally good measures, however, I will proceed on the assumption that such approximate measures can be used to provide some indication of the actual nature of relationships among scientists in a research area.[17]

However, even 'these approximate measures' cannot be taken as indicators of anything other than the flow of articulated and therefore visible information.[18]

I will now attempt to illustrate some of the theoretical points made above about the diffusion of tacit knowledge by reporting on a study of a set of experimental physicists. I make no statement about the paradigm of which their practices are a part, and I have not attempted to delineate it. I have picked on experimental physicists because the aspect of their knowledge pertaining to their doing things is more evident than would be the case for theoretical physicists. But, for the reasons touched on above, I want to treat the study as an illustration which is also relevant for theoretical sciences. Perhaps more refined observational techniques could allow direct study of the processes pertaining to the development of theoretical concepts, but I do not know how to do this at present. Nevertheless, it is important that theoretical concepts (such as 'paradigm') do not become eroded by the assumption that what cannot be operationalized cannot be discussed.

It has of course been stated many times before (for instance in the work of Menzel)[19] that scientists tend to claim (on questionnaires) that they use informal means (and even unplanned sources) for the transmission of 'technical' information. Sociologists have treated communication as though it could be exhaustively divided into the two categories 'formal' and 'informal'. Informal communication has then been treated like a more flexibly packaged version of formal communication. I want to go beyond this and stress not only the informality of some information exchange, but also its necessary capriciousness—a symptom of the lack of organization of inarticulated knowledge into visible, discrete, and measurable units.

The study I report here is of a set of scientists in different laboratories engaged for one reason or another on the same problem—namely, the building of an operating TEA laser. There is no claim that this set represents a 'group' of scientists in any sense other than a definitional one. It is true that a few members of the set saw themselves as members of a 'club', and were linked biographically and occasionally bibliographically, but the choice of a set of scientists working on the same narrowly defined problem is not supposed to imply that they communicated more with each other than with others, or were bounded in any way other than by this particular research interest. Such a set, however, has particular methodological advantages for careful research into communication among scientists. These advantages depend upon the small size of the set involved, the contemporaneity of the scientific research, and the visibility of many of the physical parameters of the technique. Thus members of every laboratory involved could be interviewed in depth, about interactions which were comparatively recent. Visible technical variations in the laser itself could be used as a lever for investigating, at a deeper level than is usually possible, the content and quality, as well as the quantity of communications among the members. In addition, a criterion of successful transfer of the technique was readily identifiable—namely, the operation of the laser.

THE TEA LASER; THE DESIGN OF THE STUDY;
THE PARAMETERS OF THE LABORATORIES INVOLVED[20]

In the late nineteen-sixties, many laboratories throughout the world were attempting to increase the power output of gas lasers by increasing their operating pressure. Early in 1970, when no-one else had achieved successful operation at pressures above about half an atmosphere, a Canadian defence research laboratory (which I will call 'Origin') announced the 'Transversely Excited Atmospheric Pressure CO_2 laser', which soon became known as the 'TEA laser'. In fact, the device had first been operated early in 1968, and a more sophisticated version had been built by the autumn of that year; but both generations of laser were classified for two years. Since 1970, a variety of models differing in design detail have been produced by many laboratories, and two or three firms are now marketing versions of the device.

The difficulty involved in constructing high pressure gas lasers is that of producing a uniform 'pumping' discharge in the gas, rather than the arc breakdowns that usually occur in electrical discharges in gases above a few torr.[21] The solutions employed in the TEA laser involve pulsed operation, with pulse times too short to allow an arc to build up, and a discharge transverse to the lasing axis which can then be of manageable voltage. The early design solution is known as the 'Pin-Bar' laser from its electrode structure, and second generation devices are known as 'Double Discharge' lasers because the main pulse is triggered by a pre-ionization discharge. The laser is, when used for most purposes, small enough so that several can be set up in the same laboratory room. The cost of the equipment might be between £500 and £2,000, mostly for mirrors and ancillary equipment such as oscilloscopes and detectors which would be found in any laser laboratory. A pin-bar laser could be built in a few days by anybody who had built one before, but a double discharge laser might take several weeks because of the complex machining operations which can be involved. Building a TEA laser is not 'Big Science'.

In the summer of 1971, I located seven British laboratories who had built, or were building, TEA lasers. This was eighteen months after the first news of the device from Origin, and some months before any device became available commercially. The laboratories were found by a snowball technique: I asked at each location for the names of others making TEA lasers. Three other methods were used to test the efficacy of this process and to search for isolates: a questionnaire sent to every physics department in the country; a check on research applications to the Science Research Council;

50

and a search of literature citing the original papers in the field. No isolates were found, and the three methods located only three, four, and one laboratory respectively; all of these had been picked up by the snowball sample. Members of these laboratories were then interviewed in some depth. Interviews generally involved considerable semi-technical discussion and a tour of the laboratory. Subsequently, it has been possible to interview at the five North American laboratories which were involved in the transfer of the technique into Britain.[22] Interviews with the British laboratories continue, from time to time, and occasionally I have found myself playing an active role in the communication network.

The seven British laboratories involved consist of one government defence research laboratory (referred to as 'Grimbledon'), one other government-run laboratory ('Whitehall'), and five university departments of physics or applied physics ('Seawich', important as a gatekeeper; 'Baird', the only Scottish university; and 'A', 'B' and 'C'). The research of four of the laboratories involves, though not exclusively, development and improvement of the laser, while in the case of the other three concern is overridingly with the provision of a beam of radiation such as the TEA laser produces, for use in other experiments[23]. The North American laboratories (in Canada and the USA) consist of two government-run establishments ('Origin' and 'X'), one university department ('Y'), and one industrial firm's research laboratories ('Z', treated as one unit on the diagram below because they interact in a complex manner with regard to their overall external information flows.)

Patent rights to some variants of the laser have been sold to two Canadian firms by Origin and X. These firms are now manufacturing the laser, and one at least is having some success in selling the device as a scientific instrument. To date, however, only one has been sold to a British laboratory, in response to a failure in a programme of construction. In spite of their commercial interest, members of both Origin and X have published papers giving details of their lasers, read papers at conferences, and received visitors from all over the world, including Britain. For their licensees they have also provided engineers' drawings, and continuing consultancy. There is no evidence (and because the Canadians are far ahead, no rationale) for industrially- or militarily-motivated secrecy regarding TEA laser design in Britain—though this may not be the case with regard to some of the applications for which the laser is being used. Devices which may be classed as 'spin-off' from laser research have in one or two cases been commercially exploited. All the laboratories involved in this study are concerned to publish articles in the scientific journals, and at the time of writing at least two of them have publications citing the original papers in the field. Others of the group have published results using the laser, but citing papers in other fields.

Though the prime concern of this paper is not to elucidate the social structure of a group of scientists, but rather to discuss the modes of transfer of real, useable knowledge among a set of scientists, it may be worthwhile to discuss the set briefly in terms of other criteria. Firstly, where they are linked bibliographically, it is into more than one subset. A bibliographical analysis would discover separate networks centred on, for instance, gas lasers, plasma physics, and spectroscopy. These networks would themselves be likely to be much larger than the TEA laser set, and would contain members with no interest in TEA lasers. I have no information about the frequency of contact between members of the TEA set compared to that between members of these other networks. Secondly, in a discussion of types of scientific specialty, Law[24] distinguishes between 'technique- or methods-based' specialties, 'theory-based' specialties and 'subject matter' specialties. It is not appropriate to assign the TEA set into one of these classes: for one thing, it is not large enough to be termed a specialty; and the TEA set seems to contain some members whose prime concern is laser building technique, and others who are part of subject matter specialties. For instance, the plasma diagnosticians are concerned with the subject of nuclear fusion. The TEA set is a set of scientists with one problem in common, and other problems not in common.

THE INVENTION AND CONSTRUCTION OF THE TEA LASER; THE DIFFUSION OF KNOWLEDGE

The process of development and construction of TEA lasers does not consist of the logical accumulation of packages of knowledge. The construction of the first device involved trial-and-error pragmatism in the face of written and verbal assurances that the principle could not possibly work. Because of this its construction seems to have taken most people by surprise. In retrospect, however, the idea of the pin-bar laser is very simple.

Serendipity was involved in the development of the double discharge laser, which grew in a quite unforeseen way out of attempts to improve the pin-bar system. This development was described as follows:

First of all we had rows of fins instead of pins, but this didn't work too well. We thought this might be because the field uniformity was too great so we put a row of trigger wires near the fins to disturb the field uniformity. Then we started finding out things. It did improve the discharge but there were delays involved. It definitely worked differently to the rationale we had when we first made it. . . .

Even today there is no clear idea how to get this thing working properly. We are even now discovering things about how to control the performance of these devices, which are unknown . . . I have four theories (for how they work) which contradict each other. . . . The crucial part (in getting a device to operate) is in the mechanical arrangements, and how you get the things all integrated together.

In the electrical characteristics of the mechanical structures. . . .

This is all the black art that goes into building radar transmitters. . . .

Further pointers to the non-systematic element in TEA laser development can be seen in the different details of design of electrode structures in different laboratories, and in the rival claims to the efficiency and power of these. Two of the laboratories which are part of the industrial firm 'Z' are copying quite different types of double discharge laser, each with a rationale for the superiority of its own model. Sometimes there are total failures in construction. One British group (not included in my sample because they have only recently entered the field) report that they built what was, so far as they could see, an exact copy of a laser successfully operated elsewhere, which failed to work. They were at a loss to explain this, and had simply given up. Another respondent reports being particularly impressed by one design of laser because he had actually seen it working. Normally, he reports, laboratories don't demonstrate their big lasers, 'because they don't work'. They will work perhaps one day in five, unpredictably.

The Role of the Literature in the TEA Laser Group

The first article submitted to a journal from Origin in late 1969 surprisingly was rejected, because it was, according to the referee's report, not particularly interesting. Subsequently, a news conference was held at Origin, and reports that the device had been built appeared in the scientific 'news' press, in such journals as *Canadian Electronics, Laser Focus* and *New Scientist.* Most British scientists first heard that the laser had been successfully operated from a small 'note' in *New Scientist*[25] referring to a 'Plywood Laser'. (One version had been constructed using a plywood box as the lasing cavity.) The first article in the formal journals appeared six months later in *Applied Physics Letters* of 15th June 1970. This article provided more detail, but, as events proved, insufficient to enable anyone to build a TEA laser. In fact, to date, no-one to whom I have spoken has succeeded in building a TEA laser using written sources (including preprints and internal reports) as the sole source of information, though several unsuccessful attempts have been made, and there is now a considerable literature on the subject.[26] Here again, it is important to distinguish between the existence of publications and their contents,[27] for there is no doubt that at least the early articles in this field were intended as priority claims and little else.[28] Some respondents stated that the early articles would have been more misleading than helpful; another commented upon the growing trend of producing 'pseudo publications', which appear to be full scientific articles, but in reality will keep back a certain amount. A member of Origin commented:

> What you publish in an article is always enough to show that you've done it, but never enough to enable anyone else to do it. If they can do it then they know as much as you do.

It cannot be stated categorically that no laboratory could ever build a TEA laser using published sources only, for it is at least possible for the device to be re-invented without even reference to the literature.[29] It was the case, however, that in the diffusion network, as far as I have explored it, written means of communication served at best to keep scientists informed of what had been done and by whom. In most cases, this information would already be known from the conference grapevine, but in the one instance where journals were really important (first informing British workers that the Canadians had succeeded in producing the TEA laser) it was a rapid publication in the 'news' press, rather than in the formal journals, which was the source. The major point is that the transmission of skills is not done through the medium of the written word.

The Actual Process of Knowledge Diffusion

The laboratories studied here (other than Origin) actually learned to build working models of TEA lasers by contact with a source laboratory either by personal visits and telephone calls or by transfer of personnel.[30] The number of visits required depended to some extent upon the degree to which appropriate expertise already existed within the learning laboratory, but many capricious elements were also involved. Attempts at building which followed visits to successful laboratories often met with failure, and were backed up with more visits and phone calls. These failures were inevitable simply because the parameters of the device were not understood by the source laboratories themselves,[31] so that even when there was a close examination of the laser in a non-secretive atmosphere, many crucial elements might be missed. For instance, a spokesman at Origin reports that it was only previous experience that enabled him to see that the success of a laser built by another laboratory depended on the inductance of their transformer, at that time thought to be a quite insignificant element. A more remarkable example concerns the machining of the electrodes of one model of double discharge laser. Here the source laboratory provided information in the form of a set of equations for the so-called 'Rogowski profiles', along with the impression that machining tolerances must be small. The difficulties involved in making the electrodes were found by one laboratory to be insuperable. In the meantime, another British laboratory had produced the shapes roughly from templates and a filing operation, and an American laboratory had simply used lengths of aluminium banister rail, both with complete success.

Network Shape

Figure 1 shows the main transfers of information which laboratories used in building their early TEA lasers. Visits between one laboratory and another are not included where they were unproductive. The dominant feature of the diffusion network is the lack of cooperation between British laboratories, other than Grimbledon. Initially a partial explanation of this might be thought to be that the six other labs did not know of each other's interest in the TEA laser. Figure 2, however, shows which British laboratories had heard of each other in connection with TEA laser building at the time of my initial survey in the summer of 1971.[32] Eight of the possible twenty-one links in Figure 2 are missing; that is, eight of the possible fifteen links between the six laboratories remaining if Grimbledon is removed. In many cases, then, universities were ignorant that others were struggling with, or had solved, the same problems as they themselves. But even when they learned the names of the others as a result of my interview, no new contacts were arranged.[33] It is also apparent that, in their contacts with Grimbledon, the question of the names of other laboratories doing this work was not a high priority; for Grimbledon could have supplied the complete list to any of them. Hence the factor of ignorance does not seem to have much value as an explanation of the lack of cooperation discovered.

Another factor might be simply that when a laboratory has developed a useful link with another, technically senior to itself (and Grimbledon was well in advance of all the others) there may be nothing to be gained from a collaborative link with a research peer. That this was by no means the whole explanation can be shown in three ways. Firstly, there was the case of the machining of the Rogowski profile electrodes mentioned above. A loss of perhaps six months and £2,000 could have been avoided by collaboration here. There is no reason to suppose that this situation is untypical, and members of most laboratories stated in the interview that they would in general be able to learn from research peers. Secondly, there is some evidence that Grimbledon became less accessible as time went on because the number of enquiries was proving to be a burden.[34] Thirdly, in some cases, universities which were late into the field did approach laboratories less advanced than Grimbledon, but were not received in a completely open manner. Hence there is evidence that cooperation between

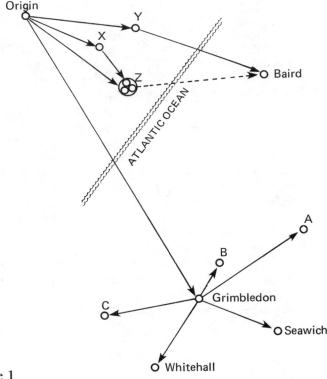

Figure 1

laboratories other than Grimbledon would have been useful, that Grimbledon would have welcomed relief from the pressure of enquiries upon itself, and that the universities were aware of these factors, both in the cases of the seekers of help and the potential sources.

Nobody reported that they were ever subject to an outright refusal of permission to visit another laboratory, and nearly everybody reported at the outset of my interview that they were 'open for anybody to look around whenever they wanted'. Tactics for maintaining secrecy—where there is no military or contractual obligation—are less forthright. Sometimes only parts of the set-up will be demonstrated. Thus one scientist reports of a visit to another laboratory:

> They showed me roughly what it looked like, but they wouldn't show me anything as to how they managed to damage mirrors. I had not a rebuff, but they were very cautious.[35]

A more subtle tactic is that of answering questions, but not actually volunteering information. This maintains the appearance of openness while many important items of information can be withheld

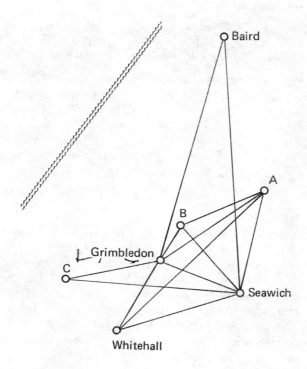

Figure 2

because their importance will not occur to the questioner.[36] One scientist put it:

> If someone comes here to look at the laser the normal approach is to answer their questions, but ... although it's in our interests to answer their questions in an information exchange, we don't give our liberty.

Another remarked succinctly:

> Let's say I've always told the truth, nothing but the truth, but not the whole truth.[37]

The major explanation for the lack of productive cooperation between the universities is to be found in their sense of competition with one another. Although this was sometimes direct competition (as in one case where an application for a research grant was turned down because of the similarity in the proposed project to that in another university), usually the feeling was more diffuse and extended to scientists working with different research goals. Several laboratories seemed to fear that all possible directions of research using the laser might be monopolized by the larger organizations. A typical comment was:

> A small laboratory like this has to be fairly careful what we say to other people who have got larger laboratories and more facilities, because they might pick up our ideas and be able to go ahead faster and we do find this kind of barrier operating. There's nothing I would like more than to be able to tell everyone everything.

The explanation of Grimbledon's unique willingness to act as the 'national gatekeeper', as it were, appears to be that it was sufficiently advanced and with sufficient resources, to be quite secure from attacks upon its prestige from British universities. Thus they were not only the laboratory best equipped to supply laser-building skills, but also the one with least to fear from doing so. This seems to account best for the persistence of the predominant star shape of the diffusion graph.

Extra-scientific Factors Involved in Information Links

As stated, I have no evidence for or against any closure or boundedness of the set or sets of communicators of which these laboratories might be construed as being part, and as the notion of social group depends on some such element,[38] the assignation is not appropriate here. Nevertheless, it is the case that all but one of the links shown on Figure 1 involved more than simply the provision of laser-building information. Of the twelve links shown, six were preceded by a history of information exchanges—about, for instance, other types of laser, electronics, or laser output detection devices. Nearly every laboratory expressed a preference for giving information only to those who had something to return. Of these six links, three also involved elements of formal organization (between government laboratories) and one (between Grimbledon and Seawich) an earlier period of colleagueship and co-authorship (at another government laboratory). Of the five remaining links all included an element of direct or indirect acquaintanceship. Seawich played an important role here as a gateway to Grimbledon.[39] One of the relations within North America was set up as a result of a member of the seeking laboratory having been best man at the wedding of a member of a different department of the source laboratory. The one relationship set up simply by the expedient of writing a letter to the source laboratory threatened to be very brittle at first, but has since been reinforced by student placements at the source.

Four of the twelve transfers of information have involved a personnel transfer—on visiting fellowships, student placements, or just periods of work at the source laboratory. Of course, the investigation of such transfers is quite outside the scope of research which depends upon frequency of contact to define relations, such as some of the work criticized earlier in the paper.

The importance of friendship relations explains in part the isolation of Baird, the Scottish laboratory, from the other British laboratories, for while they had no special friendship links across the border one of their members remarked of the American scientists from whom they obtained their information that:

> . . . these people of course had got a Ph.D. working in the same group under the same supervisor, albeit eight or ten years ago. But they're still one of our family.

In this group, then, extra-scientific factors played an important part in the setting up of scientific communication relations, and to this extent some of the findings of the researches on social groups of scientists are corroborated.

DISCUSSION AND CONCLUSION

My argument is that the nature of much scientific knowledge is such as to make it difficult to organize and hence difficult to investigate accurately with conventional sociological techniques. This argument is meant to apply to all scientific fields, but for methodological reasons one particular set (rather than a random sample) of scientists was chosen to illustrate the argument. *Ipso facto* this set is untypical of scientists in general and their area of science is untypical of science as a whole. What is more, this area is probably untypically suitable for illustrating my thesis. This is because the field is new and little of the corresponding tacit knowledge will have been learned during the scientists' apprenticeships; and because experimental work is involved, so that much tacit knowledge is embodied in visible rather than abstract objects. It may then have been difficult to produce the appropriate demonstrations using an established field, or a theoretical field, as an exemplar. Indeed, many studies, using conventional techniques[40], have suggested that theoreticians are not conscious of gaining much information from other than formal channels. Nevertheless, unless there is a serious mistake in the philosophical underpinnings of the argument, the *caveat* regarding the use of investigative techniques which treat all knowledge as homogeneously measurable must apply to theoretical science as well as experimental science,[41] for, as has been argued, a lack of consciousness of a source of knowledge is not the same as a lack of importance. The point is, as summed up by Ravetz:

> in every one of its aspects, scientific inquiry is a craft activity depending on a body of knowledge which is informal and partly tacit.[42]

I have attempted to show the complexities and uncertainties involved in the transmission of one scientific craft. These include the overt concealment of information by scientists who feel that they are in competition with others, and the effect on information transmission of personal and biographical factors which have no relation with the scientific subject matter in question, as well as the less tangible barriers which are the main concern of this paper. In the process described, publications have appeared frequently in the journals, but seemed to have had no significant information content. Even in the case of informal communication, scientists have often made use of techniques which allow an appearance of openness alongside an underlying secrecy. In cases where there has been no deliberate secrecy, a systematic transfer of knowledge has sometimes been impossible, for the source scientist has not been aware of all the relevant parameters. Thus one or both scientists involved in an interaction can be unaware of whether or not useable knowledge is being transferred. Significantly, all the laboratories which acted as sources of knowledge had completed the task of building an operating TEA laser; no-one could act as a middle man unless he was himself practised in the skill. This suggests that a participant in the flow of knowledge here was not simply a carrier of packages of information but a part of a small scientific culture.[43] In short, scientific knowledge is heterogeneous and its transmission is often capricious. This suggests that there might be a disjunction between the real sociological significance of inter-actions among scientists, and any data which can be gathered on this topic by the methods which have been conventionally used by sociologists and information scientists.[44]

I have not attempted to demonstrate that scientific specialties are not organized in social circles—though I have methodological reservations— but to suggest that the research methods used by others to demonstrate this have not yielded any structure with a significance for a sociological theory of scientific knowledge. A prerequisite of any such study must be an acknowledgement of the importance of all the elements and uses of scientific knowl-edge, not only the formal and informal elements, but the political, persuasive, and emotive, and even the intangible and unspeakable. It is possible to speak *about* that which cannot be spoken.

NOTES AND REFERENCES

An earlier draft of this paper was read at the British Sociological Association, Sociology of Science Study Group. Among others I wish to thank Colin Bell, Richard Whitley and Stephen Cotgrove for help and encouragement, and David Edge and an anonymous referee for comments which have improved the paper considerably.

1. T. S. Kuhn, *The Structure of Scientific Revolutions* (Chicago: University of Chicago Press, 1962). The concept of 'paradigm' is notoriously difficult to define, but perhaps this accounts for its fruitfulness. Kuhn himself writes:

> Given a set of necessary and sufficient conditions for identifying a theoretical entity, that entity can be eliminated from the ontology of a theory by substitution. In the absence of such rules, however, these entities are not eliminable; the theory then demands their existence. (page 197, footnote)

A complete definition of a concept is surely not a necessary prerequisite to its heuristic value.

2. Diana Crane, 'Social Structure of a Group of Scientists: A Test of the Invisible College Hypothesis', *American Sociological Review*, 34 (1969), 335—52, and *Invisible Colleges* (Chicago: University of Chicago Press, 1972); S. Crawford, 'Internal Commun-ication Among Scientists in Sleep Research', *Journal of American Society of Inform-ation Science*, 22 (1971), 301—10; J. Gaston, 'Big Science in Britain: A Sociological Study of the High Energy Physics Community' (Doctoral Dissertation, Yale University, 1969), and 'Communication and the Reward System of Science: A Study of a National Invisible College', *Sociological Review Monographs*, 18 (September 1972), 25—41.

3. A 'social circle' is a 'fuzzy edged' group whose members associate more with each other than with outsiders in respect of one or more social relation. See C. Kadushin: 'On Social Circles in Urban Life', *American Sociological Review*, 31 (1966), 786—802, and 'Power, Influence and Social Circles', *American Sociological Review*, 33 (1968), 685—99.

4.   D. J. de Solla Price, *Little Science, Big Science* (New York: Columbia University Press, 1963).

5.   The term is due to Mulkay. See M. Mulkay, *The Social Process of Innovation* (London: Macmillan, 1972). It is cited by Crane in *Invisible Colleges, op. cit.* note 2, 26.

6.   I use the term 'sociology of science' as distinct from 'sociology of scientists'. Whitley's recent article is important here. See R. Whitley, 'Black Boxism and the Sociology of Science: A Discussion of the Major Developments in the Field', *Sociological Review Monograph*, 18 (September 1972), 61—92.

7.   *Invisible Colleges, op. cit.* note 2, 1—2.

8.   *Op. cit.* note 4, 1, 4, 5.

9.   May's article showing the relative uselessness of much of the mathematical literature in one field is instructive, and many scientists will make similar comments about the literature in their own fields. See K. O. May: 'Growth and Quality of the Mathematical Literature', *Isis*, 59 (1968), 363—71.

10.   The phrase is taken from Peter Winch, *The Idea of a Social Science* (London: Routledge & Kegan Paul, 1958). Both the notion of 'tacit knowledge' (Polanyi) and the work of Kuhn are immanent in the philosophy of Wittgenstein, particularly *Philosophical Investigations* and *Remarks on the Foundations of Mathematics*. This is nicely brought out in the writings of Winch, a follower of Wittgenstein. For instance he writes:

> Imagine a biochemist making certain observations and experiments as a result of which he discovers a new germ . . . assuming that the germ theory of disease is already well established in the scientific language he speaks. Now compare with this discovery the impact made by the first . . . introduction of a concept of germ into the language of medicine. This was a much more radically new departure, involving not merely a new factual discovery within an existing way of looking at things, but a completely new way of looking at the whole problem of the causation of diseases, the adoption of new diagnostic techniques, the asking of new kinds of questions about illnesses, and so on. In short, it involved the adoption of new ways of doing things by people involved, in one way or another, in medical practice. An account of the way in which social relations in the medical profession had been influenced by this new concept would include an account of what that concept was. Conversely, the concept itself is unintelligible apart from its relation to medical practice. (*The Idea of a Social Science*, 57).

and

> 'Time' in relativity theory does not mean what it did in classical mechanics . . . (and) . . . we shall obscure the nature of this development if we think in terms of the building of new and better theories to explain one and the same set of facts; not because the facts themselves change independently, but because the scientists' criteria of relevance change. And this does not mean that earlier scientists had a wrong idea of what the facts were, they had the idea appropriate to the investigation they were conducting. ('Nature & Convention', *Proceedings of the Aristotelian Society*, 60 [May 1960], 231—52.)

11.   Winch, *op. cit.* note 10, 55—57.

12.   T. Burns, 'Models, Images and Myths', in E. G. Marquis and W. H. Gruber (eds.), *Factors in the Transfer of Technology* (Cambridge, Mass: M.I.T. Press, 1969).

13.   J. Ravetz, *Scientific Knowledge and its Social Problems* (Oxford: Oxford University Press, 1971).

14.   Bloor and Winch have clear and extended discussions of this type of point, drawn from Wittgenstein's 'Remarks on the Foundation of Mathematics'. See note 10. See also D. Bloor, 'Wittgenstein and Mannheim on the Sociology of Mathematics,' *Studies in History and Philosophy of Science*, 4, 173—91.

15.   'The Strength of Weak Ties', *American Journal of Sociology*, 78 (1973), 1360—80.

16.   The argument is not that strong links and weak links define a larger network than

strong links only—that is common sense. The argument is that weak links only define a larger network than strong links only, which is a considerable advance on common sense. In retrospect it is precisely this property of social relations which allows Kadushin's original use of the idea of 'social circle'. See Kadushin, *op. cit.* note 3, and its subsequent adoption by Crane. It is concentration on strong links which allows Crane to keep her social circles of scientists within manageable dimensions.

17. Crane, *Invisible Colleges, op. cit.* note 2, 20.

18. It being the case that some sociometric measures are sensitive to even slight inaccuracies in data collection, there are grounds for suspicion of the results of studies like Crawford's on sleep researchers. Crawford's network is defined by the question 'Who have you contacted at least three times during the last year?' Part of her analysis is in terms of cliques and isolates, though cliques may be joined or split by the addition or subtraction of only one link. Further, the choice of the number of three contacts seems entirely arbitrary. It would be interesting to know the extent to which the network topology would be changed by the substitution of, say, two or four contacts per year. The same comments apply to her analysis of the network 'core' defined by scientists who have at least three contacts per year with at least six others. See, for instance, J. C. Mitchell, 'The Concept and Use of Social Networks', in J. C. Mitchell, *Social Networks in Urban Situations* (Manchester: Manchester University Press, 1969); and P. Abell and P. Doriean, *On the Concept of Structure in Sociology* (University of Essex, 1970, unpublished mimeograph.)

19. e.g. Herbert Menzel, 'Planned & Unplanned Scientific Communication', in B. Barber and W. Hirsch (eds.), *The Sociology of Science* (Glencoe: The Free Press, 1962), 417–47.

20. Many non-sociological details are given so that the reader may judge how typical an area of science is dealt with here.

21. 1 torr = 1 mm of mercury.

22. These interviews took place in the USA and Canada, and included an interview at Origin.

23. Projects involve the use of laser radiation for damaging surfaces, producing and examining plasmas, construction of a tunable source of laser radiation, and possibly weapons and radar research.

24. J. Law, 'The Development of Specialties in Science: The Case of X-ray Protein Crystallography', *Science Studies,* 3 (1973), 275–303.

25. 22nd January, 1970.

26. Papers citing the original article have appeared at a steady rate of about thirteen per quarter.

27. See M. J. Mulkay, 'Conformity and Innovation in Science', *Sociological Review Monographs,* 18 (September 1972), 5–24, and F. Reif, 'The Competitive World of the Pure Scientist', in N. Kaplan (ed.), *Science and Society* (New York: Rand McNally & Co., 1965), 133–45.

28. This seems not to be true of more recent articles. A recent article in the *Review of Scientific Instruments* is most unusually detailed. It gives instructions for the construction of a double discharge laser which include cross sectional scale drawings, mechanical and electrical layouts, geometric coordinates, photographs and instructions for making a tool to perform the complex machining processes, and even lists of manufacturers' part numbers for the electronics, as well as for the laser body itself. One American university laboratory is attempting to build a laser following this article, but even in this case their first act was to phone the authors to make certain that the tolerances given were really correct, and as a result of this call have decided to machine the electrodes from graphite rather than aluminium, and to contract this work out. The outcome of this attempt should be interesting.

29. Grimbledon and a French laboratory were no more than six months behind

Origin in the construction of a double discharge laser, but it seems that the simpler idea of the pin-bar device had not occurred to them.

30. Of the twelve links on Figure 1 below, four involved long-term movement of persons.

31. Some versions seem to work because the trigger pulse pre-ionises the gas with a stream of electrons, and some with ultra-violet radiation.

32. Current developments have included a substantial flow of information back across the Atlantic, particularly from Grimbledon, who have perfected their own distinctive and quite successful design. In addition, the French laboratory involved in this work has had a considerable influence on Grimbledon, one other British laboratory (A) and several Transatlantic laboratories. Figure 1 shows the initial flows of information only, but in any case the situation holding between British laboratories has not changed substantially.

33. Later I deliberately fed information about useable progress in one laboratory to another. But they did not respond by trying to set up any information link.

34. One of the inventors of the device reports that this development programme was halted for a full year as a result of demands for his appearance at conferences and lectures, etc. He calculates that in the year following the announcement of the device he travelled 86,000 miles.

35. I was allowed to see the apparatus in question at this laboratory. Presumably this was because of my technical non-competitiveness.

36. Only detailed technical probing elicited the admission that this strategy was used.

37. There is probably an etiquette which governs the type of detail a visitor may ask for, and the type of record he may make of it. Memory is allowed, but photographs, for instance, are not.

38. Some of the confusion over this idea is discussed in Gaston, 'Communication and the Reward System . . .', *op. cit.* note 2, 26.

39. Seawich also introduced me to this security-conscious laboratory.

40. e.g. Gaston, 'Big Science in Britain . . .', *op. cit.* note 2.

41. I find the often-made distinction between science and technology too untidy to be useful. For an interesting discussion of this point see L. Sklair, *Organised Knowledge* (London: Hart-Davis MacGibbon, 1973), ch. 2.

42. Ravetz, *op. cit.* note 12, 103.

43. Respondents twice described failed attempts to learn to build the laser from someone who 'had all the particulars', but had not themselves built one. The point is that the unit of knowledge cannot be abstracted from the 'carrier'. The scientist, his culture and skill are an integral part of what is known.

44. I am sure this disjunction explains conclusions such as Gaston's that in a relatively slow moving scientific field, formal channels of communication would be sufficient if there were immediate publication when communications were received. Gaston, 'Big Science in Britain . . .', *op. cit.* note 2. See also Garvey and Griffith, who seem to misunderstand the significance of interpersonal contact when they write: 'Too often the only informal channel available is the inefficient and expensive one of persons seeking out a source, discovering its originator, and arranging to meet him face to face.' W. D. Garvey and B. C. Griffith, 'Scientific Communication as a Social System', *Science,* 157 (1967), 1011—16.

PART TWO

The culture of science

It is only in the last two decades that sociologists have turned to the systematic examination of scientific culture. This, of course, should have been one of the main tasks of the sociology of knowledge, a field of long standing which at one time or another attracted the interest of most of the seminal figures in the history of the social sciences. But the sociology of knowledge failed to live up to its name: it long confined itself to the realm of error and ideology, and studiously avoided anything suspected of validity.

We do possess one fine pre-war work in the sociology of knowledge tradition which considers in detail the emergence of an accepted set of scientific doctrines and techniques. Ludwik Fleck's *Genesis and Development of a Scientific Fact* (1935), recently rescued from oblivion, has rapidly become recognized as a major contribution. But that an extended study of such insight and importance was largely passed over upon its first appearance merely reinforces the point already made: there was a widespread reluctance to investigate the basis of anything considered to be genuine knowledge. Apparently it was felt that if knowledge were the product of more than reason and experience then *ipso facto* it was not knowledge at all: a successful sociological analysis of anything supposed to be knowledge could only be a successful exposé. The modern view, that the sociological study of knowledge can proceed in an entirely unrestricted, and accordingly an entirely non-pejorative way, was not so much rejected as absent from the consciousness of the times (Bloor, 1973, 1976; however, cf. Lavine, 1942).

Real interest in the nature of scientific knowledge only began to permeate sociology in the late 1960s. This was a time when idealized images of science were being called into question for a number of complex reasons, and when accounts of scientific knowledge were becoming noticeably more naturalistic and matter of fact. Sociological work at this time was both greatly encouraged by, and itself a part of, a much broader and longer-established trend, embracing a number of fields. The work of historians of science was central here. In the 1960s, inspired by a deepened professional concern with thoroughness and sound method, they were already well advanced in the production of unprecedentedly detailed accounts of many important events and episodes. This work had all the merits and all the significance of the best ethnographical fieldwork. It helped sociologists to recognize science as a form of culture like any other, and hence to realize that sociological and anthropological methods could be legitimately employed in its study (Barnes, 1974).

Equally important was the indirect effect of this historical work, which led to a thoroughgoing reassessment of the stereotypes of science sustained by historians and philosophers. Much of this reassessment turned upon the role of scientific theory. It had long been recognized that theories constituted an important part of the verbal culture of science. But theories are human inventions or constructs which go beyond the facts, and any specific body of accepted facts is formally compatible with any number of theories. Hence the existence of theories was treated as a problem rather than an interesting finding, by scholars who were keen to represent the growth of knowledge as wholly the product of reason and experience: that scientists typically adhered to one theory as *the* correct account of a set of phenomena was difficult to account for as more than a matter of custom or convention. The commonest way of circumventing this problem was to divide the verbal culture of science into two components, the factual and the theoretical, and to claim that the latter was firmly under the control of the former. Scientists were said to employ general 'rules of reason', such as any rational individual would feel bound to accept, to evaluate theories in relation to the available facts and to identify the 'best' theories (from what set of initial possibilities was something which always remained obscure).

This kind of model, which characterized judgment and evaluation in science in terms of reason and experience alone, was widely accepted for a time. But it proved difficult to reconcile with historical work which again and again showed scientists using their specific theoretical formulations as evaluative criteria in their own right, as they assessed the plausibility of other theories, the competence of experiments, and even the reliability of observations. This stimulated a re-examination of the relationship between fact and theory in scientific culture, which led to the important conclusion that a fundamental distinction between the two could not be sustained. The stereotype in which rational men kept theory firmly under the control of fact had to be replaced with one wherein statements of fact had themselves a theoretical and hence a constructed character.

This new stereotype, the merits of which have become more and more apparent as the years have gone by, became well known among sociologists through such work as that of Kuhn (1962), Toulmin (1972), Hesse (1974, 1980) and Feyerabend (1975). (An admirably succinct presentation and justification of the central point is to be found in Mary Hesse's 'Is There an Independent Observation Language?' [1970]; although Hesse's writings are the most technical and demanding of those cited above, they are extremely penetrating and amply repay the concentrated and extended study they require. Hesse's exposition of the domain and the character of theoretical discourse is also an indication of the scope and the possibilities of sociological investigation.)

From the 1960s, then, circumstances encouraged the empirical sociological study of scientific culture, and it duly got under way. As has already been described in the General Introduction, the work of the historian of science T.S. Kuhn proved a particularly valuable resource in this development because it described with unusual explicitness the character of scientific fields as sub-cultures with their own specific apparatus for socialization and social control. Indeed some important sociological studies were content simply to exemplify this point (that of Wynne, 1976, is a particularly fine example, cf. also Wynne, 1977, 1979).

The continuing fascination of Kuhn's *Structure of Scientific Revolutions* (1962) has to some extent diverted attention from important sociological themes in his later work. Here, Kuhn continually returns to the problem of how (scientific) concepts are directly attached to nature. He makes two points of great significance. First, he stresses the obvious but often forgotten fact that direct connections between concepts and phenomena cannot be adequately specified by verbal rules: they have to be *shown*. With everyday terms like 'red', or the 'kind' terms of natural history, acts of ostension are required. In the physical sciences, a complex procedure has generally to be followed, and competence in its performance acquired: the proper use of a concept is made intelligible by showing how it functions in such a procedure or *exemplar* (cf. Kuhn, 1970, 1974), which in most cases is a sequence of actions culminating in the solution of a specific concrete problem or puzzle. Thus, the use of even the most elementary scientific terms is to be thought of primarily as a competence passed on from generation to generation.

Secondly, Kuhn notes how the perceptions of *sameness* characteristic of any culture are essentially a matter of how phenomena and/or procedures are grouped under specific terms. To set two particulars under a single concept is to assert their similarity: to be able to set particulars under concepts as other competent members do is to be competent with, and to sustain the *similarity relations* characteristic of, one's culture (Kuhn, 1974).

Here is a point central to the modern sociology of knowledge. Every society uses concepts to group particulars together, as kinds of things, or kinds of property, or types of event. But particulars are never self-evidently identical to each other: people can always find resemblances between them, and differences; and hence grounds for taking them as of the same kind, and grounds for taking them as of different kinds. It follows that there are innumerable viable ways of classifying particulars and setting out what is the same as what, much as there are innumerable ways of making a map of a piece of terrain. There is no privileged single description, the only one to 'correspond to reality'. Thus, the *particular* conceptions of sameness — the particular scheme of things — favoured by a given culture is of great sociological interest as a *conventional* system specific to the culture. But that the scheme is conventional in no way implies that it is empirically inadequate or irrationally believed, any more than these things are implied of a map. The different schemes of things accepted by different cultures and sub-cultures — by aliens, ancestors, deviants and experts — are all for the most part successfully used; in this sense they are all viable, and acceptable to physical nature, even as they are also culturally specific, conventional, sustained by authority and custom, and hence of sociological interest. Hence, sociologists now feel entitled to analyse all such schemes as conventional structures, without any thought that such analysis should count as criticism or refutation. Today, there is not the slightest semantic impropriety in referring to the conventional character of knowledge.

In making a selection from Kuhn's work our primary objective has to be to convey its basic themes, and in particular its analysis of what is involved in normal science. Since many readers will wish to read the 1962 book in its entirety, and that work is in any case very much an integrated whole, we have chosen an alternative account, which illustrates Kuhn's conception of normal science as puzzle-solving in the specific context of quantitative measurement. This account has the incidental advantage of offering some useful concrete illustrations of the intimate relationship between theory and the observations produced by measurement. And it even presents a simple example of what is involved in establishing a relationship of sameness. Kuhn asks what makes the results of measurement sufficiently similar to the predictions of a scientific theory for agreement to be said to exist between the two. He suggests that there is nothing in the nature of things or in the power of reason sufficient to establish such a relation: the scientific community itself establishes and exemplifies what it regards as reasonable agreement, so that its basic assessments of the fit between theory and experience have a conventional character.

Much of the empirical research currently being produced by European sociologists of science centres upon the similarity relations established in various scientific sub-cultures (BN). Courses which concentrate upon the internal features of science will find it worthwhile to give some prominence to this material, together with related analytical work in sociology and philosophy and the relevant parts of Kuhn's work — in particular his extraordinarily insightful 'Second Thoughts on Paradigms' (1974): here is an extended and coherent literature which has sought, not entirely unsuccessfully, to get to the heart of some basic sociological problems. We ourselves, with our restricted space, must be content to stress only the basic point, which is of importance to what comes later: the knowledge of a culture always has a conventional character, and at the verbal level this is manifest in the specific relations of sameness created by its use of its concepts.

Since knowledge has a conventional character, it is important to say something of the nature of convention. It is as easy as it is incorrect to think of scientists being inducted into the conventions, the 'theoretical framework' as is sometimes said, of a physical science, and then carrying out scientific research in the way implied by the conventions. This is to treat conventions as general rules with logical implications in specific cases. It is to imagine that a scientific culture needs conventions simply as a starting point, rather as a formal system needs axioms: once the conventions, or the axioms, are provided, logical implication generates the elaborate structure which is the culture as a whole, or the formal system as a whole. On this view, conventions are necessary because 'reason' can only lead from one statement to another, and is unable to indicate premises, the starting points for inference. On this view, the scientist can be socialized into appropriate conventions or premises, and then left to do science by himself as a rational individual.

One of the several virtues of Harry Collins' contribution is that it challenges this way of understanding convention. Collins identifies a group of physicists in agreement on conventional wisdom in their field, who nonetheless disagree upon its implications. In the course of attempts to replicate an experiment which had allegedly detected gravity waves, they offered diverse specifications for a gravity wave detector and diverse standards for evaluating one detector against another. They based their specifications and standards upon existing physical knowledge, but achieved a consensus upon neither. Accordingly, they were unable to agree upon what counted as a replication of the initial experiment, and unable to settle upon a single account of the phenomenon of gravitational radiation.

The materials presented by Collins do not invite analysis as the results of careless or incompetent reasoning on the part of the scientists involved. Nonetheless, the results of any particular empirical study can always be challenged in this way. Accordingly, it is worth emphasizing that knowledge can have no logical implications in particular cases, even for ideally rational men. Again it is a matter of the sameness relation. Every particular case differs in detail from every other and can never be conclusively pronounced identical to any other, or identical in any attribute to any other. Hence, nothing can be unproblematically deduced from a rule or law, concerning any particular case, because there is always the undetermined matter of whether the case falls under the rule or the law — that is, whether it is *the same as* or *different to* those instances which have already been labelled as falling under the rule or law. Formally, this matter of similarity or difference arises at every point of use of a concept, and has to be settled at every point by the using community. Concept application is inherently open-ended (Wittgenstein, 1953; Bloor, 1973; Hesse, 1974). And Collins' excellent study can usefully be read as an illustration of this general characteristic of concept application.

Recognition of the open-ended character of concept application is of major significance. Once it is admitted, it is necessary to discount any relationship of logical implication, or sufficient determination, between the general and the particular — between, for example, laws, rules, statements of values, theories on the one hand, and actions, beliefs, inferences on the other. Thus, convention can no longer count as a determinant of actions: convention is simply what collective actions amount to. Although it may occasionally be conceptualized by single individuals as an external force, careful analysis at the collective level shows that convention, far from being a determinant of action, is its product. And to assert the conventional character of knowledge is to assert precisely this of knowledge. Verbal knowledge is our memory of previously accepted assertions of sameness, and is accordingly the product of past usage. This knowledge cannot determine our future usage or provide us with logical implications about specific future states of affairs, because the future development of relations of sameness cannot be fixed in this way. We ourselves must decide how to develop and extend existing sameness relations.

This general point, that a body of knowledge can never provide deductive implications for specific cases, is worth some stress. It is crucial to understanding the role and the basis of expertise that we recognize the inability of the expert to make unproblematic deductions about the specific cases upon which he is obliged to pronounce. And the converse is, of course, equally important: a clash between an expert's pronouncements in a specific case and what actually ensues never ever entails the falsehood of the theory used by the expert.

What, however, is to be made of the many actual instances where recommendations are presented as the inevitable, logically compelling, conclusions of a deductive chain of reasoning? If no such chain is indeed compelling, then the construction or selection of the chain, and its investment with compelling force, can only be analyzed as empirical phenomena, as part of the contingent activity of agents themselves. A careful and extensively illustrated account of how this can be done is available in David Bloor's *Knowledge and Social Imagery* (1976). In the short extract included here, Bloor identifies deductive connections as 'interpretative afterthoughts'. Far from offering a stable basis for reasoning, these connections are laid out and continually adjusted to conform with the informal *inductive* modes of inference which they are used to systematize and legitimate. Formal thought stands in an *ex post facto* relationship to informal thought, and the construction of the relationship is itself a phenomenon of sociological interest. (Note that Bloor chooses to illustrate his account with an episode drawn from mathematics, and that it does indeed serve very well. Earlier in his book Bloor has offered arguments for treating mathematics, and even logic, as forms of empirical knowledge, thus bringing their culture within the scope of the arguments adumbrated above. 'Even in mathematics, that most cerebral of all subjects, it is men who govern ideas not ideas which control men. The reason for this is simple. Ideas grow by having something actively added to them. They are constructed and manufactured in order that they may be extended. These extensions of meaning and use do not pre-exist. The future uses and expanded meanings of concepts, there entailments, are not present inside them in embryo' [Bloor 1976, p. 139].)

We are now in a position to propose that all cases of language use without exception have the character of contingent actions, and all accepted modes of language use have a conventional character in the profound sense that we have outlined. Another way of putting this is to say that the successful use of a concept, even within the culture of science, is a contingent achievement. We use this form of words to call to mind the work of ethnomethodologists, which is precisely designed to display the 'achieved' character of concept application. Ethnomethodologists are currently beginning to embark upon detailed analyses of aspects of scientific culture, and a number of such studies are likely to appear over the next few years. It is still too early to say what can be expected of work of this kind, and in particular whether it will provide any insight into the technical procedures of the culture of science. Some interesting recent work on the accounting of scientific discoveries does however give an indication of how valuable ethnomethodological techniques can be for some analytical tasks (Brannigan, 1979). And the indirect influence of ethnomethodology is clearly discernible in many of the studies now becoming available of 'the social construction of scientific knowledge' — particularly in those which take it as a matter of course that the construction of knowledge is identical with the construction of reality (Latour and Woolgar, 1979; see also BN).

In the few years since their inception modern sociological studies of scientific culture have been able to consolidate the intuition that knowledge has a conventional character, and to throw some light upon the way that it is conventional. This is a fair achievement which has had an important unifying effect upon theories of knowledge and has involved the production of a valuable body of empirical material. It is not, however, an achievement with which sociologists should rest content. The constitutively theoretical character of scientific knowledge is all too easy to exploit: it allows sociologists readily to construct any number of displays of the 'socially constructed' character of science. And the persistent scepticism directed towards sociological accounts of knowledge means that a healthy market will continue to exist for constructions of this kind. This makes it all the more important that the temptation to take such displays as ends in themselves is recognized and resisted. The claim that knowledge is a social construct does not, in itself, amount to much. It must be treated as a way of raising questions, not as an answer to a question, still less as a slogan. It merely realigns curiosity: only a numbed curiosity would actually be satisfied by it.

When we show that concept application is formally, logically and semantically open-ended, what we do is to create a problem:

what then does produce closure in particular cases? As far as logic and semantics are concerned any existing instance of concept application could have had any outcome; as a matter of fact it had one specific outcome; what then gave rise to that outcome? Only by being willing to address this kind of question can we hope for any understanding of the course of cultural change and the growth of knowledge.

One possibility currently being explored here is to adopt a thorough-going pragmatism or instrumentalism, and to try to relate the specific use of a concept at every point to objectives and interests residing in the using community (Barnes, 1977, 1981, 1982). This sociological instrumentalism invokes interests and objectives as causes of specific instances of the proper, 'reasonable' use of concepts, and not as sources of bias or distortion. Their role is to explain why one judgment of sameness is made rather than another, why A is said to be the same as B rather than C. Although some sameness relations may be more routinely acceptable than others, or more immediately perceptible than others, none are more 'natural' or more 'reasonable'. There is no specific relation which can be identified as what people would accept in the absence of goals and interests. Goals and interests merely lift one possible development of culture above other formally equivalent alternatives. The citation of objectives and interests is accordingly explanatory but non-pejorative.

Few empirical studies have so far applied this instrumentalist form of analysis to the esoteric culture of science (see Barnes and Shapin, 1979; MacKenzie and Barnes, 1979; MacKenzie, 1981), and results must be evaluated cautiously. Nonetheless, the approach deserves representation here, and Andrew Pickering's contribution serves admirably. It deals with the assimilation of a recognized new physical phenomenon, recently generated by the ever more powerful technical resources of experimental physics. Two proposals for the theoretical treatment of the phenomenon were set out and taken seriously, and within a period of two years opinion moved decisively in favour of one alternative. Yet, as Pickering shows, no formal or logical compulsion inspired this decision of the physicists: both alternatives were logically defensible. What then did, as a matter of fact, produce the decision? Pickering suggests that the chosen alternative incorporated the new phenomenon in the way which was most compatible with physicists' existing practice. But this was not a matter of disinterested logical calculation: it was not that a minimum of well-confirmed or well-corroborated results and hypotheses were discarded. It was a matter of scientists' having vested interests in a range of professional practices and in the continued use of these practices. They were convinced by the option which best maintained current practice and created opportunity for the extension of that practice. That is, they developed their culture in a way systematically related to *social* interests. And in this case it so happened that nearly all the diverse vested social interests in practice of the various groups and specialties of theoretical physics favoured the one theoretical alternative over the other. On the other hand, this citation of vested professional interests is in no way a condemnation of the scientific judgments involved in this case. The commitment of our physicists to their professional practice is what maintains our whole current system of physical knowledge. And who will say where a better interpretation of physical nature is to be found?

4

Thomas S. Kuhn

Normal measurement
and reasonable agreement

TEXTBOOK MEASUREMENT

To a very much greater extent than we ordinarily realize, our image of physical science and of measurement is conditioned by science texts. In part that influence is direct: textbooks are the sole source of most people's firsthand acquaintance with the physical sciences. Their indirect influence is, however, undoubtedly larger and more pervasive. Textbooks or their equivalent are the unique repository of the finished achievements of modern physical scientists. It is with the analysis and propagation of these achievements that most writings on the philosophy of science and most interpretations of science for the nonscientist are concerned. As many autobiographies attest, even the research scientist does not always free himself from the textbook image gained during his first exposures to science.[1]

I shall shortly indicate why the textbook mode of presentation must inevitably be misleading, but let us first examine that presentation itself. Since most participants in this conference have already been exposed to at least one textbook of physical science, I restrict attention to the schematic tripartite summary in the following figure. It displays, in the upper left, a series of theoretical and 'law-like' statements, $(x) \phi_i (x)$, which together constitute the theory of the science being described.[2] The center of Figure 1 represents the logical and mathematical equipment employed in manipulating the theory. 'Lawlike' statements from the upper left are to be imagined fed into the hopper at the top of the machine together with certain 'initial conditions' specifying the situation to which the theory is being applied. The crank is then turned; logical and mathematical operations are internally performed; and numerical predictions for the application at hand emerge from the

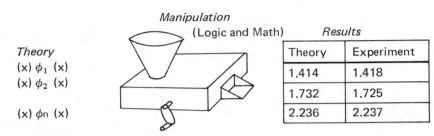

Theory	Experiment
1.414	1.418
1.732	1.725
2.236	2.237

Manipulation (Logic and Math) *Results*

Theory

$(x)\ \phi_1\ (x)$
$(x)\ \phi_2\ (x)$

$(x)\ \phi_n\ (x)$

Figure 1

chute at the front of the machine. These predictions are entered in the left-hand column of the table. The right-hand column contains the numerical results of actual measurements, placed there so that they may be compared with the predictions derived from the theory. Most texts of physics, chemistry, astronomy, and the like contain many data of this sort, though they are not always presented in tabular form. Some of you will, for example, be more familiar with equivalent graphical presentations.

The table is of particular concern, for it is there that the results of measurement appear explicitly. What may we take to be the significance of such a table and of the numbers it contains? I suppose that there are two usual answers: the first, immediate and almost universal; the other, perhaps more important, but very rarely explicit.

Most obviously the results in the table seem to function as a test of theory. If corresponding numbers in the two columns agree, the theory is acceptable; if they do not, the theory must be modified or rejected. This is the function of measurement as confirmation, here seen emerging, as it does for most readers, from the textbook formulation of a finished scientific theory. For the time being I shall assume that some such function is also regularly exemplified in normal scientific practice and can be isolated in writings whose purpose is not exclusively pedagogic. At this point we need only notice that on the question of practice, textbooks provide no evidence whatsoever. No textbook ever included a table that either intended or managed to infirm the theory the text was written to describe. Readers of current science texts accept the theories there expounded on the authority of the author and the scientific community, not because of any tables that these texts contain. If the tables are read at all, as they often are, they are read for another reason.

I shall inquire for this other reason in a moment but must first remark on the second putative function of measurement, that of

76

exploration. Numerical data like those collected in the right-hand column of our table can, it is often supposed, be useful in suggesting new scientific theories or laws. Some people seem to take for granted that numerical data are more likely to be productive of new generalizations than any other sort. It is that special productivity, rather than the function of measurement in confirmation, that probably accounts for Kelvin's dictum's being inscribed on the façade at the University of Chicago.[3]

If you cannot measure, your knowledge is meagre and unsatisfactory.

It is by no means obvious that our ideas about this function of numbers are related to the textbook schema outlined in the figure opposite, yet I see no other way to account for the special efficacy often attributed to the results of measurement. We are, I suspect, here confronted with a vestige of an admittedly outworn belief that laws and theories can be arrived at by some process like 'running the machine backwards.' Given the numerical data in the 'Experiment' column of the table, logico-mathematical manipulation (aided, all would now insist, by 'intuition') can proceed to the statement of the laws that underlie the numbers. If any process even remotely like this is involved in discovery—if, that is, laws and theories are forged directly from data by the mind—then the superiority of numerical to qualitative data is immediately apparent. The results of measurement are neutral and precise; they cannot mislead. Even more important, numbers are subject to mathematical manipulation; more than any other form of data, they can be assimilated to the semimechanical textbook schema.

I have already implied my scepticism about these two prevalent descriptions of the function of measurement. In the next two sections each of these functions will be further compared with ordinary scientific practice. But it will help first critically to pursue our examination of textbook tables. By doing so I would hope to suggest that our stereotypes about measurement do not even quite fit the textbook schema from which they seem to derive. Though the numerical tables in a textbook do not there function either for exploration or confirmation, they are there for a reason. That reason we may perhaps discover by asking what the author of a text can mean when he says that the numbers in the 'Theory' and 'Experiment' column of a table 'agree.'

At best the criterion must be agreement within the limits of accuracy of the measuring instruments employed. Since computation from theory can usually be pushed to any desired number of decimal places, exact or numerical agreement is impossible in principle. But anyone who has examined the tables in which the results of theory and experiment are compared must recognize that agreement of this more modest sort is rather rare. Almost always the application of a physical theory involves some approximation (in fact, the plane is *not* 'frictionless,' the vacuum is *not* 'perfect,' the atoms are *not* 'unaffected' by collisions), and the theory is not therefore expected to yield quite precise results. Or the construction of the instruments may involve approximations (e.g., the 'linearity' of vacuum tube characteristics) that cast doubt upon the significance of the last decimal place that can be unambiguously read from their dial. Or it may simply be recognized that, for reasons not clearly understood, the theory whose results have been tabulated or the instrument used in measurement provides only estimates. For one of these reasons or another, physical scientists rarely expect agreement quite within instrumental limits. In fact, they often distrust it when they see it. At least on a student lab report overly close agreement is usually taken as presumptive evidence of data manipulation. That no experiment gives quite the expected numerical result is sometimes called the 'fifth law of thermodynamics.'[4] The fact that, unlike some other scientific laws, it has acknowledged exceptions does not diminish its utility as a guiding principle.

It follows that what scientists seek in numerical tables is not usually 'agreement' at all, but what they often call 'reasonable agreement.' Furthermore, if we now ask for a criterion of 'reasonable agreement,' we are literally forced to look in the tables themselves. Scientific practice exhibits no consistently applied or consistently applicable external criterion. 'Reasonable agreement' varies from one part of science to another, and within any part of science it varies with time. What to Ptolemy and his immediate successors was reasonable agreement between astronomical theory and observation was to Copernicus incisive evidence that the Ptolemaic system must be wrong.[5] Between the times of Cavendish (1731–1810) and Ramsay (1852–1916), a similar change in accepted chemical criteria for 'reasonable agreement' led to the study of the noble gases.[6] These divergences are typical and they are matched by those between contemporary branches of the scientific community. In parts of spectroscopy 'reasonable agreement' means agreement in the first six or eight left-hand digits in the numbers of a table of wave lengths. In the theory of solids, by contrast, two-place agreement is often considered very good indeed. Yet there are parts of astronomy in which any search for even so limited an agreement must seem utopian. In the theoretical study of stellar magnitudes agreement to a multiplicative factor of ten is often taken to be 'reasonable.'

Notice that we have now inadvertently answered the question from which we began. We have, that is, said what 'agreement' between theory and experiment must mean if that criterion is to be drawn from the tables of a science text. But in doing so we have gone full circle. I began by asking, at least by implication, what characteristic the numbers of the table must exhibit if they are to be said to 'agree.' I now conclude that the only possible criterion is the mere fact that they appear, together with the theory from which they are derived, in a professionally accepted text. When they appear in a text, tables of numbers drawn from theory and experiments cannot demonstrate anything but 'reasonable agreement.' And even that they demonstrate only by tautology, since they alone provide the definition of 'reasonable agreement' that has been accepted by the profession. That, I think, is why the tables are there: they define 'reasonable agreement.' By studying them, the reader learns what can be expected of the theory. An acquaintance with the tables is part of an acquaintance with the theory itself. Without the tables, the theory would be essentially incomplete. With respect to measurement, it would be not so much untested as untestable. Which brings us very close to the conclusion that, once it has been embodied in a text—which for present purposes means, once it has been adopted by the profession—no theory is recognized to be testable by any quantitative tests that it has not already passed.[7]

Perhaps these conclusions are not surprising. Certainly they should not be. Textbooks are, after all, written some time after the discoveries and confirmation procedures whose *outcomes* they record. Furthermore, they are written for purposes of pedagogy. The objective of a textbook is to provide the reader, in the most economical and easily assimilable form, with a statement of what the contemporary scientific community believes it knows and of the principal uses to which that knowledge can be put. Information about how that knowledge was acquired (discovery) and about why it was accepted by the profession (confirmation) would at best be excess baggage. Though including that information would almost certainly increase the 'humanistic' values of the text and might conceivably breed more flexible and creative scientists, it would inevitably detract from the ease of learning the contemporary scientific language. To date only the last objective has been taken seriously by most writers of textbooks in the natural sciences. As a result, though texts may be the right place for philosophers to discover the logical structure of finished scientific theories, they are more likely to mislead than to help the unwary individual who asks about productive methods. One might equally appropriately go to a college language text for an authoritative characterization of the corresponding literature. Language texts, like science texts, teach how to *read* literature, not how to create or evaluate it. What signposts they supply to these latter points are most likely to point in the wrong direction.[8]

MOTIVES FOR NORMAL MEASUREMENT

These considerations dictate our next step. We must ask how measurement comes to be juxtaposed with laws and theories in science texts. Furthermore, we must go for an answer to the journal literature, the medium through which natural scientists report their own original work and in which they evaluate that done by others.[9] Recourse to this body of literature immediately casts doubt upon one implication of the standard textbook schema. Only a miniscule fraction of even the best and most creative measurements undertaken by natural scientists are motivated by a desire to discover new quantitative regularities or to confirm old ones. Almost as small a fraction turn out to have had either of these effects. There are a few that did so, and I shall have something to say about them in the next two sections. But it will help first to discover just why these exploratory and confirmatory measurements are so rare. In this section and most of the next, I therefore restrict myself to measurement's most usual function in the normal practice of science.[10]

Probably the rarest and most profound sort of genius in physical science is that displayed by men who, like Newton, Lavoisier, or Einstein, enunciate a whole new theory that brings potential order to a vast number of natural phenomena. Yet radical reformulations of this sort are extremely rare, largely because the state of science very seldom provides occasion for them. Moreover, they are not the only truly essential and creative events in the development of scientific knowledge. The new order provided by a revolutionary new theory in the natural sciences is always overwhelmingly a *potential* order. Much work and skill, together with occasional genius, are required to make it *actual*. And actual it must be made, for only through the process of actualization can occasions for new theoretical reformulations be discovered. The bulk of scientific practice is thus a complex and consuming mopping-up operation that consolidates the ground made available by the most recent theoretical breakthrough and that provides essential preparation for the breakthrough to follow. In such mopping-up operations, measurement has its overwhelmingly most common scientific function.

Just how important and difficult these consolidating operations can be is indicated by the present state of Einstein's general theory of relativity. The equations embodying that theory have proved so difficult to apply that (excluding the limiting case in which the equations reduce to those of special relativity) they have so far yielded only three predictions that can be compared with observation.[11] Men of undoubted genius have totally failed to develop others, and the problem remains worth their attention. Until it is solved, Einstein's general theory remains a largely fruitless, because unexploitable, achievement.[12]

Undoubtedly the general theory of relativity is an extreme case, but the situation it illustrates is typical. Consider, for a somewhat more extended example, the problem that engaged much of the best eighteenth-century scientific thought, that of deriving testable numerical predictions from Newton's three laws of motion and from his principle of universal gravitation. When Newton's theory was first enunciated late in the seventeenth century, only his third law (equality of action and reaction) could be directly investigated by experiment, and the relevant experiments applied only to very special cases.[13] The first direct and unequivocal demonstrations of the second law awaited the development of the Atwood machine, a subtly conceived piece of laboratory apparatus that was not invented until almost a century after the appearance of the *Principia*.[14] Direct quantitative investigations of gravitational attraction proved even more difficult and were not presented in the scientific literature until 1798.[15] Newton's first law cannot, to this day, be directly compared with the results of laboratory measurement, though developments in rocketry make it likely that we have not much longer to wait.

It is, of course, direct demonstrations, like those of Atwood, that figure most largely in natural science texts and in elementary laboratory exercises. Because they are simple and unequivocal, they have the greatest pedagogic value. That they were not and could scarcely have been available for more than a century after the publication of Newton's work makes no pedagogic difference. At most it only leads us to mistake the nature of scientific achievement.[16] But if Newton's contemporaries and successors had been forced to wait that long for quantitative evidence, apparatus capable of providing it would never have been designed. Fortunately there was another route, and much eighteenth-century scientific talent followed it. Complex mathematical manipulations, exploiting all the laws together, permitted a few other sorts of prediction capable of being compared with quantitative observation, particularly with laboratory observations of pendula and with astronomical observations of the motions of the moon and planets. But these predictions presented another and equally severe problem, that of essential approximations.[17] The suspensions of laboratory pendula are neither weightless nor perfectly elastic; air resistance damps the motion of the bob; besides, the bob itself is of finite size, and there is the question of which point of the bob should be used in computing the pendulum's length. If these three aspects of the experimental situation are neglected, only the roughest sort of quantitative agreement between theory and observation can be expected. But determining how to reduce them (only the last is fully eliminable) and what allowance to make for the residue are themselves problems of the utmost difficulty. Since Newton's day much brilliant research has been devoted to their challenge.[18]

The problems encountered when applying Newton's laws to astronomical prediction are even more revealing. Since each of the bodies in the solar system attracts and is attracted by every other, precise prediction of celestial phenomena demanded, in Newton's day, the application of his laws to the simultaneous motions and interactions of eight celestial bodies. (These were the sun, moon, and six known planets. I ignore the other planetary satellites.) The result is a mathematical problem that has never been solved exactly. To get equations that could be solved, Newton was forced to the simplifying assumption that each of the planets was attracted only by the sun, and the moon only by the earth. With this assumption, he was able to derive Kepler's famous laws, a wonderfully convincing argument for his theory. But deviation of planets from the motions predicted by Kepler's laws is quite apparent to simple quantitative telescopic observation. To discover how to treat these deviations by Newtonian theory, it was necessary to devise mathematical estimates of the 'perturbations' produced in a basically Keplerian orbit by the interplanetary forces neglected in the initial derivation of Kepler's laws. Newton's mathematical genius was displayed at its best when he produced a first crude estimate for the perturbation of the moon's motion caused by the sun. Improving his answer and developing similar approximate answers for the planets exercised the greatest mathematicians of the eighteenth and early nineteenth centuries, including Euler, Lagrange, Laplace, and Gauss.[19] Only as a result of their work was it possible to recognize the anomaly in Mercury's motion that was ultimately to be explained by Einstein's general theory. That anomaly had previously been hidden within the limits of 'reasonable agreement.'

As far as it goes, the situation illustrated by quantitative application of Newton's laws is, I think, perfectly typical. Similar examples could be produced from the history of the corpuscular, the wave, or the quantum mechanical theory of light, from the history of electromagnetic theory, quantitative chemical analysis, or any other of the numerous natural scientific theories with quantitative implications. In each of these cases, it proved immensely difficult to find many problems that permitted quantitative comparison of theory and observation. Even when such problems were found, the highest scientific talents were often required to invent apparatus, reduce perturbing effects, and estimate the allowance to be made for those that remained. This is the sort of work that most physical scientists do most of the time *insofar as their work is quantitative*. Its objective is, on the one hand, to improve the measure of 'reasonable agreement' characteristic of the theory in a given application and, on the other, to open up new areas of application and establish new measures of 'reasonable agreement' applicable to them. For anyone who finds mathematical or manipulative puzzles challenging, this can be fascinating and intensely rewarding work. And there is always the remote possibility that it will pay an additional dividend: something may go wrong.

Yet unless something does go wrong—a situation to be explored in a later section—these finer and finer investigations of the quantitative match between theory and observation cannot be described as attempts at discovery or at confirmation. The man who is successful proves his talents, but he does so by getting a result that the entire scientific community had anticipated someone would someday achieve. His success lies only in the explicit demonstration of a *previously implicit* agreement between theory and the world. No novelty in nature has been revealed. Nor can the scientist who is successful in this sort of work quite be said to have 'confirmed' the theory that guided his research. For if success in his venture 'confirms' the theory, then failure ought certainly to 'infirm' it, and nothing of the sort is true in this case. Failure to solve one of these puzzles counts only against the scientist; he has put in a great deal of time on a project whose outcome is not worth publication; the conclusion to be drawn, if any, is only that his talents were not adequate to it. If measurement ever leads to discovery or to confirmation, it does not do so in the most usual of all its applications.

THE EFFECTS OF NORMAL MEASUREMENT

There is a second significant aspect of the normal problem of measurement in natural science. So far we have considered why scientists usually measure; now we must consider the results that they get when they do so. Immediately another stereotype enforced by textbooks is called in question. In textbooks the numbers that result from measurement usually appear as the archetypes of the 'irreducible and stubborn facts' to which the scientist must, by struggle, make his theories conform. But in scientific practice, as seen through the journal literature, the scientist often seems rather to be struggling with facts, trying to force them into conformity with a theory he does not doubt. Quantitative facts cease to seem simply 'the given.' They must be fought for and with, and in this fight the theory with which they are to be compared proves the most potent weapon. Often scientists cannot get numbers that compare well with theory until they know what numbers they should be making nature yield.

Part of this problem is simply the difficulty in finding techniques and instruments that permit the comparison of theory with quantitative measurements. We have already seen that it took almost a century to invent a machine that could give a straightforward quantitative demonstration of Newton's second law. But the machine that Charles Atwood described in 1784 was not the first instrument to yield quantitative information relevant to that law. Attempts in this direction had been made ever since Galileo's description of his classic inclined plane experiment in 1638.[20] Galileo's brilliant intuition had seen in this laboratory device a way of investigating how a body moves when acted upon only by its own weight. After the experiment he announced that measurement of the distance covered in a measured time by a sphere rolling down the plane confirmed his prior thesis that the motion was uniformly accelerated. As reinterpreted by Newton, this result exemplified the second law for the special case of a uniform force. But Galileo did not report the numbers he had gotten, and a group of the best scientists in France announced their total failure to get comparable results. In print they wondered whether Galileo could himself have tried the experiment.[21]

In fact, it is almost certain that Galileo did perform the experiment. If he did, he must surely have gotten quantitative results that seemed to him in *adequate* agreement with the law ($s = \frac{1}{2}at^2$) that he had shown to be a consequence of uniform acceleration. But anyone who has noted the stopwatches or electric timers, and the long planes or heavy flywheels needed to perform this experiment in modern elementary laboratories may legitimately suspect that Galileo's results were not in *unequivocal* agreement with his law. Quite possibly the French group looking even at the same data would have wondered how they could seem to exemplify uniform acceleration. This is, of course, largely speculation. But the speculative element casts no doubt upon my present point: whatever its source, disagreement between Galileo and those who tried to repeat his experiment was entirely natural. If Galileo's generalization had not sent men to the very border of existing instrumentation, an area in which experimental scatter and disagreement about interpretation were inevitable, then no genius would have been required to make it. His example typifies one important aspect of theoretical genius in the natural sciences—it is a genius that leaps ahead of the facts, leaving the rather different talent of the experimentalist and instrumentalist to catch up. In this case catching up took a long time. The Atwood machine was designed because, in the middle of the eighteenth century, some of the best Continental scientists still wondered whether acceleration provided the proper measure of force. Though their doubts derived from more than measurement, measurement was still sufficiently equivocal to fit a variety of different quantitative conclusions.[22]

The preceding example illustrates the difficulties and displays the role of theory in reducing scatter in the results of measurement. There is, however, more to the problem. When measurement is insecure, one of the tests for reliability of existing instruments and manipulative techniques must inevitably be their ability to give results that compare favorably with existing theory. In some parts of natural science, the adequacy of experimental technique can be judged only in this way. When that occurs, one may not even speak of 'insecure' instrumentation or technique, implying that these could be improved without recourse to an external theoretical standard.

For example, when John Dalton first conceived of using chemical measurements to elaborate an atomic theory that he had initially drawn from meteorological and physical observations, he began by searching the existing chemical literature for relevant data. Soon he realized that significant illumination could be obtained from those groups of reactions in which a single pair of elements, for example, nitrogen and oxygen, entered into more than one chemical combination. If his atomic theory were right, the constituent molecules of these compounds should differ only in the ratio of the number of whole atoms of each element that they contained. The three oxides of nitrogen might, for example, have molecules N_2O, NO, and NO_2, or they might have some other similarly simple arrangement.[23] But whatever the particular arrangements, if the weight of nitrogen were the same in the samples of the three oxides, then the weights of oxygen in the three samples should be related to each other by simple whole-number proportions. Generalization of this principle to all groups of compounds formed from the same group of elements produced Dalton's law of multiple proportions.

Needless to say, Dalton's search of the literature yielded some data that, in his view, sufficiently supported the law. But—and this is the point of the illustration—much of the then extant data did not support Dalton's law at all. For example, the measurements of the French chemist Proust on the two oxides of copper yielded, for a given weight of copper, a weight ratio for oxygen of 1.47:1. On Dalton's theory the ratio ought to have been 2:1, and Proust is just the chemist who might have been expected to confirm the prediction. He was, in the first place, a fine experimentalist. Besides, he was then engaged in a major controversy involving the oxides of copper, a controversy in which he upheld a view very close to Dalton's. But, at the beginning of the nineteenth century, chemists did not know how to perform quantitative analyses that displayed multiple proportions. By 1850 they had learned, but only by letting Dalton's theory lead them. Knowing what results they should expect from chemical analyses, chemists were able to devise techniques that got them. As a result chemistry texts can now state that quantitative analysis confirms Dalton's atomism and forget that, historically, the relevant analytic techniques are based upon the very theory they are said to confirm. Before Dalton's theory was announced, measurement did not give the same results. There are self-fulfilling prophecies in the physical as well as in the social sciences.

That example seems to me quite typical of the way measurement responds to theory in many parts of the natural sciences. I am less sure that my next, and far stranger, example is equally typical, but colleagues in nuclear physics assure me that they repeatedly encounter similar irreversible shifts in the results of measurement.

Very early in the nineteenth century, P. S. de Laplace, perhaps the greatest and certainly the most famous physicist of his day, suggested that the recently observed heating of a gas when rapidly compressed might explain one of the outstanding numerical discrepancies of theoretical physics. This was the disagreement, approximately 20 per cent, between the predicted and measured values of the speed of sound in air—a discrepancy that had attracted the attention of all Europe's best mathematical physicists since Newton had first pointed it out. When Laplace's suggestion was made, it defied numerical confirmation (note the recurrence of this typical difficulty), because it demanded refined measurements of the thermal properties of gases, measurements that were beyond the capacity of apparatus designed for measurements on solids and liquids. But the French academy offered a prize for such measurements, and in 1813 the prize was won by two brilliant young experimentalists, Delaroche and Berard, men whose names are still cited in contemporary scientific literature. Laplace soon made use of these measurements in an indirect theoretical computation of the speed of sound in air, and the discrepancy between theory and measurement dropped from 20 per cent to 2.5 per cent, a recognized triumph in view of the state of measurement.[24]

But today no one can explain how this triumph can have occurred. Laplace's interpretation of Delaroche and Berard's figures made use of the caloric theory in a region where our own science is quite certain that that theory differs from directly relevant quantitative experiment by about 40 per cent. There is, however, also a 12 per cent discrepancy between the measurements of Delaroche and Berard and the results of equivalent experiments today. We are no longer able to get their quantitative result. Yet, in Laplace's perfectly straightforward and essential computation from the theory, these two discrepancies, experimental and theoretical, cancelled to give close final agreement between the predicted and measured speed of sound. We may not, I feel sure, dismiss this as the result of mere sloppiness. Both the theoretician and the experimentalists involved were men of the very highest caliber. Rather we must here see evidence of the way in which theory and experiment may guide each other in the exploration of areas new to both.

These examples may enforce the point drawn initially from the examples in the last section. Exploring the agreement between theory and experiment into new areas or to new limits of precision is a difficult, unremitting, and, for many, exciting job. Though its object is neither discovery nor confirmation, its appeal is quite sufficient to consume almost the entire time and attention of those physical scientists who do quantitative work. It demands the very best of their imagination, intuition, and vigilance. In addition —when combined with those of the last section—these examples may show something more. They may, that is, indicate why new laws of nature are so very seldom discovered simply by inspecting the results of measurements made without advance knowledge of those laws. Because most scientific laws have so few quantitative points of contact with nature, because investigations of those contact points usually demand such laborious instrumentation and approximation, and because nature itself needs to be forced to yield the appropriate results, the route from theory or law to measurement can almost never be travelled backward. Numbers gathered without some knowledge of the regularity to be expected almost never speak for themselves. Almost certainly they remain just numbers.

This does not mean that no one has ever discovered a quantitative regularity merely by measuring. Boyle's law relating gas pressure with gas volume, Hooke's law relating spring distortion with applied force, and Joule's relationship between heat generated, electrical resistance, and electric current were all the direct results of measurement. There are other examples besides. But, partly just because they are so exceptional and partly because they never occur until the scientist measuring knows *everything but* the particular form of the quantitative result he will obtain, these exceptions show just how improbable quantitative discovery by quantitative measurement is. The cases of Galileo and Dalton—men who intuited a quantitative result as the simplest expression of a qualitative conclusion and then fought nature to confirm it—are very much the more typical scientific events. In fact, even Boyle did not find his law until both he and two of his readers had suggested that precisely that law (the simplest quantitative form that yielded the observed qualitative regularity) ought to result if the numerical results were recorded.[25] Here, too, the quantitative implications of a qualitative theory led the way.

One more example may make clear at least some of the prerequisites for this exceptional sort of discovery. The experimental search for a law or laws describing the variation with distance of the forces between magnetized and between electrically charged bodies began in the seventeenth century and was actively pursued through the eighteenth. Yet only in the decades immediately preceding Coulomb's classic investigations of 1785 did measurement yield even an approximately unequivocal answer to these questions. What made the difference between success and failure seems to have been the belated assimilation of a lesson learned from a part of Newtonian theory. Simple force laws, like the inverse square law for gravitational attraction, can generally be expected only between mathematical points or bodies that approximate them. The more complex laws of attraction between gross bodies can be derived from the simpler law governing the attraction of points by summing all the forces between all the pairs of points in the two bodies. But these laws will seldom take a simple mathematical form unless the distance between the two bodies is large compared with the dimensions of the attracting bodies themselves. Under these circumstances the bodies will behave as points, and experiment may reveal the resulting simple regularity.

Consider only the historically simpler case of electrical attractions and repulsions.[26] During the first half of the eighteenth century—when electrical forces were explained as the results of effluvia emitted by the entire charged body—almost every experimental investigation of the force law involved placing a charged body a measured distance below one pan of a balance and then measuring the weight that had to be placed in the other pan to just overcome the attraction. With this arrangement of apparatus, the attraction varies in no simple way with distance. Furthermore, the complex way in which it does vary depends critically upon the size and material of the attracted pan. Many of the men who tried this technique therefore concluded by throwing up their hands; others suggested a variety of laws including both the inverse square and the inverse first power; measurement had proved totally equivocal. Yet it did not have to be so. What was needed and what was gradually acquired from more qualitative investigations during the middle decades of the century was a more 'Newtonian' approach to the analysis of electrical and magnetic phenomena.[27] As this evolved, experimentalists increasingly sought not the attraction between bodies but that between point poles and point charges. In that form the experimental problem was rapidly and unequivocally resolved.

This illustration shows once again how large an amount of theory is needed before the results of measurement can be expected to make sense. But, and this is perhaps the main point, when that much theory is available, the law is very likely to have been guessed without measurement. Coulomb's result, in particular, seems to have surprised few scientists. Though his measurements were necessary to produce a firm consensus about electrical and magnetic attractions—they had to be done; science cannot survive on guesses—many practitioners had already concluded that the law of attraction and repulsion must be inverse square. Some had done so by simple analogy to Newton's gravitational law; others by a more elaborate theoretical argument; still others from equivocal data. Coulomb's law was very much 'in the air' before its discoverer turned to the problem. If it had not been, Coulomb might not have been able to make nature yield it.

Two possible misinterpretations of my argument must now be set aside. First, if what I have said is right, nature undoubtedly responds to the theoretical predispositions with which she is approached by the measuring scientist. But that is not to say either that nature will respond to any theory at all or that she will ever respond very much. Re-examine, for a historically typical example, the relationship between the caloric and dynamical theory of heat. In their abstract structures and in the conceptual entities they presuppose, these two theories are quite different and, in fact, incompatible. But, during the years when the two vied for the allegiance of the scientific community, the theoretical predictions that could be derived from them were very nearly the same[28] If they had not been, the caloric theory would never have been a widely accepted tool of professional research nor would it have succeeded in disclosing the very problems that made transition to the dynamical theory possible. It follows that any measurement which, like that of Delaroche and Berard, 'fit' one of these theories must have 'very nearly fit' the other, and it is only within the experimental spread covered by the phrase 'very nearly' that nature proved able to respond to the theoretical predisposition of the measurer.

That response could not have occurred with 'any theory at all.' There are logically possible theories of, say, heat that no sane scientist could ever have made nature fit, and there are problems, mostly philosophical, that make it worth inventing and examining theories of that sort. But those are not our problems, because those merely 'conceivable' theories are not among the options open to the practicing scientist. His concern is with theories that seem to fit what is known about nature, and all these theories, however different their structure, will necessarily seem to yield very similar predictive results. If they can be distinguished at all by measurements, those measurements will usually strain the limits of existing experimental techniques. Furthermore, within the limits imposed by those techniques, the numerical differences at issue will very often prove to be quite small. Only under these conditions and within these limits can one expect nature to respond to preconception. On the other hand, these conditions and limits are just the ones typical in the historical situation.

If this much about my approach is clear, the second possible misunderstanding can be dealt with more easily. By insisting that a quite highly developed body of theory is ordinarily prerequisite to fruitful measurement in the physical sciences, I may seem to have implied that in these sciences theory must always lead experiment and that the latter has at best a decidedly secondary role. But that implication depends upon identifying 'experiment' with 'measurement,' an identification I have already explicitly disavowed. It is only because significant quantitative comparison of theories with nature comes at such a late stage in the development of a science that theory has seemed to have so decisive a lead. If we had been discussing the *qualitative* experimentation that dominates the earlier developmental stages of a physical science and that continues to play a role later on, the balance would be quite different. Perhaps, even then, we would not wish to say that experiment is prior to theory (though experience surely is), but we would certainly find vastly more symmetry and continuity in the ongoing dialogue between the two. Only some of my conclusions about the role of measurement in physical science can be readily extrapolated to experimentation at large.

NOTES AND REFERENCES

1. This phenomenon is examined in more detail in my monograph, *The Structure of Scientific Revolutions,* to appear when completed as vol. 2, no. 2, in the *International Encyclopedia of Unified Science.* Many other aspects of the textbook image of science, its sources and its strengths, are also examined in that place.

2. Obviously not all the statements required to constitute most theories are of this particular logical form, but the complexities have no relevance to the points made here. R. B. Braithwaite, *Scientific Explanation* (Cambridge, 1953) includes a useful, though very general, description of the logical structure of scientific theories.

3. Professor Frank Knight, for example, suggests that to social scientists the 'practical meaning [of Kelvin's statement] tends to be: "If you cannot measure, measure anyhow" ' (*Eleven Twenty-Six,* p. 169).

4. The first three laws of thermodynamics are well known outside the trade. The "fourth law" states that no piece of experimental apparatus works the first time it is set up. We shall examine evidence for the fifth law below.

5. T. S. Kuhn, *The Copernician Revolution* (Cambridge, Mass., 1957), pp. 72–76, 135–43.

6. William Ramsay, *The Gases of the Atmosphere: The History of Their Discovery* (London, 1896), chaps. 4 and 5.

7. To pursue this point would carry us far beyond the subject of this paper, but it should be pursued because, if I am right, it relates to the important contemporary controversy over the distinction between analytic and synthetic truth. To the extent that a scientific theory must be accompanied by a statement of the evidence for it in order to have empirical meaning, the full theory (which includes the evidence) must be analytically true. For a statement of the philosophical problem of analyticity see W. V. Quine, 'Two Dogmas of Empiricism' and other essays in *From a Logical Point of View* (Cambridge, Mass., 1953). For a stimulating, but loose, discussion of the occasionally analytic status of scientific laws, see N. R. Hanson, *Patterns of Discovery* (Cambridge, 1958), pp. 93–118. A new discussion of the philosophical problem, including copious references to the controversial literature, is Alan Pasch, *Experience and the Analytic: A Reconsideration of Empiricism* (Chicago, 1958).

8. The monograph cited in note 1 will argue that the misdirection supplied by science texts is both systematic and functional. It is by no means clear that a more accurate image of the scientific processes would enhance the research efficiency of physical scientists.

9. It is, of course, somewhat anachronistic to apply the terms 'journal literature' and 'textbooks' in the whole of the period I have been asked to discuss. But I am concerned to emphasize a pattern of professional communication whose origins at least can be found in the seventeenth century and which has increased in rigor ever since. There was a time (different in different sciences) when the pattern of communication in a science was much the same as that still visible in the humanities and many of the social sciences, but in all the physical sciences this pattern is at least a century gone, and in many of

them it disappeared even earlier than that. Now all publication of research results occurs in journals read only by the profession. Books are exclusively textbooks, compendia, popularizations, or philosophical reflections, and writing them is a somewhat suspect, because non-professional, activity. Needless to say this sharp and rigid separation between articles and books, research and non-research writings, greatly increases the strength of what I have called the textbook image.

10.   Here and elsewhere in this paper I ignore the very large amount of measurement done simply to gather factual information. I think of such measurements as specific gravities, wave lengths, spring constants, boiling points, etc., undertaken in order to determine parameters that must be inserted into scientific theories but whose numerical outcome those theories do not (or did not in the relevant period) predict. This sort of measurement is not without interest, but I think it widely understood. In any case, considering it would too greatly extend the limits of this paper.

11.   These are the deflection of light in the sun's gravitational field, the precession of the perihelion of Mercury, and the red shift of light from distant stars. Only the first two are actually quantitative predictions in the present state of the theory.

12.   The difficulties in producing concrete applications of the general theory of relativity need not prevent scientists from attempting to exploit the scientific viewpoint embodied in that theory. But, perhaps unfortunately, it seems to be doing so. Unlike the special theory, general relativity is today very little studied by students of physics. Within fifty years we may conceivably have totally lost sight of this aspect of Einstein's contribution.

13.   The most relevant and widely employed experiments were performed with pendula. Determination of the recoil when two pendulum bobs collided seems to have been the main conceptual and experimental tool used in the seventeenth century to determine what dynamical 'action' and 'reaction' were. See A. Wolf, *A History of Science, Technology, and Philosophy in the Sixteenth and Seventeenth Centuries,* new ed. prepared by D. McKie (London, 1950), pp. 155, 231–35; and R. Dugas, *La mécanique au xviiᵉ siècle* (Neuchâtel, 1954), pp. 283–98; and *Sir Isaac Newton's Mathematical Principles of Natural Philosophy and His System of the World,* ed. F. Cajori (Berkeley, 1934), pp. 21–28. Wolf (p. 155) describes the third law as 'the only *physical* law of the three.'

14.   See the excellent description of this apparatus and the discussion of Atwood's reasons for building it in Hanson, *Patterns of Discovery,* pp. 100–102 and notes to these pages.

15.   A. Wolf, *A History of Science, Technology, and Philosophy in the Eighteenth Century,* 2d ed. revised by D. McKie (London, 1952), pp. 111–13. There are some precursors of Cavendish's measurements of 1798, but it is only after Cavendish that measurement begins to yield consistent results.

16.   Modern laboratory apparatus designed to help students study Galileo's law of free fall provides a classic, though perhaps quite necessary, example of the way pedagogy misdirects the historical imagination about the relation between creative science and measurement. None of the apparatus now used could possibly have been built in the seventeenth century. One of the best and most widely disseminated pieces of equipment, for example, allows a heavy bob to fall between a pair of parallel vertical rails. These rails are electrically charged every 1/100th of a second, and the spark that then passes through the bob from rail to rail records the bob's position on a chemically treated tape. Other pieces of apparatus involve electric timers, etc. For the historical difficulties of making measurements relevant to this law, see below.

17.   All the applications of Newton's laws involve approximations of some sort, but in the following examples the approximations have a quantitative importance that they do not possess in those that precede.

18.   Wolf (*Eighteenth Century,* pp. 75–81) provides a good preliminary description of this work.

19.   Ibid., pp. 96—101. William Whewell, *History of the Inductive Sciences,* rev. ed., 3 vols. (London, 1847), 2:213—71.

20.   For a modern English version of the original see Galileo Galilei, *Dialogues Concerning Two New Sciences,* trans. Henry Crew and A. De Salvio (Evanston and Chicago, 1946), pp. 171—72.

21.   This whole story and more is brilliantly set forth in A. Koyré, 'An Experiment in Measurement,' *Proceedings of the American Philosophical Society* 97 (1953): 222—37.

22.   Hanson, *Patterns of Discovery,* p. 101.

23.   This is not, of course, Dalton's original notation. In fact, I am somewhat modernizing and simplifying this whole account. It can be reconstructed more fully from A. N. Meldrum, 'The Development of the Atomic Theory: (1) Berthollet's Doctrine of Variable Proportions,' *Manchester Memoirs* 54 (1910): 1—16; and '(6) The Reception accorded to the Theory advocated by Dalton,' ibid. 55 (1911): 1—10; L. K. Nash, *The Atomic Molecular Theory,* Harvard Case Histories in Experimental Science, case 4 (Cambridge, Mass., 1950); and 'The Origins of Dalton's Chemical Atomic Theory,' *Isis,* 47 (1956): 110—16. See also the useful discussions of atomic weight scattered through J. R. Partington, *A Short History of Chemistry,* 2d ed. (London, 1951).

24.   T. S. Kuhn, 'The Caloric Theory of Adiabatic Compression,' *Isis* 49 (1958): 132—40.

25.   Marie Boas, *Robert Boyle and Seventeenth-Century Chemistry* (Cambridge, 1958), p. 44.

26.   Much relevant material will be found in Duane Roller and Duane H. D. Roller, *The Development of the Concept of Electric Charge: Electricity from the Greeks to Coulomb,* Harvard Case Histories in Experimental Science, case 8 (Cambridge, Mass., 1954), and in Wolf, *Eighteenth Century,* pp. 239—50, 268—71.

27.   A fuller account would have to describe both the earlier and the later approaches as 'Newtonian.' The conception that electric force results from effluvia is partly Cartesian but, in the eighteenth century its *locus-classicus* was the aether theory developed in Newton's *Opticks.* Coulomb's approach and that of several of his contemporaries depends far more directly on the mathematical theory in Newton's *Principia.* For the differences between these books, their influence in the eighteenth century, and their impact on the development of electrical theory, see I. B. Cohen, *Franklin and Newton: An Inquiry into Speculative Newtonian Experimental Science and Franklin's Work in Electricity as an Example Thereof* (Philadelphia, 1956).

28.   Kuhn, 'The Caloric Theory of Adiabatic Compression.'

5

H.M. Collins

The replication of
experiments in physics[1]

INTRODUCTION

The epistemological stance which informs this paper is well expressed in the following passage taken from McHugh's *On the Failure of Positivism*.[2]

> We must accept that there are no adequate grounds for establishing criteria of truth except the grounds that are employed to grant or concede it—truth is conceivable only as a socially organized upshot of contingent courses of linguistic, conceptual and social behaviour (p. 329).

This view is difficult to hold consistently, for much of our knowledge seems so 'solid' as to require a justification in terms other than those which describe human actions. This, I think, is because when we consider the grounds of knowledge, we do it within an environment filled with objects of knowledge which are already established. To speak figuratively, it is as though epistemologists are concerned with the characteristics of ships (knowledge) in bottles (validity) while living in a world where all ships are already in bottles with the glue dried and the strings cut. A ship *within* a bottle is a natural object in this world, and because there is no way to reverse the process, it is not easy to accept that the ship was ever just a bundle of sticks.[3] Most perceptions of the grounds of knowledge are structured in ways derived from this perspective. Theories, histories and epistemologies are often subject in some degree to what one might call the 'ethnocentricism of now'.[4]

My contention is that because of the institutionalization of the production of scientific truth, it is possible to make a partial escape from the cultural determinism of current knowledge, in studies of science. It is actually possible to locate this process in scientific laboratories, in letters, conferences and conversations. It is possible

94

then to perform a kind of automatic phenomenological bracketing for ideas and facts, by looking at them while they are being formed, before they have become 'set' as part of anyone's natural (scientific) world. In short, the contemporaneous study of contemporary scientific developments, I suggest can provide an entry to a sociology of knowledge which is less subject than usual to some philosophical and methodological problems. It should be expected that this will generate a picture of science in which the figurative 'ships' are still being built by human actors, to be subsequently erected in their bottles by a trick invented and worked by human actors also—a picture of science which is much more relativistic.[5]

The rest of the paper illustrates this point by reference to a study of a set of scientists who are currently engaged in the search for a new 'natural' phenomenon—gravitational radiation. The initial focus is on the replication of experiments in the field and associated transfers of knowledge between the scientists. The models of scientific knowledge and its transfer which give rise to the interpretation offered, were originally set up following previous research into scientific communication, in particular a study of an area of laser research.[6] These two models will be explained first.

TWO MODELS OF SCIENTIFIC KNOWLEDGE AND ITS TRANSFER

The algorithmical model

The first model assumes knowledge is reducible to something like the programme of a digital computer. Thus it is implied that there is a finite series of unambiguous instructions which can be formulated, transferred, and when correctly followed will enable a scientist to copy another's experiment exactly. Such a series of instructions is known as an 'algorithm', hence this model is the 'algorithmical model'.

According to the algorithmical model we would expect a scientist wishing to replicate an experiment to search his available information sources for the algorithm, follow it, produce an exact copy of the original apparatus, and *ipso facto* identical results. Where this process does not take place the model implies that the explanation might be found in the incompleteness of the algorithm in available information sources, and it follows that the sociologist should look for the causes of this incompleteness. And indeed, the implicit use of such an approach has led to the discovery of competitive secretiveness, incompleteness of the journal article and so on.[7]

The enculturational model

The second model of the transfer of knowledge is generated by noting that the meaning of an *'exact copy of the original'* is itself problematical. The suggestion that two copies are the 'same' presupposes a cultural limitation on the list of variables which might be measured for each experiment.

The implication of this point for the algorithmical model can perhaps be seen in the following passage taken from Kurt Vonnegut's *'Slaughterhouse 5'*, where the hero Billy Pilgrim finds himself on the strange planet 'Tralfamador'.[8]

> One of the biggest moral bombshells handed to Billy by the Tralfamadorians, incidentally, had to do with sex on Earth. They said their flying saucer crews had identified no fewer than seven sexes on Earth, each essential to reproduction. Again Billy couldn't possibly imagine what five of those seven sexes had to do with the making of a baby, since they were sexually active only in the fourth dimension.
>
> The Tralfamadorians tried to give Billy clues that would help him imagine sex in the invisible dimension. They told him that there could be no Earthling babies without male homosexuals. There *could* be babies without female homosexuals. There couldn't be babies without women over sixty-five years old. There *could* be babies without men over sixty-five years old. There couldn't be babies without other babies who had lived an hour or less after birth. And so on.
>
> It was gibberish to Billy (p. 78).

The moral is that there is no way for us to know whether or not we possess *the* algorithm for sexual reproduction. The additional parameters suggested by the Tralfamadorians are quite compatible with our knowledge of sex. (Indeed they have the advantage of explaining otherwise puzzling failures of conception, if these are taken as occasions when one or other of the five unknown sexes is not performing properly.) What is more, there is an infinity of other parameters which it would be possible to posit as relevant and which would be compatible with our knowledge. So, there is an infinity of other possible algorithms, and no *formal* way of selecting the right one (if indeed there is one).[9] We, however, being within the culture of Western Earth 'know' which parameters are legitimately acceptable, and which are gibberish in the explanation of conception. *Ipso-facto* we know when cases are anomalous (virgin births, unexplained sterility).

Thinking in these terms, the problem becomes one of explaining the successful copies of experiments rather than the failures. The model which seems most appropriate is one which involves the transmission of a *culture* which legitimates and limits the parameters requiring control in the experimental situation, *without necessarily formulating, enumerating or understanding them,* and which *ipso-facto* generates the set of anomalous experiments (failures which can't be explained by uncontrolled legitimate parameters).

An important corollary of this model is that *the transfer of knowledge is not directly monitorable.*[10]

The enculturational model fits many of the findings of previous studies better than the algorithmical model.[11] For instance, in the laser study, it was found that a scientist who was to learn to build a successful copy of a laser nearly always needed to spend some time in close interaction with another who had already built one. Written sources, including internal laboratory reports, were never (in this study) sufficient, nor was contact with a 'middle man' who had seen and talked about the laser but had not built a copy himself. Furthermore, even prolonged interaction with a successful scientist was not a *sufficient* condition for the transfer of knowledge. Indeed, there were cases where one scientist, with the best of intentions, tried to teach another the skill required, but the second found his apparently isomorphous model an inexplicable failure. In such cases neither scientist knew that the appropriate knowledge had not been transferred until *after* the second scientist failed.[12]

Both model and findings point to the conclusion that *neither an observer nor a scientist himself* can tell whether or not that scientist has the knowledge appropriate for replicating another's experiment by any means such as the observation of the scientist's apparatus, or the monitoring of channels of information, or the construction of an inventory of the scientist's knowledge before he has tried to make use of it. The only way to know if any culture has been absorbed is in successful interaction with native members. In the case of scientists, successful interaction means producing results acceptable to the scientific community, for instance, doing an experiment which works. *A fortiori* this is the only indicator of successful knowledge transfer between scientists available to a sociologist.[13]

To repeat, the only way that a scientist can know that he has the knowledge required to replicate an experiment properly, is to make a replica which does what is counted as 'working'. This is also the only way for a sociologist to know this. In most scientific fields such an indicator is readily visible, but the conclusion will be seen to have great significance for the understanding of communication among the scientists in the field to be described.

THE DETECTION OF GRAVITATIONAL RADIATION

The sociological method followed in this area (as in the other areas which I am investigating—including the laser area), involves the collection of names of workers by snowball sampling starting from a familiar laboratory, followed by depth interviews. An important feature of the interviews is the extent to which they are built around detailed *technical* discussion of the experiment and scientists' interactions rather than straightforward sociometric questioning. By the end of 1972 I had interviewed members of all the laboratories in Britain and the U.S.A. engaged in building or running artificial antennae for the detection of gravitational radiation. (I only interviewed some of those who had run experiments using the Earth as an antenna—nearly, *but not quite all* such experiments were uncontentious failures anyway because of the high level of noise in the Earth's crust.) Of the thirteen experimental establishments visited, nine were university departments of physics (including the three British ones); one was a department of geophysics, and the others were the research laboratories of three large American firms.[14]

Most of the scientists interviewed in this field agreed that the existence of gravitational waves (the gravitational equivalent of electromagnetic radiation) is predicted by Einstein's general theory. They all agreed that the amount of energy required was far too great for the generation of detectable gravitational radiation on Earth within the foreseeable future. Also they were agreed that astronomical catastrophes such as the collapse of a star, should produce gravitational radiation. Up to about 1969 however, only one scientist had actually thought it worthwhile to try and detect this cosmic radiation because the resulting flux passing by the Earth was expected to be so small as to make terrestrial detection almost impossible. One British scientist speaking of the first experimenter in the field remarked: 'It has always struck me as a remarkable human phenomenon that anyone could address himself to this extraordinarily difficult problem . . . ploughing a lonely furrow with very little financial support'.

To that extent, then, the gravitational radiation field is not ordinary science, which is not to say immediately that it is not normal science. Indeed, in so far as the finding of gravitational waves would be a confirmation of the general theory of relativity, this field could be seen as doing no more than filling out the relativistic paradigm. Further, the measurements required while being of an unprecedented accuracy are seen *by some* as being extensions of techniques that have been often used before. What is extraordinary however is that the originator not only claims to have detected gravitational radiation, but claims to have detected it in such great quantity as to make it very difficult to reconcile with current cosmological theories.

The first researcher (I will refer to him as 'O') commenced work in 1957–8, and he reported positive results some ten to twelve years later in 1969.[15] The original experiment involves suspending an aluminium cylinder or bar in a vacuum, insulating it as far as possible from all known disturbing forces, and observing the residual disturbances in the cylinder. The output trace from the cylinder will show fluctuations due to, for instance, seismic and electromagnetic disturbances, Brownian motion in the atoms of the aluminium, noise in the electronics, etc. When these are subtracted, by reference to the control equipment or by statistical techniques, certain residual peaks remain, and these are attributed to the effects on the bar of gravitational radiation. (The effects are minute: a change of length in the region 2×10^{-14} cms.—one-tenth of the radius of an electron—in a cylinder one and a half meters long may need to be observed.) Other claims which seem to support the extra-terrestrial origin of the disturbances are that they can be seen simultaneously on two or more detectors separated by hundreds or thousands of miles,[16] and more recently evidence has accrued that peaks in activity are correlated with sidereal time rather than solar time, suggesting that the galactic centre rather than the solar system is the source of the radiation.[17]

It may not be long before the scientific community decides that the claims of the originator are completely spurious, or on the other hand, revolutionary. When that happens, and a new natural element in the scientific world has been constructed, the following section of my paper will look quaint. That is what is particularly interesting about writing it *now* before the solid existence of the facts clouds the look of contingency about their origins. Even at the time of writing, this is beginning to happen. Two or three of the secondary laboratories have had their instruments working long enough to consider that they are entitled to claim non-confirmation of the original findings, and none have claimed to have confirmed them. The *New Scientist* has recently headlined a short piece 'Another Nail in the Coffin of Gravity Waves'. But in fact, the counter evidence seems to amount to little in formal terms.

SCIENTISTS' ACTIVITIES IN THE FIELD

In other fields scientists often deliberately block the flow of communication to others because of the competitive pressures they experience. My original hypothesis was that communication would be more free in the Gravitational Radiation field because the scientists would feel themselves to be members of a revolutionary 'outgroup', and this would produce solidarity. This does not seem to be the case however. On the one hand various groups of secondary experimenters feel themselves to be in competition with each other to be 'first to be second' with observations of the effect, or in competition with everyone, 'O' included, to be first to be able to claim that the observations are incorrect. This competitive pressure definitely makes some scientists reticent about giving details of their experiments to others. Also many of the scientists claim that 'O' has not been forthcoming with his data and experimental designs, though this is denied at source. It would not be appropriate here to go into the evidence for and against this latter issue, or the reasons posited. In any case, while this may be a matter of historical interest, it is not of immediate sociological relevance. What is significant is that detailed communication with 'O' is not important to the secondary scientists in the field, because *they are not interested* in building what they perceive as isomorphic models of the detecting apparatus.[18] With the exception of scientists working on simultaneity experiments, and one case of administratively enforced cooperation, all the scientists I spoke to are building detectors which differ in obvious ways, not only from each other's, but also from that of the originator.

Now common sense and previous research would lead one to expect that scientists competing to be the first to replicate an experiment, would go to some lengths to discover the details of the original in order to copy it as exactly as possible. For instance, in the previously mentioned laser study, scientists' first efforts were always directed towards building what they perceived as exact copies of successful models. Workers in the gravity wave field explained their lack of interest in producing carbon copies of the originator's apparatus in several ways. Some said that an exact copy could gain them no prestige. If it confirmed the first researcher's findings, it would do nothing for *them,* but would win the Nobel prize for *him,* while on the other hand, if it disconfirmed the results there would be nothing positive to show for their work. But, if their apparatus was better in some way, in the case of positive results, they would be ahead of the field, and in the case of negative findings they could be seen as being better experimenters than the first researcher. Another type of explanation is typified in the remarks of three different scientists:

> I would feel kind of ridiculous building an exact copy because people would say—well, it's exactly that. Just a copy, no ideas in it.

> We're like anyone else, we like to do things first or better or by ourselves. It's more satisfying.

> . . . . . in some respects the least creative and the least interesting thing to do is just to build a carbon copy of somebody's apparatus . . . most people in this field are an expert in some part of it. They are either an electronics expert or they are an astrophysicist or they are a low temperature expert. . . . With regard to that part of the experiment in which they are an expert, they usually come to the conclusion that they can do it better. . . . So that when Smith and Jones build their apparatus they are going to put their own electronics into it, because Jones is one of the world's top experts in electronics.

> So what you copy is the part that you are not an expert on and the part you are an expert on you build something you think is better.

He added, significantly for the argument that will follow:

> But it may or may not be true that it is better. . . . This is a problem.

Other possible explanations include the methodological rationale, that a new effect is more powerfully demonstrated by a variety of methods than by one method.[19] (Two scientists said that they did not want to build the same pitfalls or artefacts into their apparatus as the originator.) Again, if the effect is established more refined devices might be more appropriate as gravitational telescopes, to give good time resolution (and hence directional resolution), and to see more distant sources.

No doubt these reasons and rationales (and no doubt it is possible to find others) go some way to explain the actions of the scientists in this field, and indeed, may explain them completely. However, I would like to offer an alternative and possibly more fundamental interpretation. This interpretation presents itself initially in the form of a problem of sociological methodology.

It was argued above that within the context of the enculturational model of the transmission of scientific knowledge, the only indicator available for the sociologist that such a transfer had taken place, was that a secondary scientist could produce results acceptable to the scientific community. However, in the gravity wave field, the nature of 'result acceptable to the scientific community' has not yet been established. *What counts as a 'working gravity wave detector' is still a contentious matter.* Now, if the scientists in the field do not agree about what constitutes a 'good gravity wave detection experiment', the sociologist is left with no index to show when the knowledge required to build a 'good gravity wave detector' has passed from one scientist to another. For the same reason *the scientists themselves* do not know what constitutes such knowledge, and so long as they are not sure that the originator's device is detecting gravity waves, that is, so long as they are not sure that *he has* built a 'good gravity wave detector,' there is no special reason for them to try to build one isomorphic with his. What is more, if their own devices do not 'detect gravity waves' this does not necessarily identify them as ineffective scientists (as would the building of a non-functioning laser for instance), for they *may* legitimately deny that the gravity waves are there to be found.

I will illustrate the claim that scientists do not agree about what constitutes a good experiment in this field with sets of material from interviews.

The *first* set of extracts shows the variation in scientists' opinion regarding the value of others' experimental set-ups and reported results. In each case, three scientists at different establishments are reporting on the experiment of a fourth.

Experiment W

Scientist (a)   . . . that's why the W thing though it's very compli-cated has certain attributes so that if they see something, it's a little more believable . . . They've really put some thought into it. . . .

Scientist (b)   They hope to get very high sensitivity but I don't believe them frankly. There are more subtle ways round it than brute force. . . .

Scientist (c)   I think that the group at . . W . . . . . are just out of their minds.

Experiment X

Scientist (i)   . . . he is at a very small place . . . [but] . . . I have looked at his data, and he certainly has some interesting data.

Scientist (ii)   I am not really impressed with his experimental capabilities so I would question anything he has done more than I would question other people's.

Scientist (iii)   That experiment is a bunch of shit!

Experiment Y

Scientist (1)   Y's results do seem quite impressive. They are sort of very business-like and look quite authoritative. . . .

Scientist (2)   My best estimate of his sensitivity, and he and I are good friends . . . is . . . [low] . . . and he has just got no chance [of detecting gravity waves].

Scientist (3)   If you do as Y has done and you just give your figures to some . . . girls and ask them to work that out, well, you don't know anything. You don't know whether those girls were talking to their boyfriends at the time.

Experiment Z

Scientist (I)   Z's experiment is quite interesting, and shouldn't be ruled out just because the . . . group can't repeat it.

Scientist (II)   I am very unimpressed with the Z affair.

Scientist (III) Then there's Z. Now the Z thing is an out and out fraud!

The *second* set of extracts shows that scientists perceived differently the importance of minor variations in bar type detectors, and perceived differently which are to be counted as copies of which others.

(i)   You can pick up a good text book and it will tell you how to build a gravity wave detector. . . . At least based on the theory that we have now. Looking at someone else's apparatus is a waste of time anyway. Basically it's all nineteenth-century technology and could all have been done a hundred years ago except for some odds and ends. The theory is no different from electromagnetic radiation. . . .

From this point of view, all the detectors built should be capable of seeing the radiation if it is there, for there are no particular problems associated with building them.

(ii) The thing that really puzzles me is that apart from the split bar antenna (the distinctive British version of the device) everybody else is just doing carbon copies. That's the really disappointing thing. Nobody's really doing research, they're just being copy cats. I thought the scientific community was hotter than that.

This scientist (whose apparatus was least like the originator's) perceived all the others as though they *were* carbon copies. To him the differences between the detectors was not significant. All either were, or were not good detectors.

On the other hand, the following two remarks show that the *differences* rather than the similarities between detectors, may be perceived as of more significance in respect of seeing the radiation.

(iii) . . . it's very difficult to make a carbon copy. You can make a near one, but if it turns out that what's critical is the way he glued his transducers, and he forgets to tell you that the technician always puts a copy of Physical Review on top of them for weight, well, it could make all the difference.

(iv) Inevitably in an experiment like this there are going to be a lot of negative results when people first go on the air because the effect is that small, any small difference in the apparatus can make a big difference in the observations. ... I mean when you build an experiment there are lots of things about experiments that are not communicated in articles and so on. There are so-called standard techniques but those techniques, it may be necessary to do them in a certain way.

Finally, the originator sees the dissimilarities between the detectors as the dominant feature, and feels that these differences make the secondary detectors less effective than his own device.

(v)   Well, I think it is very unfortunate because I did these experiments and I published all relevant information on the technology, and it seemed to me that one other person should repeat my experiments with my technology, and then having done it as well as I could do it they should do it better. ... It is an international disgrace that the experiment hasn't been repeated by anyone with that sensitivity.

The *third* set of evidence shows variation in scientists' perception of the value of various parts of the originator's experimental procedures. 'O' had been reporting results and had been heavily criticised for some time before the majority of secondary scientists began setting up their own experiments. In answering criticisms and improving his results 'O' produced a series of experimental elaborations, one or other of which convinced different secondary scientists to take the findings seriously enough to start work themselves. The first of these elaborations was the demonstration of coincident signals from two or more detectors separated by large distances. Some scientists found this convincing. Thus one scientist said:

> I wrote to him specifically asking about quadruple and triple coincidences because this to me is the chief criterion. The chances of three detectors or four detectors going off together is very remote.

On the other hand some scientists believe that the coincidences could quite easily be produced by the electronics, chance, or some other artefact. Thus:

> ... from talking it turns out that the bar in ... and the bar in ... didn't have independent electronics at all. ... There was some very important common contents to both signals. I said ... no wonder you see coincidences. So all in all I wrote the whole thing off again.

However, 'O' then ran an experiment where the signal from one of the bars was passed through a time delay, and showed that under these circumstances the coincidences disappeared! (suggesting of course that they were not an artefact of electronics or chance). Several respondents made remarks such as '... the time delay

104

experiment is very convincing', whereas others did not find it so. 'O''s discovery of the correlation of peaks in gravity wave activity with sidereal time was the outstanding fact requiring explanation for some scientists thus:

> ...I couldn't care less about the delay line experiment. You could invent other mechanisms which would cause the coincidences to go away. ... The sidereal correlation to me is the only thing of that whole bunch of stuff that makes me stand up and worry about it.... If that sidereal correlation disappears, you can take that whole ... experiment and stuff it someplace.

Against this, two scientists remarked:

> The thing that finally convinced a lot of us ... was when he reported that a computer had analysed his data and found the same thing.

and:

> The most convincing thing is that he has put it in a computer. ...

But, another said:

> You know he's claimed to have people write computer programmes for him 'hands off'. I don't know what that means. ... One thing that me and a lot of people are unhappy about, is the way he's analysed the data, and the fact that he's done it in a computer doesn't make that much difference. ...

Fourthly, we may look at some of the reasons that bring scientists to believe or doubt experimental findings in this field, and we can produce a list of factors which were mentioned during interviews. (Different scientists were affected by different combinations of these factors, and I am sure the list is far from exhaustive.) A major source of disbelief in 'O''s findings were theoretical calculations of available energy sources in the galaxy, but 'non-scientific' sources of attitudes to 'O' and other scientists included:

1. Faith in experimental capabilities and honesty, based on a previous working partnership.
2. Personality and intelligence of experimenters.
3. Reputation of running a huge lab.
4. Whether or not the scientist worked in industry or academia.
5. Previous history of failures.
6. 'Inside information'.
7. Style and presentation of results.
8. Psychological approach to experiment.
9. Size and prestige of university of origin.
10. Integration into various scientific networks.
11. Nationality.

As one scientist put it, in outlining the source of his disbelief in 'O'"s results:

> You see, all this has very little to do with science. In the end we're going to get down to his experiment and you'll find that I can't pick it apart as carefully as I'd like.

There is then, no set of 'scientific' criteria which can establish the validity of findings in this field.

Fifthly and finally, the following set of interview extracts shows that scientists are engaged in other than formal methods of argument and persuasion. In the first place, the following quotations should be read as showing lack of consensus over formal criteria.

1. We had a . . . summer school in . . . which lasted for two weeks. . . . When we first went there, there was a certain amount of excitement in the air because it appeared from the discussion that 'P' was saying that he had looked for 30 days and had not seen any coincidences at the level of sensitivity of 'O' and 'O' was not feeling very good about it. They had many hours of talk together, and officially when the conference left, they agreed that 'O' was more sensitive than 'P'. . . . So officially there wasn't any disagreement. Unofficially I don't know. . . .

2. Well there are two professors from . . . going round spreading rumours that 'Q' is repeating my experiment with one degree of magnitude greater sensitivity. This he has denied in a telephone conversation with 'R'. Where the truth lies, I don't know. . . .

3. . . . when I heard about these results and the fact that this second-hand report cameback that discredited 'O' I called up 'O' and told him what the rumours were that were going around so that he would know, and I said that I doubted if that was really the true picture, and he told me quite a lot more, and that they really didn't have anything near his sensitivity. . . .

4. Then there was this rumour that 'S' had seen 'T'"s paper and decided that the statistics were absolute junk, and refused to publish it, and 'T' wanted him to publish it and he wouldn't publish it, and then when 'S' went on vacation, the other editor 'V' let it go past because he decided it was making too much noise.

5. . . . in fact there is a samisdat kind of thing circulating around which 'S' wrote which he didn't want to publish pointing out all these kind of inconsistencies.

6. There was an attempt by 'U' and a few others to get me involved and intrigued into a 'Young Turks' rebellion. . . . I did not get involved in that. . . . 'U' is a negativist in his attitude toward things quite often, and he set out to topple 'O' for a while. . . .

There are sort of two camps and 'U' is the main activist of the camp that is out to topple 'O'.

We see then that a methodological entailment of the enculturational model of knowledge transfer must lead to re-formulation of the sociological problem in this field. It is no use asking how scientists *learn to* replicate the originator's experiment for there is no agreed criterion of replication.

NEGOTIATING THE CHARACTER OF PHENOMENA

The most fruitful way of interpreting the activity of scientists in this field, I suggest, is not as attempts to competently replicate, or competently test 'O's findings, but rather as *negotiations about the meaning of a competent experiment* in the field. *Ipso-facto,* they are negotiating the character of gravitational radiation and building the culture of that part of science which may become known as 'gravitational wave observation'. In general terms, to make a claim for the existence and character of a phenomenon is to make a demand for a particular organization of conceptual and perceptual categories so that events which take place at different locations and times and under different circumstances, are seen as the 'same'[20] —i.e. manifestations of that phenomenon. And the *experimental* replication is, as I interpret it, the *scientifically institutionalized counterpart of this.* When a scientist claims that an experiment has been properly replicated he is claiming that these two sets of events, the original and the replication, should be treated as the *same*. Further, he is claiming that all experiments which are to be included in the set of 'competent experiments' in the field, must be seen as manifesting the phenomenon. It follows that inclusion or exclusion of various experiments from the set of competent experiments settles the character of the phenomenon.

A simple example may make this more clear. Consider the notion of 'heat'. Because there is general agreement on the character of the phenomenon heat, a variety of entirely different (heteromorphic) activities are seen by scientists as the 'same', forming the set of competent temperature measuring experiments. For instance, dipping a glass tube filled with mercury into a liquid, and dipping two dissimilar metals linked by a voltmeter (a thermocouple) into it, may be the same experiment—measuring the temperature of the liquid. In seeing these experiments as the same and competent many of the characteristics of heat are implied, for instance, it is neither spontaneously generated by glass, nor say 'repelled by voltmeters'. Also it is not affected by the number of women over 65 years old at the time of the experiment for this does not need to be recorded for the experiment to be competent. On the other hand, the length of time that the thermometers remain within the liquid is taken as important because a property of heat is that it takes time to be transferred. Hence experiments in which, say, a glass thermometer is dipped rapidly into liquid and immediately withdrawn, are not counted as competent. If such experiments *were* to be counted as competent *the properties of heat would change.* It would flow instantly and be 'temporarily destroyed by short contact with glass but not by long contact!' The concepts of heat and temperature link different experimental activities into one set and exclude others which might appear the same to an observer without those concepts. Further, they provide the category of 'anomalous' or 'inexplicably bad' experiment when unexpected results are produced. They provide this category because *they limit* the type of factor which can be invoked to explain an unexpected result. (If there were no limit, *any* result could be explained.) Thus, it is culturally inadmissible to invoke a deficiency in the number of women over 65 to explain a malfunctioning thermometer.[21] (Scientists, of course, have a battery of such 'explanations' which seem to have a tension–relieving function. For instance, 'gremlins', 'fifth law of thermodynamics', 'transient effects' and in medical science auto-suggestion!)

In the case of temperature measurement (and laser building) the cultural milieu which legitimates the ascription of the labels 'competent experiment' and 'bad or anomalous experiment', is an established one, but as we have seen, this is not so in the case of the detection of gravity waves. There is as yet in this field no general agreement about what is an expected result so no clarity about the boundaries of the set of 'unexpected results'.

We can now see 'O's' series of experimental elaborations and other scientists' denial of the importance of these, as negotiations over whether 'O's' experiments are 'competent', and we can now see the fifth set of interview extracts given above as reports of other kinds of negotiating tactic operating in the field. If 'O's' experiments become seen as competent, then of course, expected results will be manifestations of gravity waves (residual peaks) on apparatus like 'O's', and any experiment which does not show such manifestations will be classed as not competent. Furthermore, gravity waves will have a character such that they can be detected on apparatus such as 'O's' and not on certain other apparatus. Those who do not believe that 'O' has discovered gravity waves, will class the experiments differently. These attitudes can be summarized in the following table:

ATTITUDES OF SCIENTISTS TO GRAVITY WAVE EXPERIMENTS
Believe in high fluxes of Gravity Waves

		YES	NO
Find Residual Peaks	YES	Competent	Not Competent
	NO	Not Competent	Competent

(These categories are not entirely unambiguous, nor are they completely independent. It would be foolish to deny that some scientists might come to believe in gravity waves if their apparatus shows clear residual peaks. At the time of my fieldwork however, scientists' attitudes fitted the table as far as could be seen, but only 'O' fitted the YES—YES box. All boxes had at least one member.)

THE BOUNDARIES OF THE CULTURE

We have argued that the culture of a scientific field limits and legitimates the content of arguments invoked to explain differences in findings between experiments, and we have argued that this culture is relatively undeveloped in the case of gravitational radiation. An analysis of the actual arguments used by scientists in their claims of competence or otherwise for the experiments, favours this interpretation. Most arguments which have appeared in print seem to fit comfortably with normal physics.[22] For instance, in relation to various experiments the following variables have been cited in arguments favouring one interpretation or another.
1.   Means of extracting energy from the bar.[23]
2.   Material of which bar is constructed.[24]
3.   Type of electronic processing of signals.[25]
4.   Statistical techniques for extracting signal from noise.[26]
5.   Calibration of apparatus.[27]
6.   Estimates of frequency of chance spurious signals from bar, and frequency of random sensitive and non-sensitive states.[28]
7.   Frequency of radiation/sensitive frequency of bar.[29]
8.   Length of bursts of radiation.[30]
9.   Proximity of source of radiation.[31]
10. Band-width of radiation.[32]
11. Other accepted phenomena which might cause spurious effects which are not controlled or shielded for.[33]

Some other invocations however, seem to show that the cultural milieu is not limited to orthodoxy.[34]

Thus it has been suggested that:

12. If radiation from the centre of the galaxy were being focussed in some way toward the Earth, this would explain the large fluxes of radiation.

13. In some way pulses of gravitational radiation are triggering the release of energy stored in the bar. This is a developed version of the generalized notion that gravity waves are coupled to material more strongly than had been thought.

14. Explanation of the results of these experiments might require reference to a 'fifth force'. (Some force in addition to the currently known magnetic, gravitational, strong and weak forces.)

15. The gravity wave findings are solely products of mistakes, deliberate lies, or self-deception.

16. Finally, the explanation might require reference to psychic forces.[35]

Of these latter items only the first two have received mention in print under the author's name (so far as I know). One scientist has postulated the 'fifth force' (item 14) under a pseudonym, and at least one other has given it serious thought:

> I've been talking and thinking about that and I have done a couple of tests of possible candidates for fifth forces and found no results. I just don't know.

Most scientists went to great length to point out to me that they did not believe gravity wave findings were a deliberate hoax or lie, though a self-deception hypothesis was less rare. Most scientists, but not all, were uncomfortable when presented with their colleague's suggestion that psychokinesis could be responsible for the results. Typically, the response was initially a grin, followed when pressed by an admission that it was a logical possibility, and that they were open-minded to this suggestion.

As one put it:

> ... when you are groping in a new scientific field you are prepared to go to almost any extreme, and if [the results are] eventually explained by ESP I'm not going to roll over and die. I have an open mind.

Scientists then were in general unwilling to put their names to this latter class of explanatory hypotheses, though they were prepared to consider them, and this I would suggest is a consequence of the difficulty of reconciling its members with the practice of modern physics.[36] Arguments *ad hominem*, are in effect an abandonment of the idea of replicability, for the identity of the experimenter is then *always* a relevant variable. Once the idea of self-deception has been legitimated it can be applied to all experiments *and 'replications'* into the future. The acceptance of psychokinesis would similarly make it possible to attribute all experimental results to the wishes of the experimenter—or anyone else for that matter. On the other hand, the notions of 'fifth force' and stored energy seem to relate to nothing that has gone before. Perhaps this occasional invocation of *ad hoc* flights of imagination, or elements form other cultures in the negotiation of a new phenomenon, is a general symptom of the type of science which is difficult to reconcile with the dominant 'paradigm', and sociologists should look for these symptoms if they wish to find other than 'normal' science.

SUMMARY

The 'Enculturational Model' of scientific knowledge fits the findings of previous research on the transfer of knowledge between scientists. A consequence of this model is that the only criterion that the knowledge required to repeat an experiment 'properly' has been transferred, is that the experiment 'works'. Scientists (in the laser field anyway) may try to gain this knowledge by personal contact with successful scientists, and by repeating the experiment in a way which they perceive as isomorphous with a successful one.

In the gravity wave field scientists were not concerned to conduct experiments isomorphous with that of the originator of the research. There are many possible ways of explaining this, but a convincing interpretation of their actions is that in the absence of general agreement of what is to count as a 'working experiment' in this field, secondary experiments which do not show the same results as the original experiments may still be seen as 'competent' so there is no special impetus to copy the original experiment. Scientists' actions may then be seen as negotiations about which set of experiments in the field should be counted as the set of competent experiments. In deciding this issue, they are deciding the character of gravity waves.[37]

The perspectives put forward in this paper stem from thinking of scientific knowledge as a cultural artefact, and in these terms the gravity wave scientists are still developing the terms and objects of their culture. At present this culture is sufficiently open to allow the postulation of several 'unorthodox' explanatory hypotheses.

I would like to return now to the quotation with which I started and the implication that a close look at contemporaneous science would reveal a 'relativistic' aspect. What is actually seen in the case of gravitational radiation research, is a set of varied experiments and a set of scientists disagreeing about how they are to be inter-preted, or as I have put it, negotiating the character of gravity waves. As far as can be seen from this distance, the originator can argue indefinitely that experiments which claim to be disconfirm-ations of his results, are not good replications. Critics of his find-ings will argue that they *have* produced good experiments. As far as can be seen there is nothing outside of 'courses of linguistic, conceptual and social behaviour' which can affect the outcome of these arguments, and yet this outcome decides the immediate fate of high fluxes of gravity waves—it settles the design of the ship in the bottle.

NOTES AND REFERENCES

1.   Earlier versions of this paper were read at the universities of Bath and Essex, and presented at a conference on the 'Comparative Sociology of Sciences' at the Civil Service College, London on 7th December 1973. I am grateful for the many useful comments made at these meetings, and for the careful reading and helpful suggestions of several of my colleagues at Bath, especially Graham Cox.

2.   See J. D. Douglas, *Understanding Everyday Life,* R.K.P., 1971. McHugh's writing is largely influenced by the ethnomethodologists, and the influence of this school should be visible in the paper which follows. More important however (at least in a biographical sense), is the influence of the writings of Winch and Wittgenstein. My excuse for begin-ning the paper with an ethnomethodological quotation rather than one from Wittgenstein (apart from its appositeness) is that I am not able to disentangle the implications for the sociology of science of the two approaches. It seems that I am not alone in this, for an anonymous but authoritative source has recently written:

> Wittgenstein here (in *Remarks on the Foundations of Mathematics*) tends to beat the 'ethnos' at their own game. See McHugh, P., Raffel, S., Foss, D., Blum, A., *On the Beginning of Social Inquiry,* R.K.P., 1974, p. 104.

Other papers which are of immediate relevance to the stance taken here include: D. Bloor, 'Wittgenstein and Mannheim on the Sociology of Mathematics', (*Studies in the History and Philosophy of Science,* 4 (1973), 1973—91), where Bloor shows how Witt-genstein's 'Remarks . . .' can open the way for a sociology of mathematical knowledge; 'Evaluation', which is chapter four in McHugh *et al,* 'On the Beginning . . .', *op. cit.,* and two papers which are as yet unpublished, and which demand an interpretive approach to the sociology of science: A. McAlpine and A. Bitz, 'Some Methodological Problems in the Comparative Sociology of Science' (read at a conference on the comparative sociology of science, London, 7th December 1973); J. Law and D. French, 'Normative and Interpretive Sociologies of Science'. See also parts of H. M. Collins, 'The TEA Set: Tacit Knowledge and Scientific Networks', *Science Studies,* 4, (1974), 165—86.

3.   Wittgenstein's description of a strange world where the inhabitants sell wood at a price determined by the area of the base of the pile, irrespective of its height and shape,

Escher's drawings and Magritte's paintings help to relieve one of this certainty. (See L. Wittgenstein, *Remarks on the Foundations of Mathematics*, **I**, 149.)

4.   Further, this feeling of solidness about some knowledge may in some cases lead to an idea of progress, current knowledge seeming to be better than superseded knowledge, because it has a prima-facie claim, being a natural object. Perhaps some of Kuhn's ambiguities may be put down to this. Curiously, even Collingwood believed in scientific progress. He writes, 'The interest of science, in relation to the conception of progress, seems to be that this is the simplest and most obvious case in which progress exists and is verifiable'. (R. G. Collingwood, *The Idea of History*, O.U.P., 1946. Paperback edition p. 332.)

5.   Thomas Kuhn's now famous book (T. S. Kuhn, *The Structure of Scientific Revolutions*, University of Chicago Press, 1962) has given rise to a new interest in the sociological explanation of science (this is not to say that some of Kuhn's analysis was not anticipated, e.g. by Polanyi, but only that his work was seminal) in which 'correct' science is seen as being as fit a subject for such explanation as 'false' science. (See D. Bloor, *op. cit.*, and B. Barnes, 'Sociological Explanation and Natural Science: A Kuhnian Reappraisal', *Arch. Europ. Sociol.*, XIII (1972), 373—391. R. Whitley, 'Black Boxism and the Sociology of Science' in *Sociological Review Monograph*, 18 September, 1972.) The overall programme is to show how scientific concepts are related to the societies (or political interests) in which they are embedded. What is needed if these demands are to be met, is some discussion of the mechanism of the construction of scientific cognitions which shows where 'society gets in'. Simply to show that particular elements in science are congruent with particular interests or cultures is not sufficient. It has to be shown that 'the scientific method'—the actual practices of scientists—could yield one result rather than another in different social circumstances.

6.   See Collins, 'The TEA Set . . .', *op. cit.*

7.   For instance, see 'The Competitive World of the Pure Scientist', F. Reif, in *Science and Society*, (ed.) N. Kaplan, Rand McNally and Co., 1965, W. D. Garvey and B. C. Griffith, 'Scientific Communication as a Social System', *Science*, 157 (1967), 1011—16. M.J. Mulkay, 'Conformity and Innovation in Science', *Sociological Review Monograph* 18 (September 1972), 5—24.

8.   Panther edition, 1970.

9.   This seems to be a formulation of the problem of induction.

10.   That is probably why the algorithmical model of knowledge is the one that has almost exclusively informed information science, and certain schools in the sociology of science. This has resulted in what I consider to be a mass of rather trivial work based on certain easily observed indicators such as the 'Science Citation Index'.

11.   For example, see Gruber and Marquis, *Factors in the Transfer of Technology*, M.I.T. Press (1969).

12.   Readers familiar with the literature of the sociology of science will have recognised in the points made under 'Enculturational Model' some which are often associated with the term 'Tacit Knowledge'. (See M. Polanyi, *Personal Knowledge*, R.K.P. (1958); J. R. Ravetz, *Scientific Knowledge and its Social Problems*, O.U.P. (1971); Jevons, *Scientific Education*, Allen and Unwin, 1969.) This term *could* be used for the non-explicated rules which are followed 'as a matter of course' in Wittgensteinian language games, and the idea seems to be similar to the 'taken for granted reality' of the phenomenological school. Further major differences in the content of tacit knowledge can be seen as differences in Kuhnian paradigms. (The link between Wittgenstein and Kuhn is discussed in R. Trigg, *Reason and Commitment*, Cambridge, 1973.) However, a problem with the term is that it is usually used in a restricted context, to refer only to elements of the 'technical' knowledge required of a scientist, that is, knowledge pertaining to artisanship rather than cerebration. (This is *in spite* of Ravetz's comment that '*in every*

one of its aspects scientific activity is a craft activity depending on a body of knowledge which is informal and partly tacit' (*op. cit.,* 103) my emphasis.) The point I am making is more general; it refers to the transfer of culture, not just practical expertise.

13.   This is offered as a general conclusion for established scientific fields. Experimental replications which don't work at first are taken as 'anomalous cases' etc., or as signifying lack of knowledge, but continued failure will be taken to signify a lack of 'experimental ability' which may cause the scientist to 'give up' or disqualify him from continuing as an experimentalist. 'Good experimental scientists' are those whose 'normal' experiments generally do work. Colin Bell has discussed replications in sociological research. (See *Futures,* April, 1974.)

14.   The following points may be of interest:

1. Only one of the three firms has any reputation for supporting 'pure' research. Members of the firms interviewed suggested that management viewed their work as mainly pure research which might generate good publicity for the firm, or might lead to practical payoff in the future. Certainly, firms with experimental expertise in the field would be in a good position to benefit commercially should a new science of 'gravitational astronomy' grow from the research.

2. A possible long-term application of gravity waves speculated upon by one university physicist is communication. Matter is virtually transparent to gravity waves (according to received theory) while being relatively opaque to radio waves.

3. Most scientists (not 'O') expended only a part of their research effort on gravity wave research.

4. The cost of a gravity wave experiment varied between about $10,000 and $100,000 for the hardware.

15.   I will not refer to the source papers in the physics journals as I want to preserve the anonymity of the scientists in these contemporaneous fields as far as possible.

16.   Several pairs of laboratories are co-operating on coincidence experiments.

17.   A correlation of peaks with solar time might suggest that some phenomenon associated with terrestrial night and day, might be involved, and this would be of no startling scientific interest.

18.   Some recent experiments which *have not yet produced results* are much more like the original. At this stage, no-one would comment upon whether they were the 'same'. This may appear significant in the light of the discussion which follows.

19.   This was suggested by Ted Benton and Ray Dolby separately.

20.   The discussions in P. Winch, *The Idea of a Social Science,* R.K.P., 1958, on pages 24—29 and 83—86 are helpful in this context.

21.   One might describe such an argument as one which 'breaks the rules' of the game of temperature measurement. It is a part of the Wittgensteinian concept of following a rule that occasions when the rule is broken, can be recognized.

22.   Most of these points may be found in journals such as *Nature* and *Physics Today.*

23.   Alternatives are for instance, piezo-electric crystals measuring strain in the surface of the bar—or sandwiched between parts of the vibrating mass in several different ways— or the variation in the value of a capacitor whose plates are linked to the end of the bar —or a weak couple between the large bar and a much smaller mass.

24.   Some materials are more efficient, but one experiment used highly efficient pure aluminium which was reported to 'creep' to a considerable extent producing an unacceptable level of 'noise'.

25.   The electronic circuits could be producing the 'signals' themselves, or swamping them in their own noise, or contributing to the appearance of simultaneity in signals from different detectors, or acting as receivers for spurious non-gravitational disturbances.

26.   Decisions have to be made regarding the criterion which separates out 'signals' from noise. In the most crude systems, decisions are made by looking at a 'print-out'

which will show a number of 'peaks'. Peaks above a certain predetermined level will be counted as 'signals'. This selection process **may** be done by panels of judges, by the experimenter himself, or by other scientists. Alternatively, computers may be used to do this analysis or an analysis using more sophisticated statistical techniques.

27.   Some scientists complained that 'O' had not given details of the method of calibration of his bar. This affects the arguments on sensitivity.

28.   Occasionally there will be high peaks of noise in the bar. The frequency of 'real signals' is obtained by subtracting the expected spurious signals from the total observed. There is some argument around these points more especially concerning the number of spurious coincident signals which should be observed from different detectors. The issue is made more complex by the fact that it is not expected that the antennae will detect all the incoming gravitational signals, but only those which arrive when the detector is in a random 'sensitive state'. Should a signal 'arrive' at a point when the noise in the detector is generating a 'negative' peak, a net zero output might be registered. This means that coincidences can be registered only when both detectors involved in the experiment chance to be in a sensitive state. Hence, if there are more detectors involved in a coincidence experiment, there is less chance of registering a coincidence.

29.   Sensitivity of most of the antennae is a function of the frequency of the incoming radiation. Bar type devices are most sensitive at their primary resonating frequency. This frequency is a function of their size, shape and material of construction. The original experiment's central frequency was 1661 Hertz (cycles per second), but other experiments have not taken this as their main search frequency. One or two other experimental designs are non-resonant, and they claim to be able to pick up signals across a broad frequency band. There is some question about the sensitivity of these latter.

30.   Some detectors are more sensitive to short bursts of radiation than others. This is an effect independent of the total amount of energy in the burst.

31.   If the source of radiation were near, the theoretically embarrassing estimates of total energy involved would be avoided. (See footnote 32.)

32.   The hypothesis of a narrow bandwidth has been invoked to explain the large amount of radiation without producing disturbing cosmological consequences. Large amounts of radiation in only one narrow frequency range would be acceptable because it would not imply that large *absolute* amounts of energy were being generated and therefore would not imply a rapid conversion of the mass of the universe into energy.

33.   Other candidates which have been put forward include: neutrino fluxes, electric storms, sun spots, and most recently, currents in the ionosphere. An invocation of this type enables some members to fill the YES—NO box in the table, i.e. they find residual peaks but do not count these as manifestations of gravity waves.

34.   Cutting the list at this point is to some extent arbitrary—it depends upon *my* judgement of what is acceptable in physics. Doubtless, some physicists would put some of points 9, 10, 12 and 13 in the opposite category.

35.   This suggestion attributes the residual peaks to (for instance) the desires of the experimenter operating through psychokinesis—the power of mind over matter. A rumour circulates that the originator had consulted with J. B. Rhine (the figurehead of scientific research into the paranormal), though both of the parties deny this. Another experimenter had seen a signal on his detector for the first time immediately after a telephone conversation with the originator and toyed with the idea that some psychokinetic effect had been involved. Another two or three experimenters had taken an interest in research on 'extra sensory perception' and related effects. Some members of the parapsychological association are most interested in the work, and are delighted in so far as they believe that some of the scientists involved are considering the psychokinesis hypothesis seriously. (This was reported to me while I was engaged on research into parapsychology. At the time I had no idea that there was any connection between the two fields, and far from being a response of any question of mine, the report came as

a total surprise to me.) Finally, an experiment is being planned with the collaboration of one of the secondary experimenters to test the ability of a gifted psychic on the apparatus. (Reported by the physicist involved.)

36.   The arguments at the beginning of the paper about the cultural determinism of current knowledge, can now be seen to operate. Many of the reasons invoked to explain the differences between the results of gravitational radiation detection would never be put forward with reference to any well established phenomenon. Indeed, it is probable that before long most of these arguments (especially those that don't appear in print), will be totally forgotten. Probably these arguments will be reinterpreted as 'cranky', 'not serious', 'did I really say that', 'I couldn't really have thought that', and so on. So far as the history of the subject is concerned they will cease to exist, and indeed, be literally inconceivable.

37.   Ravetz (*op. cit.*, p. 156/157) in an interesting discussion, makes the point that 'criteria of adequacy' in 'immature' fields are weak, leading to 'pitfalls' and 'a situation where conclusions of problems are illusory, and facts cannot exist'. 'Strong criteria of adequacy' would seem to correspond to what I mean by established culture, and 'weak criteria of adequacy' to open culture. Ravetz's dicussion however, does not allow that scientists might compete over the definition of appropriate criteria of adequacy, and does not show how different sets of criteria can lead to the discovery of different phenomena (facts). He writes as though there is one appropriate set of criteria, which will develop as a successful field matures.

I have drawn heavily on Lakatos (e.g. 'Falsification and the Methodology of Scientific Research Programmes' in Lakatos and Musgrave [eds], *Criticism and the Growth of Knowledge*, (C.U.P., 1970) in thinking about the way that scientists may support their views against criticism, by setting up sub-hypotheses. Indeed, the argument: 'X's experiment is not a replica of mine' is a general form of sub-hypotheses. What Lakatos does not provide is any discussion of the cultural limitation on the sub-hypotheses that may be legitimately proposed.

This paper is not intended to criticize in any way those scientists who kindly gave up their time to speak to me about their fascinating work. Indeed, it is only since third parties have read the final draft of the paper that it has occurred to me that an intended criticism might be read into it. Quotations taken out of context, e.g. that experiment 'is an out-and-out fraud', 'the group at W are just out of their minds' etc., should be read as dramatic and rhetorical expressions of lack of conviction with others' experiments, not literal ascriptions of fraudulent intent or insanity. Sometimes what counts as evidence for the sociologist of science may look like gossip to the scientist, but this is an unfortunate contingency only. As regards the overall relativistic position developed, I would hope that this would no more be seen as 'an attack on science' than would a solipsistic position be seen as an 'attack on sense perception'. Creative science is a marvellous and beautiful exercise of intellect, ingenuity and skill. It would be absurdly arrogant for any sociologist to criticise its internal operation and thus claim to know how to do science better than the practitioners themselves.

6

David Bloor

Formal and informal thought

In the course of a rather dry disagreement that Mill was having with Bishop Whately he dropped some disturbing and exciting hints about the nature of formal reasoning. The context is unpromising. Mill is debating with Whately the question: does the syllogism contain a *'petitio principii'*? The issue can be stated very simply by looking at the following syllogistic argument:

> All men are mortal
> The Duke of Wellington is a man
> Therefore the Duke of Wellington is mortal

If we are in a position to assert the first premise, that all men are mortal, then we must already know that the Duke is mortal. So what are we doing when we conclude or infer his mortality at the end of the syllogism? Surely the syllogism begs the question or reasons in a circle? Mill believes that there is indeed a circle here. Part of the subsequent account of reasoning that he gives to justify this view is well known, but some of its most suggestive features pass unnoticed.

LORD MANSFIELD'S ADVICE

The familiar part of Mill's theory is that reasoning proceeds, as he puts it, from particulars to particulars. Bearing in mind that the Duke of Wellington was alive when Mill wrote, then the inference to his mortality was by inductive generalisation and the association of ideas. Experience yields reliable inductive generalisations concerning death and these are naturally extrapolated to cover cases which appear relevantly similar to those occurring in the past. The case of the Iron Duke is assimilated to the previous cases signalised by the generalisation. Mill says that the real process of inferring consists in the move from particular past cases to particular new cases. The thought process involved does not therefore really depend on, or proceed via, the generalisation that all men are mortal. It gets along without the help of the major premise of the syllogism. As Mill put it: 'Not only may we reason from particulars to particulars without passing through generals, but we perpetually do so reason' (II, III, 3).

If the general premise of a syllogism is not involved in our acts of reasoning then what status is to be assigned to it? This is where Mill drops his hints. General propositions for Mill are merely 'registers' of inferences that we have already made. The reasoning, he insists, lies in the specific acts of assimilating the new cases to the old ones, 'not in interpreting the record of that act'. In the same discussion Mill refers to the generalisation that all men are mortal as a 'memorandum'. The inference to the mortality of any specific person does not, says Mill, follow from the memorandum itself but rather from those very same past cases which were themselves the basis of the memorandum.

Why call the major premise of a syllogism a record, a register or a memorandum? To talk of premises and principles in this way conveys two ideas. First, it suggests that they are derivative or mere epiphenomena. Second, whilst indicating that they are not central to the act of reasoning itself, it hints that they do perform some other positive function, albeit not the one that is usually attributed to them. Mill's language here suggests a book-keeping or bureaucratic role, a means of documenting and filing what has happened.

Mill neatly epitomises and extends this account by his story of Lord Mansfield's advice to a judge. This was to give decisions boldly because they would probably be right, but on no account to give reasons for them, for these would almost infallibly be wrong. Lord Mansfield knew, says Mill, that the assigning of reasons would be an afterthought. The judge in fact would be guided by his past experience, and it would be absurd to suppose that the bad reason would be the source of the good decisions.

If reasons do not produce conclusions, but are mere afterthoughts, then what relation do they bear to them? Mill sees the connection between general principles and the cases that fall under them as something which has to be created. An interpretive bridge has to be built. Thus: 'This is a question, as the Germans express it, of hermeneutics. The operation is not a process of inference, but a process of interpretation' (II, III, 4).

Mill treats the syllogism in a similar way. Its formal structures are connected to actual inferences by an interpretive process. It is 'a mode in which our reasonings may always be represented'. That is to say: formal logic represents a mode of display; an imposed discipline; a contrived and more or less artificial surface structure. This display must itself be the product of a special intellectual effort and must itself involve some form of reasoning. What is striking is the order of causality and priority that this account reveals. The central idea is that formal principles of reason are the tools of informal principles of reasoning. Deductive logic is the creature of our inductive propensities; it is the product of interpretive afterthoughts. I shall refer to this idea as the priority of the informal over the formal.

How does the priority of the informal over the formal express itself? The answer is two-fold. First, informal thought may use formal thought. It may seek to strengthen and justify its predetermined conclusions by casting them in a deductive mould. Second, informal thought may seek to criticise, evade, outwit or circumvent formal principles. In other words the application of formal principles is always a potential subject for informal negotiation. This negotiation is what Mill referred to as an interpretive or hermeneutic process. It concerns the link which must always be forged between any rule and any case which allegedly falls under that rule.

The relation between the formal principles of logic and informal reasoning is clearly a delicate one. Informal thought seems at one and the same time to acknowledge the existence and the potency of formal thought — why else would it exploit it? — and yet it has a will of its own. If Mill is right it goes its own way, moving inductively from particular to particular, governed by associative links. How can it do both of these things at once?

Consider the syllogism: all A is B, C is an A, therefore C is B. This is a compelling pattern of reasoning. It emerges out of our learning of simple properties of physical containment. We have an informal tendency to reason as follows: if a coin is put in a matchbox and the matchbox is put in a cigar box, then we go to the cigar box to retrieve the coin. This is the prototype of the syllogism. The simple situation provides a model for the general pattern which comes to be counted as formal, logical and necessary. Formal principles, like the syllogism above, thus harness a natural proclivity to draw conclusions. For this reason it can represent a valuable ally or an important enemy in any case that is being made. It may therefore become important to subsume a problematic case under this pattern or to keep them apart, depending on the informal purposes.

In order to evade the force of an inference it is obviously necessary to challenge the application of the premises, or the concepts in the premises, to the case in hand. Perhaps the item designated by the letter C is not really an A; perhaps not all things counted as As really are Bs. In general, distinctions must be drawn; boundaries re-allocated; similarities and differences indicated and exploited; new interpretations developed, and so forth. This form of negotiation does not call the syllogistic rule itself into question. After all, the rule is embedded in our experience of the physical world, so some range of application will have to be granted it, and tomorrow we may want to appeal to it ourselves. What can be negotiated is any particular application.

Informal thinking therefore has a positive use for formal principles as well as a need to circumvent them. Whilst some informal purposes will be exerting pressure to modify or elaborate logical structures and meanings, others will be banking on their stability and maintenance. Informal thinking is both conservative and innovatory.

The idea that logical authority is moral authority may be in danger of neglecting these more dynamic elements in logical thought: competing definitions; opposing pressures; contested patterns of inference; problematic cases. To forget these would be to assume that logical authority always works by being taken for granted. The present point is that it also works by being taken into account: by being a component in our informal calculations. Authority which is sustained by being taken for granted may be said to be in static equilibrium to contrast it with the image of dynamic equilibrium. This static acceptance may be a more stable and compelling form of authority but even this stability can be disturbed.

There is no reason why a sociological theory should not allow for both phenomena. Indeed the coexistence of these alternative styles of constraint is a central feature of all aspects of social behaviour. In some people and in some circumstances moral or legal precepts, for example, may be internalised as emotionally charged values which control behaviour. In other cases these precepts may be apprehended simply as pieces of information: things to be born in mind when planning behaviour and predicting the responses of others. The concurrence of these two modes of social influence in mathematics — and the theoretical problem of untangling them — can only serve to strengthen its similarity with other aspects of behaviour.

The negotiated application of formal principles of inference explains certain important examples of variation in logical or mathematical behaviour. Of course, the more formalised the logical principles at issue the more explicit and conscious is the negotiation process; the less explicit the principle, the more tacit the negotiation. I shall illustrate the negotiated character of logical principles with three examples. The first will concern the negotiated overthrow of a self-evident logical truth. The second example concerns the much discussed question of whether the Azande tribesmen have a different logic to us. The third case will be the negotiation of a proof in mathematics. This will be based on the brilliant historical study of Euler's theorem made by I. Lakatos (1963—4). Here Lakatos offers something of great value to the sociologist, much more than might be guessed from his methodological remarks that I discussed earlier. [Eds: We only reproduce the first of these three examples here.]

PARADOXES OF THE INFINITE

Consider again the syllogism: all As are B, C is an A, therefore C is a B. It was argued that this reasoning is based on the experience of containment and enclosure. If anyone is in doubt as to how or why the syllogism is correct he need only look at the diagrammatical form into which it can be cast, and which is equivalent to it (see Figure 1). The diagram connects the syllogism to an important common sense principle, namely that the whole is greater than the part.

Figure 1 The whole is greater than the part.

It is tempting to assume that because experiences of enclosure are ubiquitous they will uniformly and without exception impress this principle on all minds. It is not surprising that those who believe in the universality of logic cite such principles of evidence. Thus Stark (1958) says:

> So far as purely formal propositions are concerned, there simply is no problem of relativity. An example of such a proposition is the assertion that the whole is greater than the part. In spite of all that the super-relativists have argued, there can be no society in which this sentence would not hold good, because its truth springs immediately from the definition of its terms and hence is absolutely independent of any concrete extra-mental conditioning (p. 163).

Stark is not saying that this truth is innate. He allows that it comes from experience, but so direct is the connection with experience that nothing can insinuate itself between the mind and its immediate apprehension of this necessity. Experiences of this kind are universal and so the self-same judgments arise. Always and everywhere the whole is greater than the part.

It is certainly correct to say that this idea is available in all cultures. It is a feature of our experience which can always be appealed to, and so some application will always be found. But this does not mean that any particular application is compelling or that its truth is immediate, or that there is no problem of relativity. Indeed this case is particularly interesting because it shows the opposite of what Stark thinks. There is a body of mathematics called transfinite arithmetic which is successfully based on an explicit rejection of the principle that the whole is greater than the part. Properly understood this example therefore shows that apparently self-evident truths backed by compelling physical models can be subverted and renegotiated.

Consider the sequence of integers: 1, 2, 3, 4, 5, 6, 7, ... Select from this endless sequence another endless sequence consisting of only the even numbers, 2, 4, 6, and so on. It is possible to associate these two sequences thus:

$$1 \quad 2 \quad 3 \quad 4 \quad 5 \quad 6 \quad 7 \quad \ldots$$

$$2 \quad 4 \quad 6 \quad 8 \quad 10 \quad 12 \quad 14 \quad \ldots$$

In common sense terms the even numbers can be counted. More technically it can be said that the even numbers are put in a one—one correspondence with the integers. This one—one correspondence will never break down. For every integer there will always be a unique even number to pair with it. Likewise for every even number there will be a unique integer. Suppose it is now said that sets of objects which have a one—one correspondence between their members have the same number of members. This seems intuitively reasonable, but means that there are the same number of even numbers as there are integers. The even numbers, however, are a selection, a mere part, a subset of all the integers. Therefore the part is as great as the whole and the whole is not greater than the part.

The inexhaustible supply of integers can be expressed by saying that there is an infinite number of them. Infinite aggregates thus have the property that a part can be put in one—one correlation with the whole. This property of infinite aggregates was known many years before the development of transfinite arithmetic. It was taken as evidence that the very idea of aggregates of an infinite size was paradoxical, self-contradictory and logically defective. Cauchy for example denied their existence on this basis (Boyer (1959), p. 296). However, what had once been grounds for dismissing infinite sets came to be accepted as their very definition. Thus Dedekind (1901, p. 63) says: 'A system S is said to be infinite when it is similar to a proper part of itself', where 'similar' in this definition is what has been called one—one correspondence.

How can a contradiction be transmuted into a definition, how is such a renegotiation possible? What has happened is that the model of physical enclosure which underlay the conviction that wholes are greater than their parts has given away to another dominant image or model: that of objects being placed in one—one correspondence with each other. This, too, is a situation which it is easy to exemplify and experience in a direct and concrete way. Once this alternative model has become the centre of attention then the simple routine of aligning even numbers with integers becomes a natural basis for concluding that the part (the even numbers) is as great as the whole (all of the integers). Informal thought has subverted an apparently compelling principle by pressing the claims of a new, informal model. A new range of experience has been located and exploited. If compelling logical principles consist in a socially sanctioned selection from our experience then they can always be opposed by appealing to other features of that experience. Formal principles only feel special and privileged because of selective attention. Given new concerns and purposes and new preoccupations and ambitions then the conditions exist for a readjustment.

The conclusion is that there is no absolute sense in which anyone must accept the principle that the whole is greater than the part. The very meanings of the words do not compel any given conclusion because they cannot compel the decision that any new case must be assimilated to the old cases of this rule. At most, prior applications of this model create a presumption that new and similar cases will also fall under the same rule. But presumption is not compulsion, and judgments of similarity are inductive not deductive processes. If it is proper to speak of compulsion then the compelling character of a rule resides merely in the habit or tradition that some models be used rather than others. If we are compelled in logic it will be in the same way that we are compelled to accept certain behaviour as right and certain behaviour as wrong. It will be because we take a form of life for granted. Wittgenstein expressed it neatly when he said in the 'Remarks' (1956): 'Isn't it like this: so long as one thinks it can't be otherwise, one draws logical conclusions.' (I, 155). Nevertheless Wittgenstein believes it is right to say that we are compelled by the laws of inference: in the same way as we are compelled by any other laws in human society.

REFERENCES

Boyer, C.B. (1959). *The History of the Calculus and its Conceptual Development.* New York: Dover Publications.

Dedekind, R. (1901). *Essays on the Theory of Numbers.* Translated by W.W. Berman. New York: Dover Publications (1963).

Lakatos, I. (1963–4). 'Proofs and Refutations', *British Journal for the Philosophy of Science,* 14 : 1–25, 120–39, 221–43, 296–352.

Mill, J.S. (1848). *A System of Logic: Ratiocinative and Inductive.* London: Longmans.

Stark, W. (1958). *The Sociology of Knowledge.* London: Routledge and Kegan Paul.

Wittgenstein, L. (1956). *Remarks on the Foundations of Mathematics.* Oxford: Basil Blackwell.

7

Andrew Pickering

Interests and analogies

INTRODUCTION

This paper follows a recent debate in high energy (elementary particle) physics from its initiation to its eventual resolution.[1] The aim of the paper is not, however, simply to present a case-study in the history of modern physics. Its primary objective is to recommend a simple model of cognitive organization and cognitive change in science which I shall call the 'interest' model.

The interest model has two key concepts: that of an 'exemplar', and, of course, that of an 'interest'. Let us take 'exemplars' first. Kuhn first introduced the term in the 'Postscript' to his *Structure of Scientific Revolutions,* and my analysis extends the discussion he gave there.[2] For Kuhn, an exemplar is a shared example, upon which scientists base their work at the research front. He emphasizes that such a shared example involves the concrete demonstration in some practical situation of the utility of a cultural product — a new experimental technique, a new theoretical model, or whatever — and that it is precisely through such demonstrations, through *practice* involving exemplars, that new concepts are related to the natural world and acquire their meanings. Particular research networks within a given scientific community are to be seen as engaged in the articulation — the working-out in practice — of particular exemplars. As Kuhn put it:

> More than other sorts of components of the disciplinary matrix, differences between sets of exemplars provide the community fine-structure of science.[3]

What is involved in the articulation of an exemplar? An exemplar constitutes the embodiment of an analogy: some new or 'unsolved' aspect of a scientific field is organized and perceived through a particular exemplary achievement as being the 'same' as some already well-understood aspect of the field.[4] In its original statement, the exemplar makes the analogy explicit in a crude

125

form, and its articulation consists of the elaboration of the crude achievement by a process of modelling: more sophisticated perspectives are constructed at the research front by importation of conceptual resources from the area with which the analogical connection has been made. Thus, exemplars function in scientific investigation by linking problem-areas on the research front to an existing body of accepted practice. And the use of exemplars at the research front involves the importation into the front, at the point of use, of cultural resources and practices already routinely used within the field at some other point.

Note that although an exemplar is available, in principle, for the use of any scientist or group of scientists, only a specific sub-set of scientists are able to use it at any time. To use an exemplar requires competence, and this is typically only possessed by those in the area where the exemplar is part of routine scientific accomplishment, and by those engaged in articulating the exemplar at the research front. To those competent in its use an exemplar means more than it does to others: the continuation of their current practical activity depends upon its continued acceptance, and their standing as the originators of valued knowledge and the possessors of valued skills may be intimately bound up with the standing of the exemplar. One can speak of the group or groups having expertise relevant to the articulation of some exemplar as having an 'investment' in that expertise, and, as a corollary, as having an 'interest' in the deployment of their expertise in the articulation of the exemplar. An 'interest', then, is a particular constructive cognitive orientation towards the field of discourse. As a shorthand description one can refer to an exemplar being so constructed that it 'intersects with the interests' of some particular group or groups.[5]

One final point remains, and must be stressed. Given the complex and fragmented structure of science, the even greater complexity of its history, and the way in which individual scientists acquire and develop competences, it is often the case that an exemplar deployed at the research front refers back not just to one specific body of practice but to several, frequently quite distinct, such bodies. It is important to remember that an exemplar is not a monolithic, unitary entity. It is possible to use it in many different ways, referring back to different routine accomplishments and patterns of competence. Individual scientists can set about articulating an exemplar along one of several dimensions, according to their prior investments and interests; and, in so doing may well acquire new expertise and interests relevant to other dimensions of the exemplar.

Theoretical development in science consists, then, in the construction and elaboration of exemplars. This is the process through which concepts become entrenched in scientific practice. And a specific development of scientific theory, a specific sequence of constructions and elaborations of exemplars, is preferred to alternative developments because of its relationship to the existing patterns of interest or investment in the relevant scientific community. Certainly, in the case to be considered here, I shall argue that one of two competing theoretical models was preferentially incorporated into the practice of the physics community because of their different relationship to existing interests. The two models became differentially entrenched in scientific practice precisely to the extent that the elaborations of each model intersected with existing interests. The result was that one model became central to the practice of several research networks, whereas the other was abandoned by all but a few isolated individuals.

Another way of putting this is to say that, at the communal level in physics, one theoretical construction became real and the other did not. This matter of the *interpretation* of practice is not the primary focus of the present paper, but it is important nonetheless. And since there is a relationship between the entrenchment into practice of a theoretical construct and its subsequent interpretation as real, the point is worth brief mention. Natural scientists are often favourably disposed to a realist account of their practice: their interpretative discourse often centres upon which of their theoretical concepts can properly be identified with reality — as corresponding to the contents of the natural world. When a concept becomes thoroughly entrenched in the practice of several diverse research groups it acquires impersonal qualities: it ceases to be intelligible as the 'property' of any individual or group. This allows the concept the more easily to be perceived in isolation from its social origins, and hence the more easily to be represented as in correspondence with an asocial natural world. This, I shall suggest, is how the reality of theoretical constructs actually is socially accomplished.

THE 'NEW PARTICLES': CHARM AND COLOUR

We can now turn from the abstract sketch of the interest model to its application to an historical example. The example concerns a major development which took place in high energy physics (HEP) between 1974 and 1976.[6] During this period several highly unusual elementary particles — the 'new particles' in the parlance of the time — were discovered, and a great deal of theoretical effort was directed towards an explanation of their properties. HEP theorists initially articulated a wide range of theoretical models of the new particles, but by mid-1976 a consensus had arisen that one of these models — known as 'charm' — was right, and its competitors — principally a model known as 'colour' — were abandoned. The aim of the following sections of this paper will be to use the interest model to illuminate the way in which the reality of charm was accomplished, in comparison with the failure of the alternative models to achieve that status. I want first to give a broad outline and some general discussion of the developments at issue, and then to turn to a more detailed analysis in terms of the interest model. I should, however, clarify one point in advance. Several recent studies of the social construction of scientific knowledge have focussed on debates over experimental techniques: Collins, for instance, has characterized his study of the gravitational radiation controversy as a study of negotiations over what constitutes a 'working' gravitational wave detector.[7] The study discussed in this paper concerns negotiations of a different kind. Throughout the period considered here, members of the HEP community treated experimental data as facts: no important negotiations over experimental techniques took place, but instead negotiations concerned how well the various theoretical models could cope with the data in the context of the existing traditions of HEP.[8] With this proviso in mind, then, let me now turn to an outline of the actual events.

On 11 November 1974 two groups of HEP experimenters, one from the East Coast and one from the West Coast of the USA, announced their independent discoveries of a new and highly unusual elementary particle.[9] One group named it the 'J', the other the 'psi', and by common consent it has become known as the 'J-psi'. I will return to the question of why it was unusual later; for the moment it is sufficient to note that its properties were not easily reconciled with expectations then current within the community. Within days of the discovery announcement the journals were flooded with a variety of theoretical speculations on the nature of the J-psi. Ten days after the announcement of the J-psi a further experiement revealed the existence of another unusual particle — the 'psi-prime'[10] — and detailed measurements were soon made of the properties of both the J-psi and psi-prime. In the light of these observations, many of the early speculations on the nature of the J-psi quickly came to be seen as untenable, and in early 1975 only two theoretical models survived as serious contenders for the explanation of the new particles. These models were known as 'charm' and 'colour', and it is the way in which the choice was made between them that I want to discuss here.[11]

The new particles had immediately become a major focus for experimental effort, and in response to the accumulating data the charm and colour models were progressively refined and articulated. In essence, one can say that each model came to be embodied as an ever more complex string of statements about the interactions of elementary particles. The individual statements relating to a particular model were quite distinct, and carried quite different associations, from those of its rival, but nonetheless the protagonists of both models were sufficiently ingenious to avoid contradiction with the data: each new datum was accommodated by an appropriate adjustment somewhere in the string of statements — the adjustments, of course, differing according to the model used. Thus, at any particular time, the two models were able to give quite different explanations of the same set of observations.

This circumstance offers further support, should such be needed, to the idea that there is no such thing as a crucial experiment in science; and, from a somewhat wider perspective, constitutes a neat exemplification of the Duhem-Quine thesis, the latter being the holistic proposition that scientific knowledge is an inter-connected network and that unexpected experimental observations can, in principle at least, be accommodated by appropriate adjustments anywhere within the system. Contemporary philosophers of science have approached the problems posed by Duhem and Quine in a variety of ways, but in this paper I look at how they are handled in practice by a real scientific community.

In outline, consensus developed as follows. In late 1974 there was, as I have mentioned, no obvious preferred explanation for the new particles, although charm and colour were the most frequently discussed candidates.[12] In the first six months of 1975 little new data of any significance appeared, but the colour explanation came to be seen as increasingly unlikely, although several individuals and groups continued to defend it in the literature and in conference talks. In July 1975 the discovery of the first of what proved to be a new family of related particles was announced, and this led to the wholesale abandonment of colour. From this time onwards only a handful of theorists continued to advocate colour in the literature and at conferences. In the spring of 1976 a second family of related particles were discovered, which were almost universally interpreted as charmed particles, and by summer 1976 'charm [was] ready for the textbooks',[13] as one leading theorist put it. Nonetheless, the recalcitrant colour theorists interpreted this second family as coloured particles and maintained their position. It was only in the summer of 1977, when detailed measurements had been reported on members of this second family, that they concluded that 'there was no way'[14] that colour could be defended, and abandoned it.

There were thus, as I will elaborate below, three distinct 'epochs' in the establishment of charm. In the first, from November 1974 to June 1975, a loose consensus formed in favour of charm: in the second, from July 1975 to May 1976, there was a relatively solid consensus favouring charm, and alternative models — principally colour — received little attention; and in the third, from June 1976 onwards, charm was regarded by the HEP community as a whole as established.[15] The changes in consensus going from the first to the second epoch and from the second to the third were quite clearly associated with important qualitative changes in the data — the successive discoveries of new families of particles related to the J-psi and psi-prime — and there is a great temptation to treat this as the entire analysis: 'the data decided', one wants to say — but this would be a mistake. It would be a mistake because it ignores the consequences of the Duhem-Quine thesis, i.e. it ignores the fact that the colour theorists were able to hold their model immune from falsification until well into the third epoch, and, in principle, for ever.

How, then, was the choice made by the community between the competing models? In the remainder of this paper I will argue and illustrate the following thesis: the proponents of the two models made differing theoretical responses to the emerging data, and the interaction of these responses with the pre-existing matrix of interests supported by the HEP community determined the triumph of charm and the demise of colour. What I want to show is that, as they were developed, the charm model came to intersect with the interests of an ever increasing number of the sub-cultures of HEP, and the colour model did not. One consequence of this style of analysis is that it is inevitably technical, and it should be borne in mind in what follows that the discussion of technical points will necessarily be over-simplified and abbreviated, not least because of the very success of charm's protagonists in intertwining their chosen model with all the most promising (and popular) theoretical lines in HEP in the early 1970s.

First some technical preliminaries. Three distinct forces are recognised as important in the world of elementary particles; in order of decreasing strength these are known as the strong, electromagnetic and weak interactions; (the fourth, gravitational, force is too weak to have any perceptible effect). Most particles, for instance the proton and neutron, experience the strong interaction, and such particles are known as 'hadrons'. A subdivision of the family of hadrons, which will be significant in this discussion, is made on the basis of their intrinsic angular momentum or 'spin': particles having integral spin are known as 'mesons', while particles having half-integral spin are known as 'baryons'. A few particles, for instance the electron, experience only the weak and electro-magnetic interactions, and these are known as 'leptons'.

This paper will be mainly concerned with conjectures as to the properties of hadrons, since both the charm and colour explanations of the J-psi and its subsequently discovered relatives asserted that that was what they were. As soon as the J-psi was discovered it was clear that if it was a hadron, it was highly unusual. Quite simply, it lived too long. At the time it was the heaviest particle known (more than three times more massive than the proton), and it was well established that massive hadrons are unstable, decaying rapidly to lighter hadrons with a lifetime of the order of 10^{-23} s; a lifetime, furthermore, which decreases with increasing mass of the parent. The J-psi was observed to live around 2000 times longer than any straightforward estimate of its expected lifetime, and the immediate theoretical problem was to find an explanation for this.

Both charm and colour gave explanations which depended upon variants of the 'quark' model. In this model hadrons are not seen as truly elementary particles, but rather as composites of more fundamental entities known as quarks. Mesons are pictured as made up of a quark plus an antiquark, while baryons are pictured as comprising three quarks. Quarks are thought to come in several different types, and the various species of quark are distinguished by assigning them different 'quantum numbers', in just the same way as is routinely done for the hadrons of which they are cons-stituents. Some of these quantum numbers can be understood in terms of their macroscopic equivalents — one can think of the 'spin' of an elementary particle in analogy with the angular mom-entum of a spinning top — but others cannot, and are best thought of as book-keeping devices used to explain the various conservation laws which are observed to hold in the interactions of elementary particles. Quarks are thought to carry two distinct sets of these book-keeping quantum numbers, quaintly known as 'flavours' and 'colours', and these were at the heart of the competing explana-tions for the new particles.

Let me give the charm explanation first. In the pre-psi era there were three established flavours of quarks — 'up', 'down' and 'strange'. The charm model asserted that a fourth flavour — 'charm' — existed, and that the J-psi was a meson composed of a charmed quark plus a charmed antiquark. Because the book-keeping rules for charm were those of simple addition, the net charm of the J-psi was supposed to be zero (the quark having charm + 1, the antiquark charm − 1). The charm explanation of the long lifetime of the J-psi was based on an empirical rule known as the 'Zweig rule'. The Zweig rule stated that decays of hadrons were inhibited whenever they involved the mutual annihilation of constituent quarks. This rule had been observed to hold in the few previous instances where it appeared to be applicable, and since the only way the J-psi could decay to the old hadrons was through mutual annihilation of its constituent charmed quarks, it clearly offered at least a qualitative understanding of its longevity.

Thus the charm explanation depended upon the existence of a new quark which carried a new flavour. The colour explanation, on the other hand, required no new quarks, exploiting instead the freedom of the colour quantum numbers. Quarks are supposed to come in three colours — say, red, yellow, and blue. Colour labels combine together like vectors, and the old hadrons were all supposed to be colourless — each red quark, for instance, having its colour cancelled by a red antiquark, and so on. The colour model asserted that, for the first time, in the J-psi colour was visible; that in the J-psi the colour quantum numbers of the constituent quarks did not cancel out. Thus the J-psi was supposed to be a red particle (or blue, or yellow — which particular colour was immaterial). Hadronic colour was supposed to be conserved by the strong interactions; hence the J-psi could not decay to old colourless hadrons via the strong interactions; and hence its anomalously long life span.

These, then, are the essences of the two models, and we can now proceed to look at their respective fortunes.

THE FIRST EPOCH

I will discuss the three epochs in chronological order. In the first epoch, when only the J-psi and psi-prime had been discovered, the two models faced superficially equal obstacles to acceptance. Nonetheless, the charm model outstripped its rival in popularity, and the explanation for this lies in the differing extent to which the articulations of the two models intersected with the interests of various subcultures.

The immediate obstacle to the acceptance of charm was that the Zweig rule went some way towards explaining the lifetime of the J-psi and psi-prime, but not far enough. A straightforward extrapolation of the suppression of decays, from the few old hadrons to which the rule was applicable to the new particles, led one to expect that the J-psi would decay around 40 times faster than it actually did. There were, of course, two ways of looking at this discrepancy: as one group of theorists put it, 'this is either a serious problem or an important result, depending on one's point of view'.[16] Theorists predisposed to charm saw it as the latter, and those to colour as the former, but it is highly significant that there was a third, ostensibly neutral, group which was inclined to side with charm. This was the school of 'hadrodynamicists', theorists who had invented, elaborated and used the Zweig rule throughout the 1960s. These theorists had invested time and energy in the Zweig rule; they had the expertise required to see the world constructively from its perspective, and one can straightforwardly say that they had an interest in the elaboration of the rule in the fresh context offered by the new particles. As I have mentioned, the Zweig rule was supported by relatively few observations, and if one accepted charm then a striking new datum was added to the set — namely, that the rule became more effective at higher energies — and there was new work to be done. Thus the most striking mismatch between the charm model and the data could be construed as a puzzle which intersected with the interest of an existing research network, and normal science could be, and was, done on it.[17]

Furthermore, the proponents of charm themselves provided a possible set of tools for attacking this particular puzzle. To understand why they were in a position to do this some background is called for. The existence of charm was volubly advocated by a group of theorists centred on Harvard even before the J-psi was discovered. These physicists were the spearhead of what might be called the 'gauge theory revolution'. Gauge theory is a particular example of a quantum field theory, and as such is seen as a possible explanatory base for the entire subject matter of HEP. Quantum field theory in general had been in the doldrums from the late 1950s onwards, facing intransigent mathematical problems, until it was rescued by important theoretical developments in the early 1970s.[18] As a result of these developments those few physicists who had retained an active interest in field theory came to realize that gauge theory in particular had many attractive features. Harvard was the most influential centre of field theoretic expertise, and throughout the 1970s theorists there (and, later, elsewhere) devoted a great deal of effort to the construction of models of elementary particle interactions consistent with gauge theoretical ideas. Again, this stream of practice is readily understood in terms of the prior interests of the theorists involved. Furthermore, as we shall see, several of these models came to constitute exemplars for the work of theorists not initially having an interest in gauge theory. What must be stressed at the outset is that gauge theorists did not confine their attention exclusively to charm, and that, since many physicists do not dissect their knowledge in the way I am presently attempting, successes for gauge theory anywhere fed back as support for charm (and vice versa).[19]

During the period under consideration various gauge-theoretic exemplars were constructed, and the birth of any new stream of practice based on a gauge-theoretic exemplar increased the proportion of the community for whom gauge theory constituted a central item of discourse. Gauge theory would enter the vocabulary of the theorists elaborating the model, the experimenters involved in the production of relevant and potentially relevant data, and so on. This enlargement of the 'gauge-theory speaking' community continually acted to further the 'out-thereness' of gauge theory, offering an immediate boost to the 'reality' of any new gauge-theoretic model. There was, in a very real sense, a revolution in progress in the period I am considering — and, as its most visible specific focus, charm both marked a watershed in the revolution and was reinforced by it.

To return to the story then, by mid-1974 gauge theorists had become convinced, for reasons which it is unnecessary to go into here, that for their approach to make sense — for their long-term goals in theoretical development to be accomplished — charm must exist. When the J-psi was discovered, they immediately interpreted it in terms of charm, and also gave a distinctive explanation of the super-Zweig rule required to explain its longevity.

This explanation was based on a simple model motivated by the gauge theory candidate for the theory of the strong interactions, now known as 'quantum chromodynamics', or 'QCD' for short. Two facets of the QCD model, or the 'charmonium' model as it was called, are relevant to the present discussion. Firstly, it made concrete predictions for the existence of yet more long-lived new particles, and I will return to this in my analysis of the second epoch. Secondly, it constituted an exemplar for work outside the charmonium programme itself. Let me explain what I mean by this. The charmonium model asserted that the charmed quark and anti-quark which compose the J-psi interact in a simple way consistent with the general structure of QCD. As such it was an example of a long tradition of 'spectroscopic' quark models. It differed from its predecessors in specifying the model parameters in terms of the expectations of QCD. There was, of course, no reason to limit the application of the model to the new particles, and one group of Harvard physicists published an extensive analysis of the spectroscopy of the old hadrons using an appropriate formulation of the new model, in which they showed how existing puzzles could be seen in a new light, how new puzzles could be created and solved, and so on.[20] The new model showed the existing network of quark spectroscopists how they could enrich their approach and generate new soluble puzzles, and — in the background — the rising tide of gauge theory guaranteed that such work would be of interest to an increasing fraction of the community, and not simply within the specialist network. It intersected with the interests of quark spectroscopists — they accordingly took it up, and this work fed back, by enlarging the 'charm-speaking community', as support for the reality of charm.[21]

Here we can see clearly both the analogical and multidimensional aspects of exemplars. The analogical element of the new spectroscopic quark model resided in its treatment of the quark interactions within hadrons: these were treated in essentially the same way as the interactions of charged particles within well-understood atomic systems (more on this in the discussion of the second epoch). The multi-dimensionality of the exemplar arose from the juxtaposition of at least four identifiable elements: the atomic analogy just mentioned; the gauge-theory idiom in which it was couched; the contemporary practice associated with the spectroscopic quark model; and the set of relevant data, which was redefined by the model itself. Each of these dimensions intersected with the interests of some group or groups, although, as it happened, only the dimension relating to contemporary quark modelling required any degree of sophistication, and hence it was quark-modellers (rather than, say, atomic physicists) who set about elaborating the new model.

To summarize, then: what charm theorists did in the first epoch was: (i) generate a puzzle — the super-Zweig rule — which intersected with the interests of hadrodynamicists; (ii) supply their own solution to this puzzle — the charmonium model — which had predictive power (and hence was welcomed by experimenters), and which intersected with the interests of hadron spectroscopists by enriching both their puzzles and their techniques; and (iii) point out that in a wider context charm supported, and was supported by, the gauge theory revolution. If we now turn to the activities of colour theorists in the same period we will be able to gain some inkling of why the consensus moved strongly in favour of charm. There was, in fact, considerable diversity amongst the various colour theorists, but I shall attempt to circumvent this problem by dealing only with the most tenaciously defended variant of the model — that proposed by Gordon Feldman (of Johns Hopkins University) and Paul Matthews (then of Imperial College, London, now Vice-Chancellor of the University of Bath).[22]

As I mentioned above, at the heart of the colour model was the proposition that the J-psi and psi-prime were manifestations of a new conserved quantum number called colour. This immediately explained the longevity of the new particles, and was, furthermore, historically dignified: the flavour known as 'strangeness' had been invented and accepted in similar circumstances in the 1950s. Thus, the colour model was, if anything, faced with fewer problems than charm when first proposed. However, it soon acquired a comparable obstacle to acceptance. Early in 1975, detailed experiment on the psi and psi-prime had revealed that photons (quanta of the electromagnetic field) were only to be found infrequently amongst their decay products. This was a considerable setback for the colour model since, straightforwardly articulated, the model predicted that decays involving photons would be the predominant mode.[23]

This embarrassment for colour modellers became known as the 'radiative damping' problem, and Feldman and Matthews responded by invoking what they called the 'Feynman rule'. If one follows up the reference they give for this rule one finds (in a paper co-authored by Richard Feynman) the somewhat apologetic introduction of an arbitrary suppression factor, designed simply to improve the fit to the data of a phenomenological quark model. It was this arbitrary factor which Feldman and Matthews invoked to circumvent the radiative damping problem.[24]

Here we can see important similarities and differences between the charm and colour programmes. The two programmes were similar in that each encountered unexpected difficulties with the data: the problem of the anomalous lifetime of the new particles for charm, and the radiative damping problem for colour. Protagonists in each camp were obliged to resort to philosophically *ad hoc* special pleading to overcome these problems. But the difference between the two programmes lay in the character of this special pleading. Charm theorists construed the lifetime problem as a manifestation of a super-Zweig rule — and thus related it to an existing body of practice and interests. And they went even further by supplying their own interpretation of the rule, which was constructed in such a way as to appeal to a whole constellation of interests. In contrast, Feldman and Matthews resorted to a solution of the radiative damping problem which was not only philosophically *ad hoc*, but also — and much more damagingly — *ad hoc* in a *sociological* sense: the Feynman rule, which depended solely upon an arbitrary suppression factor, drew upon no existing bodies of practice, intersected with no established interests, and thus made no converts for the colour camp.[25]

In terms of the generation of either interesting puzzles or new exemplars charm had a distinct and unarguable edge over colour. Also, although I will not discuss it in detail here, the wider context of the gauge theory revolution favoured charm: in essence this was because the charm model as articulated by most theorists was based on fractionally charged quarks, while those theorists who related colour to their long-term interest in gauge theories did so on the basis of the unfashionable idea of integrally charged quarks.[26]

THE SECOND EPOCH

For all of these reasons consensus within the HEP community moved strongly in favour of charm and against colour during the first epoch, but nonetheless the charm consensus remained tentative; at an international conference held in June 1975, Gordon Feldman recalls that around 10-15 per cent of the submitted papers favoured colour rather than charm, and there were no less than four independent review talks on colour given, as compared with only one on charm.[27] However, a dramatic change soon took place. In July 1975 the discovery of what proved to be the first of a whole new family of related long-lived particles (which I will refer to as the 'chi' particles) was announced, and at another international conference held in August 1975 (at which time evidence had been announced for three such chi particles), a marked change in the balance of power was evident; Haim Harari, who gave the theoretical review of the new particles, only briefly discussed colour and concluded that 'in spite of the ingenuity that went into some of these models their gross features are incompatible with the experimental observations';[28] and James Bjorken, in his conference summary talk, described charm and the charmonium model as 'standards of reference [which] an experimentalist will naturally use to interpret his data'.[29] The July 1975 announcement of the first chi particle thus marked the beginning of the second epoch, and from that point onwards Feldman and Matthews were left as the only public advocates of colour.

This transformation from the loose consensus of epoch one to the more solid consensus of epoch two is extremely interesting.

The first point to be made is that, as I mentioned earlier, the charm model did predict the existence of a new family of particles. The 'charmonium' model, as the Harvard theorists called it, required the existence of five new long-lived particles less massive than the psi-prime, into which the psi-prime would occasionally decay with specified characteristics. Thus, at first glance, one is tempted to assert that the HEP community acted in accordance with some model of scientific rationality akin to that implicit in Lakatos' idea of a 'research programme' in favouring charm in the second epoch; the charmonium model, one might want to say, involved a progressive problem-shift which received increasing empirical support throughout the second epoch. In more straightforward language, one might rephrase this as that it was obvious to all concerned that the charmonium model was right. This was, I think, how the situation appeared to the majority of the HEP community, but, even so, some explanation of why certain explanations are seen to be obvious is called for, and this point requires some emphasis before I give my own analysis of 'obviousness'.

The difficulty which arises in any simple explanation of the success of the charmonium model is connected with the lack of explicit criteria of 'confirmation'. Why, for instance, in August 1975, were the three chi particles then established seen as empirical support rather than as refutation for the charmonium model which predicted five? This question is not without point, since, as we shall see, the colour model, too, led one to expect the existence of particles intermediate between the psi and psi-prime. One might answer the question with the assertion that the community was guessing that a further two particles would be discovered. This is true, but the question then remains as to which factors disposed physicists to make that particular guess. Furthermore, by the summer of 1976 two further chi particles had been tentatively identified — making up the required total of five — and hailed as confirmation of the model.[30] However, these latter two particles had properties which were in quite clear disagreement with the straightforward predictions of the model, and the 'it's obvious' explanation remains problematic even taking these into account.

This objection might seem a mere quibble, in the light of the evident weight given by the HEP community to the correct prediction of five chi particles, but to indicate that the quibble must be taken seriously let me refer briefly to later developments. In the period subsequent to 1976 the charmonium model was worked out by theorists in increasing detail and with increasing sophistication, and these theorists came to conclude that at least one of the chi particles — the least massive — had properties inconsistent with the predictions of the model. They concluded that, if this particle actually was a member of the same family as the others, then something must be radically wrong with the model. At the same time a high-precision experiment was underway at Stanford to investigate the properties of the chi particle. This 'Crystal Ball' experiment, as it was known, failed to find any evidence of the lightest chi particle at the previously reported mass — but did find evidence for the 'same' particle at a considerably higher mass, much closer to the prediction of the model.[31]

This is not the place to elaborate on these later developments, but it is clear that, during the second epoch, the partially worked-out charmonium model was instrumental in maintaining the reality of some rather tenuous data, even while the data was seen as supporting the reality of the model. Hence one can only conclude that the dominance of charm in the second epoch stands in need of some explanation, and that an explanation in terms of, say, 'research programmes' and 'novel facts' would run into severe difficulties — since at least some of the 'novel facts' in question were stabilized only by the model they were apparently confirming.[32]

How then is one to understand the success of the model? The explanation I will give is in terms of the intersection of interests. The forces which had been at work in the first epoch — the active streams of work on the Zweig rule and the new QCD-related quark spectroscopy, and the increasing momentum of the gauge theory revolution — could only be reinforced by any sign of confirmation of the charmonium model, and vice versa. And to these forces were added new ones.

To unravel the new sources of support for charm in the second epoch it is necessary first to outline the theoretical basis of the charmonium model. The model interpreted both the psis and the chis as bound states of charmed quarks, and it departed radically from earlier quark models in its assumption that charmed quarks are much heavier than the old up, down and strange quarks — so heavy, in fact, that within hadrons such as the psis and chis the constituent charmed quarks move very slowly (non-relativistically) in comparison with the highly relativistic motion entailed by quark models of the old hadrons. Now, non-relativistic motion is much more tractable theoretically then relativistic motion. In particular, the charmonium model depended upon an image of a charmed quark non-relativistically orbiting its antiparticle under the influence of a central potential, in exact analogy with the atomic system of an electron orbiting a positron (the antiparticle of the electron). The latter system is known as positronium, and is one of the textbook applications of quantum mechanics to atomic physics, being almost identical to the hydrogen atom in its formal treatment. The name charmonium was coined to make explicit the parallel between the envisaged structure of the new particles and the simplest and best understood atomic structure; and the psis and chis were pictured in the charmonium model as energy levels of the charmonium system in exact analogy with the energy levels of the positronium spectrum. The charmonium model illustrates perfectly the role of an exemplar as the concrete embodiment of an analogy relating a new field of research back to an established body of practice.

The model was intuitively transparent to any trained physicist, and this had two important consequences. Firstly, whenever the model encountered a mismatch with reality the resources were available to essentially anyone to attempt to fix it up, and for others to appreciate such work. In order to calculate the charmonium spectrum one had first to specify the potential in which the quarks moved. The earliest models chose simple potentials, and as data accumulated and discrepancies were noted with the predictions, more sophisticated potentials — involving 'hyperfine' splittings — were introduced to explain them in complete analogy to the detailed treatment of atomic spectra. Given the undergraduate training of physicists and the crude statement of the charmonium model, the pattern of extension of the model to the puzzles posed by the chi particles was obvious — and as data accumulated on the chi particles it became food for an active stream of normal science.[33]

This stream of normal science was one clear source of fresh support for charm during the second epoch. Another source was less direct but more fundamental: namely, the charmonium model made quarks themselves 'real'. To fully appreciate this argument requires some knowledge of the history of the quark concept, which I can only outline here. Since their invention in 1964 it had never been clear to physicists what to make of quarks: were they real objects, or were they no more than mnemonics for observed regularities? Had it proved possible to isolate a single quark and study its properties this question would have been resolved, but an extensive programme of experiment had failed to do so; quarks — if they existed — remained obstinately confined within hadrons. Theoretical models based on quarks had been enormously successful in explaining various features of the data throughout the 1960s and 1970s, but nonetheless all of these models were based on assumptions or approximations which made little sense if taken too seriously. The constituent quark model of hadron masses, for

instance, was based on approximations which could not be justified in the context of the theory of special relativity — and relativity has always carried more weight in HEP then quarks.

Now, what the charmonium model did was to show that *if*, by
some lucky chance, there happened to exist a species of quark
which was very heavy, *then* it was possible to construct models of
hadrons containing them which were not based on 'nonsensical'
assumptions. No fundamental principles had to be ignored in the
charmonium model: if charmed quarks were heavy, then the
charmonium spectrum made just as much sense as the atomic
spectra on which physicists are weaned. And furthermore, if the
charmonium model worked — if it fitted the data — the quarks
were just as real as electrons and positrons. Of course, this
reasoning would strictly speaking only apply to the heavy, charmed
quarks, but since these were to be seen as simply a new flavour of
the old, light quarks, these, in their turn would be seen as real.
There was thus great scope for positive feedback here. If the
charmonium model worked, quarks would become real and doubts
over the significance of the considerable amount of work being
done on the basis of the quark model would be reduced; the entire
community would be disposed to see puzzles thrown up by any
variant of the model as valid and worthy of attention, and so on.
It was a classic bootstrap situation in which the more one believed
in and utilised the quark concept the more one was inclined to see
any indication from the data as validation of the charmonium
model, and the more one believed the charmonium model to have
been confirmed, the more one was inclined to believe in and utilize
quarks.

The massive intersection of interests is, I believe, the only possible explanation for the dramatic loss of support for colour in the
second epoch. It is particularly noteworthy that those physicists
supporting colour on the basis of models which could encompass
both colour and charm at once abandoned the former for the latter.
Also, in the third epoch, when charm had finally become institutionalized, all of the major retrospective review talks concentrated on the quark-charm-gauge theory nexus.

The activities of Feldman and Matthews in defence of colour
during the second epoch only highlight the importance of the
intersection of interests. Their response to the discovery of the chi
states was completely *ad hoc* in my sociological sense. They pointed
out that the existence of the chis did not falsify colour, but beyond that they had little to say. The colour model predicted the
existence of coloured versions of all the old colourless hadrons.
Feldman and Matthews had interpreted the J-psi and psi-prime

as coloured versions of particles called 'rho mesons', and similarly they were at liberty to interpret the chi particles as coloured 'pions' and so on. Unfortunately they were unable to discuss what the masses of the chis should be, how they related to the psis, how long they should live – in short, all of the questions experiment could investigate.[34] This was because their style of explanation was based on group-theoretical symmetry arguments and, for reasons I will not go into here, could not easily be adapted to the dynamical perspectives mentioned above which made charmonium so attractive. Thus the discovery of the chi states did not lead the colour model to intersect with any new interests – it simply remained silent on the topic – and it was stripped of almost all of its adherents by the success of charm.

THE THIRD EPOCH

So much, then, for the second epoch. It is now time to go on to the third and final epoch in which charm was enthroned as the real thing. Despite the success of the charmonium model in the second epoch, it could not be said that it demonstrated that charm, as required by gauge theories, existed. The charmonium model only required that some new species of heavy quark existed, not necessarily carrying the specific quantum numbers associated with charm. As a dynamical model anathema to the philosophy of colour its success was quite sufficient to bring about the demise of the latter, but to establish the existence of charm something more was required.

That something more was the observation of 'naked charm' – particles made up of a single charmed quark plus an old non-charmed quark and hence carrying non-zero total charm, in contrast to psis and chis in which the charm of quarks and antiquarks cancelled out. To cut a long story short, the first manifestation of naked charm was found in April 1976, in the shape of a long-lived meson known as the 'D'. The decays of this particle were seen to be highly unusual; and unusual, furthermore, in exactly the way the protagonists of charm had long predicted.[35] This was enough to institutionalize charm.[36] In view of the forces at work which I have discussed earlier, and the ability of charm to provide an acceptable framework for normal science on this most recent new family of particles, this surely calls for no further explanation, and none will be given.

From the point of view of this paper the most interesting event in the third epoch was the abandonment of colour by Feldman and Matthews. Interestingly enough, this did not come about at once – indeed, as soon as the discovery of the D had been announced they published a paper entitled 'The Discovery of Coloured Kaons?'[37] – but by the summer of 1977 they had been worn down by the accumulating data. Their reasons for abandoning colour centred on rather technical issues, but I will attempt to give a brief sketch. In the 'coloured kaons' paper they had given definite predictions, which differed markedly from the equivalent predictions of charm, for a more massive particle related to the D. Unfortunately, subsequent experiment categorically supported charm and refuted colour. Since Feldman and Matthews were hardly in a position to challenge the fundamentals of the experimental method, they sorted through the resources of their model searching for a new candidate having the required quantum numbers to partner the D in agreement with experiment. Here they collapsed. Their only suitable candidate explained the existing observations, but also predicted phenomena which were not observed. They could have fixed this up by the invention of some mechanism analogous to that used to explain away the old radiative damping problem, but this would have been hopelessly *ad hoc*. Gordon Feldman certainly saw it as a totally meaningless manoeuvre and, as he put it, 'I made sure we were wrong . . . there was no way . . . these particles are presumably not . . . colour.[14] Colour could have been preserved from falsification, but the cause was manifestly lost – it would have served no purpose.

The point to be stressed here is that by this time it made essentially no difference to the reality of charm whether Feldman and Matthews maintained their defence of colour or not. Their sociologically *ad hoc* avoidance of falsification had failed to give birth to any new streams of practice, and they had been reduced to a discourse-community of two. Quite literally, they were reduced to talking to themselves. As Professor Feldman put it: 'just to be able to talk to other physicists I have to be able to speak about it [i.e. charm]'.[14] In the eyes of the HEP community, the colour model was irrevocably attached to its authors, Feldman and Matthews, rather than residing 'out there' in the real world.

This completes my account of the establishment of charm and the demise of its principal rival, and the interpretation of this episode in terms of the 'interest model'. The concepts of 'exemplars' and 'interests' have been used to explain why the existence of charm came to be embodied in the practice of an increasing number of groups of HEP theorists, and why a similar development failed to occur in the case of colour. Through a series of exemplary achievements – giving solutions to its initial problems which were fruitful in a whole variety of directions, and tying these solutions into a gauge-theoretic framework — and the sociologically *ad hoc* failure of colour to emulate these, charm became entrenched in the practice of several sub-cultures of theoretical HEP, and in the process became real.

NOTES AND REFERENCES

1.   More extensive documentation and analysis of this case-study is to be found in A. Pickering. 'The Role of Interests in High-Energy Physics: The Choice between Charm and Colour', in K.D. Knorr, R. Krohn and R. Whitley (eds), *The Social Process of Scientific Investigation. Sociology of the Sciences, Volume IV, 1980*. (Dordrecht and Boston: Reidel, 1981), 107—138.
2.   Kuhn, T.S. *The Structure of Scientific Revolutions*, Chicago and London: Chicago University Press. 2nd ed. 1970. 174—210.
3.   *ibid.,* 187.
4.   The relation between exemplars and analogies has been emphasised by Margaret Masterman, 'The Nature of a Paradigm', in I. Lakatos and A. Musgrave (eds), *Criticism and the Growth of Knowledge*, Cambridge, 1970.
5.   As Kuhn has remarked (*op.cit.,* note 2, 187—204) the utility of concepts such as exemplars (and, by implication, expertise, investments, interests, and so on) derives from the existence of an inarticulated, tacit component of scientific practice (see M. Polanyi, *Personal Knowledge,* London: Routledge and Kegan Paul, 1973; H.M. Collins, 'The TEA Set: Tacit Knowledge and Scientific Networks', *Science Studies,* 4. 1974. 165—186., and 'Building a TEA Laser: the Caprices of Communication', *Social Studies of Science,* 5 1975. 441—50).
6.   The empirical data on which this paper is based derive from an extensive review of the HEP literature (technical and popular), and correspondence and interviews with around 60 leading physicists, including the main protagonists of the 'charm' and 'colour' models (see below).
7.   Collins, H.M. 'The Seven Sexes: a Study in the Sociology of Phenomenon, or the Replication of Experiments in Physics', *Sociology,* 6. 1976. 141—184.
8.   HEP theorists and experimentalists constitute distinct professional groups. The accomplishment of the reality of objects clearly requires the involvement of both groups, and the simplest way in which this can come about is through each group regarding the products of the other as 'fact'. This is by no means always the case in HEP, but, since much of the data on the new particles was produced using *standard* techniques, solidly rooted in practice, there was little incentive for theorists to query the empirical material at their disposal. (For a minor qualification to this, see the discussion of the 'chi' particles given below.)
9.   Aubert, J.J. *et al.,* 'Experimental Observation of a Heavy Particle J', *Physical Review Letters,* 33. 1974. 1404—5, and J.—E. Augustin, 'Discovery of a Narrow Resonance in e^+e^- Annihilation', *Phys. Rev. Lett.,* 33. 1974. 1406—7.
10.   Abrams, G.S. *et al.,* 'Discovery of a Second Narrow Resonance', *Phys. Rev. Lett.,* 33. 1974. 1453—4.
11.   An analysis of the failure of other proposed models of the new particles would go along exactly the same lines as that to be given for colour.
12.   The three major rapid publication HEP journals are *Physical Review Letters, Physics Letters* and *Lettere al Nuovo Cimento.* Between them they published 62 theoretical papers submitted before the end of January 1975 on the J-psi and the psi-prime. Of these articles 24 favoured charm, 7 colour, 11 the so-called 'IVB hypothesis' which was

quickly abandoned, and 14 unconventional models (the remaining 6 concerned the application of conventional techniques to the data).

13.   Ellis, J. 'Charm, Après Charm and Beyond', Lectures presented at the Cargèse Summer Institute on High Energy Physics, Cargèse, Corsica, 5—22 July 1977 (CERN preprint TH. 2365 [1977] p.1).

14.   Feldman, G. interview, 10.5.78.

15.   I will not document these changes in consensus in detail here. Since the new particles generated a great deal of excitement within the HEP community they were intensively reported in the scientific press, and one can obtain an overview of the evolution of consensus from the pattern of this reporting. Thus *Science* carried one report (186. 1974. 909—11) in epoch one, which indicated that there was no theoretical consensus on the nature of the new particles, and two reports (189. 1975. 443—5 and 191. 1975. 452, 492) in epoch two which discussed only charm. The most detailed contemporary popular reporting was in the *New Scientist;* from November 1974 to March 1975 a variety of theoretical explanations were noted, but in the thirteen articles subsequently published during the first two epochs, charm was the only explanation to be discussed.

16.   M.K. Gaillard, B.W. Lee and J.L. Rosner, 'Search for Charm', *Reviews of Modern Physics,* 47. 1975. 277—310. p.300.

17.   To give some sort of quantitative feel for this, let me note that *Physics Letters* (the major European rapid publication journal) published 14 papers on the Zweig rule *per se* (i.e. treating it as a fact to which the new particles were relevant rather than as part of a putative explanation of the new particles) during epochs one and two (submission dates of these papers range from 12.5.75 to 7.6.76). I have been unable to locate any analogous papers, in any journal, devoted to the parallel problem engendered by colour (see below).

18.   The important breakthrough was the demonstration by G. 't Hooft, then a graduate student at the University of Utrecht, that gauge theories possessed a most desirable mathematical property known as 'renormalizability' (G. 't Hooft, 'Renormalization of Massless Yang-Mills Fields', *Nuclear Physics,* B33. 1971. 173—199.).

19.   The most important independent source of support for gauge theories during this period came from the accumulating data on 'weak neutral currents', predicted by gauge theory models and first discovered in 1973. For popular accounts of the relevance of these, see 'Neutral Currents: New Hope for a Unified Field Theory', *Science,* 182. 1973. 372—374, and 'The Detection of Neutral Weak Currents', *Scientific American,* 231, Dec. 1974. 108—119.

20.   De Rujula, A., George, H. and Glashow, S.L. 'Hadron Masses in a Gauge Theory', *Physical Review,* D12. 1975. 147—162.

21.   For instance, O.W. Greenberg of the University of Maryland was, in the 1960s, one of the pioneers of both colour and quark spectroscopy. During epoch one he propounded colour as an explanation of the new particles. Discussing his reasons for subsequently abandoning colour in favour of charm (interview, 11.5.78) he remarked that 'QCD has made a lot of impressive successes', and cited a recent fit of baryon masses, based on the exemplar discussed here, as a 'great advance'.

22.   The lack of unanimity amongst colour modellers was clearly significant in their failure to achieve objective status for their viewpoint: each variant of the model was readily seen to be proper to the individual or small group proposing it.

23.   This problem had become apparent by January 1975. See O.W. Greenberg's review of colour models, 'Electron-Positron Annihilation to Hadrons and Color Symmetries of Elementary Particles', in A. Perlmutter and S. Widmayer (eds), *Theories and Experiments in High-Energy Physics,* New York: Plenum Press, 1975.

24.   Feldman, G. and Matthews, P.T., 'Has $\psi 4$ Already Been Observed at Stanford Linear Accelerator Center?', *Phys. Rev. Lett.,* 35. 1975. 344—6.

25.   One could labour this point further, but probably the following comment of Feynman and his colleagues on introducing the suppression factor in their original paper will suffice: '. . . in a most unsatisfactory way, [we] have included . . . an adjustment factor F . . . This is frankly just empirical fitting' (R.P. Feynman, M. Kislinger and F. Ravndal, 'Current Matrix Elements from a Relativistic Quark Model', *Phys. Rev.*, D3. 1971. p.2707).

26.   In the original formulation, given in 1964 by M. Gell-Mann and G. Zweig, quarks were supposed to carry fractional electric charges (either one- or two-thirds of the charge of the electron). Building upon the exemplary achievement of these two physicists, the main line of quark theory, including gauge theory formulations, had continued to assume fractionally charged quarks. With the introduction of the colour quantum number it became possible to envisage integrally charged quarks.

27.   Zichichi, A. (ed.), *Proceedings of the EPS International Conferencee*, Palermo, Italy, 23-28 June 1975, Bologna: Editrice Compositori, 1976.

28.   Kirk, W.T. (ed.), *Proceedings of the International Symposium on Lepton and Photon Interactions at High Energies*, Stanford, 21-27 August 1975, p.323.

29.   *ibid.*, p.991.

30.   See, for instance, the review given by B.H. Wiik, 'Plenary Report on New Particle Production in e^+e^- Colliding Beams', in *Proceedings of the 18th International Conference on High Energy Physics*, Tbilisi, USSR, July 1976, pp.N75—85.

31.   I am extremely grateful to Professor James Bjorken of the Stanford Linear Accelerator Center who first made me aware of these developments.

32.   For several instances of unreflective reference to 'novel facts' in connection with Lakatos' methodology of research programmes, see the essays contained in *Method and Appraisal in the Physical Sciences*, C. Howson (ed.), Cambridge, London, New York and Melbourne: Cambridge University Press, 1976.

33.   The first and simplest charmonium model was constructed by two Harvard theorists, H.D. Politzer and T. Appelquist. Although they had given seminars on the model, predicting the existence of very long-lived particles even before the discovery of the J-psi, they did not submit their work for publication until the discovery had been announced; (their first paper on the model was 'Heavy Quarks and e^+e^- Annihilation', *Phys. Rev. Lett.*, 34. 1975. 43—45). The model was quickly adopted by theorists all over the world, leading to a stream of increasingly sophisticated phenomenology of the new particles which continues to the present.

34.   Feldman and Matthews stated their position on the chi particles in 'A Case for Colour', *Nuovo Cimento*, 31A. 1976. 447—486., and 'Colour Symmetry and the Psi Particles', *Proceedings of the Royal Society, London*, A355. 1977. 621—627.

35.   Goldhaber, G., Pierre, F.M. *et al*, 'Observation in e^+e^- Annihilation of a Narrow State at 1865 MeV/c^2 Decaying to K π and K π π π', *Phys. Rev. Lett.*, 37. 1976. 255—259.

36.   See the *Proceedings of the 18th International Conference on High Energy Physics*, Tbilisi, USSR, July 1976 — for instance, the reviews given by R. Schwitters (pp.B34—39), B.H. Wiik (pp.N75—85) and A. De Rujula (pp.N111—127). See also the popular scientific press of the time, for instance 'Charmed Quarks Look Better Than Ever', *Science*, 192. 1976. 1219—1220; Iliopoulos Wins His Bet', *Nature*, 262. 1976. 537—538; and 'Naked Charm Revealed at Stanford', *New Scientist*, p.413 (20.5.76).

37.   Feldman, G. and Matthews, P.T., 'The Discovery of Coloured Kaons?', *Phys. Lett.*, 64B. 1976. 353—358. This was the last paper which Feldman and Matthews were to publish on colour. My account of subsequent developments in their position is based on interview and correspondence with G. Feldman.

PART THREE

The interaction of science and technology

As was noted in the General Introduction, our concern is with science purely and simply as a phenomenon; that is, with activities generally accepted and described as scientific, with knowledge and culture generally so described, with contexts generally so described. Our concern with technology is of just the same kind. Sociological studies must attempt to take science and technology as they find them, and not themselves stipulate what scientific and technological activities ought to consist in. The adoption of this perspective does not, however, solve the problem of the identification of science and technology, since there is no unchallenged social consensus upon what either actually consists in, or upon where the one ends and the other begins. Conceptions of the nature of science and technology, how they are demarcated, and how they are related to each other, are historically variable and endlessly controversial. Otto Mayr's contribution conveys a fine sense of this variability. He points out that all the diverse constructions of the science-technology relationship are themselves in need of description and explanation (a simpler task than that of describing and explaining the relationship 'itself'). And he stresses the major ideological significance of different constructions of the relationship. Mayr's brief contribution reveals remarkable sociological insight, scarcely putting a foot wrong in its treatment of notoriously difficult conceptual and methodological problems.

To accept Mayr's argument is to accept that any general model of the science-technology relationship must inevitably be inadequate, and problematic in its application. Such a model may, nonetheless, retain some utility for specific purposes. There are, in our society, activities and roles ordered around the extension of knowledge and competence without any regard to practical application. There are other roles and activities concerned solely to increase and improve the stock of existing practically useful techniques, processes and artefacts. These two distinct kinds of role currently

147

serve as paradigm cases when science and technology are referred to: this reflects the widespread implicit understanding that science and technology are partially separate communities with partially separate cultures. And it is legitimate to enquire into the form of the relationship between these two cultures. It has to be recognized that all conceptions or models of the relationship will have limitations, will indeed be strictly invalid, and will offer a constant temptation to false inference and overgeneralization. Nonetheless, with cautious use such models can have considerable pragmatic value.

Two such models demand attention here. First, there is the traditional hierarchical model which treats technology as applied science (cf. Table 1). On this model, the production of new knowledge is the concern of science: scientists creatively construct new hypotheses and theories, and rigorously evaluate them against observations and experimental results drawn from nature. Technology is then the routine activity of working out and realizing the 'implications' of scientific theories. It is a humdrum, uncreative activity crucially dependent upon basic science.

This model has enjoyed widespread acceptance over a long period and still enjoys a degree of support (cf. Layton, 1977; also the quotations at the start of Gibbons' and Johnson's contribution here). But its credibility is mainly related to its utility as a legitimating device, and is only tenuously related to evidence and considered argument. The model treats every technological development as based upon some previous scientific advance, and 'since Francis Bacon this had been a fundamental tenet for the social licence of basic science' (Layton, 1977: 206).

Hierarchical conceptions have suffused some massive, expensive and remarkably inept empirical investigations of the relationship between science and technology, two of which, Project Hindsight and Project TRACES, have over the years metamorphosed from sources of information into empirical phenomena in their own right, exerting a certain morbid fascination over those concerned with the history of science policy (Kreilkamp, 1971; Layton, 1977). The assumption of a hierarchical relationship has also become incorporated into one of the more widely disseminated stereotypes of the exploitation of science, the so-called 'lag' theory. To apply the 'lag' theory one takes a technological innovation and works backwards along the lines of cultural change which terminate in it, noting particularly those points at which a scientific discovery or some part of the knowledge of pure science was involved. One then selects one such discovery, preferably the most recent discovery encountered, and accounts the later technological innovation as the logical consequence of that discovery. The innovation thus redounds to the credit of basic science, and the 'lag' between the discovery and the innovation which it 'implies' serves as a measure of the competence of technologists.

At the present time advocacy of the hierarchical model is in a steep decline, and among those involved in the serious study of science and technology it may already be defunct. There is general agreement that it travesties technological activity, and an increasingly widespread recognition that it offers a seriously inadequate image of science also. As far as technology is concerned, our understanding has been transformed by recent case studies and historical investigations which set new standards in thoroughness, methodological awareness and exegetical sophistication. It is no longer plausible to conceive of technology as applied science.

One thing which practically any modern study of technological innovation suffices to show is that far from applying, and hence depending upon, the culture of natural science, technologists possess their own distinct cultural resources, which provide the principal basis for their innovative activity. New kinds of instruments are predominantly developments of older kinds, new materials of old materials, and similarly with machinery, procedures, processes, and designs. Technologists are the recipients of an inherited culture, a good part of which is non-verbal (Ferguson, 1977). Their primary activity involves the development and articulation of that received culture in the course of solving their practical problems: technologists apply technology. Hence the state of science sets no necessary constraints on the possibilities of technology, nor, conversely, does a scientific advance automatically indicate a corresponding leap forward in technology. In most cases therefore one must seek to account for a specific innovation as a possible development of the existing technological culture which is realized in response to the pull of external demand, or possibly in response to the needs generated by the interdependence of different aspects of technology upon each other (Hughes, 1976).

It remains true, of course, that technologists do, at times, make use of the findings and theories of basic science, and that this has great significance. But it is equally the case that scientists make occasional use of the ideas and artefacts of technology, and that this is of considerable importance in understanding the growth of science (see Kuhn, 1961, for a particularly interesting illustration). Moreover, the way in which technologists use or 'apply' science is profoundly misrepresented by the hierarchical model.

Case studies of technological innovation, even the most clearly 'science-based' innovation, do not offer accounts of routine activity wherein the 'implications' of science are deduced and realized. Innovation is characteristically shot through with uncertainty: even the risks of failure it entails cannot be reliably calculated (Schon, 1967). Existing knowledge is always liable to prove insufficient: additional unexpected features invariably appear in every new artefact or material or process, throwing new difficulties in the path of further advance, or possibly proving advantageous and allowing the successful outcome of a project which would otherwise have been doomed to failure. Projects may set off in one direction and end 'successfully' at an entirely unlooked for destination as the result of uncertainty of this kind. Where a scientific theory is used, there may be disagreement or obscurity as to what precisely its 'implications' are. And those 'implications' may be disconfirmed by practice, so that theory has to be reconsidered, and new and different 'implications' 'deduced' from it. Or the theory may itself be reconstructed, replaced or supplemented as the technological activity proceeds. (Changes of this kind occurred, for example, as a result of the 'anomalous results' mentioned by Gibbons and Johnson (182) as arising in the course of work on semiconductors: far from this work merely realizing the technological implications of a physical theory, it simultaneously advanced both technology and physical theory.)

All this, of course, merely confirms what would be expected on the basis of the analysis in the previous section. There, it was argued that knowledge cannot have logical implications in a particular case, since how the case falls under the relevant concepts — what the case is the same as — is an open-ended endlessly problematic question. If this is so then technologists are no more able than are scientists to draw specific logical implications from a body of knowledge. Science must rather stand as a resource for possible inventive/creative use by both scientists and technologists, and so too indeed must the knowledge of technology itself. Technology must be conceived of as an inventive activity predominantly developing the scope and extending the significance of technological knowledge via its active, creative application, and to a lesser extent, doing the same with whatever parts of science it takes up and incorporates. Conversely, science develops and extends old scientific knowledge, together with whatever of technology it takes up and incorporates — an account of science which involves only a minor elaboration of that set out in the previous section.

We are now in possession of the basic elements of the modern understanding of the science/technology relation (cf. Table 1.) This characterizes science and technology as distinguishable subcultures each with their own bodies of lore and competence. And

Table 1 Two alternative conceptions of Science (S), Technology (T) and the form of their relationship

BASIC MODEL	Hierarchical Technology as $\quad$ S Applied Science $\quad\downarrow$ $\qquad\qquad\qquad$ T	Symmetrical S $\longleftrightarrow$ T
FORM OF COGNITION	S creative/constructive T routine/deductive	S creative/constructive T creative/constructive
PRIMARY BASIS OF COGNITION	S nature $\;\big\}$ $\qquad\qquad\quad$ determinants $\qquad\qquad\quad$ of cognition T science $\,\big\}$	S existing science $\big\}$ resources $\qquad\qquad\qquad\qquad\quad$ for T existing $\qquad\quad\big\}$ cognition $\quad$ technology
RESULTS	S discoveries T inventions and $\quad$ applications	S inventions T inventions
MAJOR CONSTRAINTS ON RESULTS	S state of nature T state of science	S no single major constraint T no single major constraint
EVALUATION OF RESULTS	S evaluates discoveries in an unchanging context-independent way. T is evaluated according to its ability to infer the implications of S. Success in T is proper use of S; failure in T is incompetent use of S.	S and T, both being inventive, both involve evaluation in terms of contingent ends. No *a priori* reason why activity in T should not be evaluated by reference to ends relevant to agents in S, or vice versa.
COGNITIVE FORM OF THE RELATIONSHIP	T deduces the implications of S and gives them physical representation. No cognitive feedback from T to S.	T makes occasional creative use of S. S makes occasional creative use of T. S and T cultural resources.
RESULT OF RELATIONSHIP	Predictable	Unpredictable
PRIMARY MEDIATING AGENCY	Words	People

it identifies the general form of the relationship between the sub-cultures as a symmetrical one, allowing two-way *interactions* — which is not to say that the sub-cultures necessarily have equal effects upon each other, but merely that the form of the relationship is identical from whichever side it is viewed, and that nothing in the relationship itself favours one direction of causal connection over its opposite.

Note that there is nothing at all profound in this simple inter-active model of the science/technology relationship. It is of real value as a corrective to earlier models, and to the many miscon-ceived modes of thought which still survive largely as a result of the curious evaluative distinctions we continue to make between science and technology. But the model does little more than set out the general form apparent in the relationship between any two sub-cultures. Wherever and however one makes an analytic division which separates a culture into two parts, the relationship between the parts will have the basic form set out above: given what is generally understood by a culture, this basic form is simply the product of the analytical division into sub-cultures. This, however, is all to the good in the present case. Recall what Mayr had to say of the problems involved in identifying science and technology, the difficulty of demarcating the boundary between them, and the consequent limitations set upon the scope and validity of any specific conception of the science/technology relation. It should now be apparent that however science and technology are demarcated, provided only that they are demarca-ted as sub-cultures, an interactive, symmetrical model of the relationship between the entities thus defined will be appropriate. An interactive model remains relevant over a wide range of charac-terizations of science and technology: it is possible to disagree about what science and technology consist in and yet agree upon the form of their relationship.

Derek de Solla Price sets out an interactive model based upon precise stipulative definitions of science and technology. The present value of his contribution is, however, to a great extent independent of the merits or otherwise of those definitions: his discussion usefully supplements and extends what has already been said in a number of ways. First, Price supports an interactive model with evidence and argument quite different from that set out above, and likely to resonate with a different range of pre-suppositions. Secondly, his particular model is in some ways more elaborate than the 'bare-bones' account just presented. Price characterizes the actual interaction of science and technology at the present time as weak, unsystematic and difficult to predict, yet nonetheless as an important determinant of the broad rate of advance of both: science and technology are parallel structures which march in step with each other as a result of their weak symbiotic interaction. (Price accepts the full force of the metaphor of symbiosis. He regards the weak interaction with science as *essential* to the long-term vitality of technology, although in a revealing paragraph he makes it clear that this claim is a matter of faith and conviction only, and cannot be adequately justified (171—2)). Finally, and most importantly, Price turns to the question of how the science/technology relation is mediated, what the primary mechanism of interaction consists in.

His answer is in perfect tune with the conclusions of the previous two sections, that knowledge does not reside in the written word and cannot adequately be conveyed via the written word. The research front in both science and technology is, says Price, 'basically contained in people' (172). Accordingly transfer between research fronts is primarily accomplished by personal mobility (just as transfer within research fronts is accomplished, as we saw earlier). Price notes how the educational system operates as an institutionalized system for the transfer of knowledge by mobility. And he also stresses the value of high rates of mobility of personnel in scientific and technological occupations, a point supported by a number of studies of innovation (Shimshoni, 1970; Gibbons and Johnston, 1974; cf. also Freeman, 1974). The postulate that the individual practitioner is the basic unit in the transfer of knowledge and competence is near to being orthodox doctrine at the present time.

The task of conveying a proper understanding of how scientific and technological skills are combined and used in specific contexts is one we cannot hope to accomplish here. The detailed consideration of extended case studies is essential if such an understanding is to be obtained, and apart from the technical demands they make, such studies lose their value if they are condensed or presented only in part. On the other hand, one cannot pretend to know in general what one does not know in particular: to discuss the science/technology relationship without any reference at all to actual cases would be indefensible. We have settled upon an uncomfortable compromise by including Gibbons' and Johnson's brief but excellent review of the history of the development of the transistor. Without becoming unduly technical this piece does illustrate some of the general themes previously discussed, and makes apparent something of the complexity of processes of technological innovation. Moreover, anyone who is moved to a more thorough and extended examination of the events which

Gibbons and Johnson describe will find a plethora of materials available both in the scientific and the secondary literature (Braun and MacDonald, 1978).

8

Otto Mayr

The science-technology relationship

In today's industrial civilization, the words 'science' and 'technology' refer to two complexes of activities, persons, institutions, and values, and they are heavily charged with sentiment. Since they often appear side by side, it is natural to ask about their relationship or, rather, about the relationship of the various phenomena that are labelled by them. This relationship has eluded all who have tried to grasp it, and we may simply speak of 'the problem of the science-technology relationship.' [. . .]

Naturally, historians have tried to bring [their] particular skills and attitudes to bear on the problem of the science-technology relationship. How well have they done?

Let us first survey briefly a number of efforts representative of the empirical-inductive approach that explore the science-technology relationship specifically on the level of intellectual exchange, that is, that study interactions between the theoretical understanding of nature and the inventions and techniques of industry. The following selection is arbitrary but, I hope, not untypical.

A review, for example, of the invention of the 'Silent Otto' internal combustion engine in the 1870s concluded that the inventor, N.A. Otto, an ingenious former travelling salesman, was supported by a team of extremely gifted engineers, some of them with excellent academic training; nevertheless, the theoretical premises on which the successful engine was based later on proved false.[1] A study of the introduction of the loading coil into telephone technology, besides clearing up an important priority question, gives a classic example of a technological innovation — by the way, one of huge financial significance — that was the direct result of a discovery in mathematical physics.[2] An analysis of John Ericsson's caloric engine showed how the engineer, by applying the currently conventional theories of heat, suffered a sensational failure because he was violating the Second Law of Thermodynamics, which was then just being formulated.[3] An inquiry into the origins of the industrial production of sal ammoniac in the 18th century showed that the theoretical understanding of that substance grew and matured simultaneously with several industrial processes for synthesizing it; a connection between the two developments, which is likely to have existed, could not be documented.[4] A later paper by the same author investigated a crash program by the French government, in the 1780s and 1790s, to increase the production of saltpeter: the program, presided over by none other than Antoine Lavoisier, was successful in spite of its failure to develop methods of producing saltpeter artificially: Lavoisier had simply rationalized the age-old natural methods and improved some technical details.[5] An analysis of the attempts of various celebrated nineteenth-century physicists to solve the problem of speed regulation of prime movers revealed that their results failed to gain practical acceptance simply because indeed they had no practical advantages to offer.[6] A study of the introduction of the high-speed steam engine in the 1860s and 1870s, finally, showed how some little-known American mechanical engineers produced a method of balancing steam engines, together with a remarkable theoretical explanation which was not only readily accepted in America but was received in Europe as almost a sensation.[7]

This selection could doubtless be extended, although the literature in this category is not very ample. One is likely to read case studies of this sort with mixed feelings. As far as they go, they seem sound and uncontroversial. They conform to the rules of conventional historical workmanship, and, while one could quibble about details, their conclusions, within their limited scope, appear reasonable.

On the other hand, where do such efforts lead? They do not yield data that fall into simple patterns. They are empirical, but they do not result in points on a graph through which one could draw a smooth curve. They frustrate attempts to induce generalization from them, and for the synthetic historian they must be more embarrassment than help. Moreover, if such a small sample showed so little convergence, would not a greater number of such case studies have even more widely scattered results?

The apparent failure of the empirical method characteristic of their craft has driven many historians to embrace the customary method of philosophers, namely, to try to establish the relationship by means of theoretical reasoning and logical deduction. That is, we begin by formulating logical definitions of 'science' and 'technology' and go on to build models of their mutual relationship; or, in reverse order, we first postulate a model and then define 'science' and 'technology' to fit it. In either case, it is only after we have arrived at a hypothetical model that we look for historical evidence to substantiate it.

The models of the science-technology relationship that one encounters most frequently have one feature in common: they portray science and technology as two distinct entities that are opposite and mutually exclusive, implying that things pertaining to science and technology can be separated like black and white beans. We hear, for example, that science and technology relate to each other like dancing partners or like mirror images and that science is the juice squeezed out of the lemon of technology. Often it is said that science and technology were joined in marriage in the 19th century, a metaphor that concedes, at least, that the relationship was not constant throughout history.

Sometimes the two are placed into a hierarchical relation, in which one of them originates everything that is new and significant, while the other merely receives, reacts, carries out. The role of initiator is usually, but not always, given to science. It can also be heard, although not so often, that scientific advance is only a consequence of technological practice.

Models or, rather, metaphors like those above state the opposite nature and the mutual exclusiveness of science and technology as sharply as possible. Other models are used that are more conciliatory and less rigorous, that allow for some overlap and common territory between the two. Sometimes, for example, science and technology are pictured as spread out along a spectrum, with 'pure' science on one end, the traditional crafts on the other, and 'applied science' and the 'engineering sciences' somewhere in between. Still another model presents science and technology (or theory and practice) as poles of a magnet, implying that they are not separate but, rather, both part of one whole.

Among this offering of models everyone will have his preferences. But, to evaluate them more objectively, we must ask: what is the difference, the characteristic distinction, between science and technology? If the two are indeed mutually exclusive opposites, what is the criterion that separates them?

The terms, 'science' and 'technology' refer to phenomena on many different levels. They refer to bodies of knowledge, to activities, to the goals and motivations behind such activities, to forms of education, to social and professional institutions, etc. If we could draw boundaries between science and technology on each of these different levels, we would discover that the resulting maps would not be congruent. If we can make out boundaries at all between what we call science and technology, they are usually arbitrary. Take, for example, the level 'bodies of knowledge.'

Traditionally we regard physics as a science and the manufacture of diesel engines as a technology. But what is thermodynamics, when textbooks are available in all shades of emphasis, ranging from purely practical concerns to the most esoteric theory? How should we classify subjects like aerodynamics, semiconductor physics, or even medicine?

Nor can we tell a scientist from a technologist by watching him at work. Both may employ the same mathematics, both may work in laboratories of like appearance, both may be seen getting their hands dirty in manual labour.

If we go on to consider actual historical figures, the distinction between science and technology becomes no clearer. At all times, 'pure' scientists were rare, while many scientists of high reputation were active contributors to technology (Archimedes, Galileo, Kepler, Huygens, Hooke, Leibniz, Euler, Gauss, Kelvin); and many engineers became famous in science (Hero of Alexandria, Leonardo da Vinci, Stevin, Guericke, Watt, Carnot). Nowadays, practitioners seem to be more clearly identified by an academic degree or by a job title, but, if we look at their actual work, the labels again turn out to be arbitrary. Many, perhaps most, present-day 'scientists' turn out to work for technological goals, whereas academic engineers occasionally are occupied with research that has no technological applications in mind at all.

On the level of social organization, the distinctions between science and technology are just as arbitrary. If a school, academy, or professional organization carries the words 'science' or 'technology' in its name, this is more an indicator of how that concept rates on the current scale of values than of the actual interests and activities of its members. More often than not, however, science seems to have enjoyed a higher social status than technology, and professional organizations have been effective instruments of acquiring and defending such status. There is some clubbish exclusiveness about 'scientists.' Thomas Edison, for example, who perhaps did more for the systematic application of 'science' to industrial purposes than any other historical individual, was an outsider to them, fond of ridiculing formal science and boasting of how little he owed to it. Is that perhaps why the new *Dictionary of Scientific Biography* — a totally admirable enterprise, to be sure — deals with Edison[8] on roughly one page, while devoting 45 pages to Euclid?[9]

One level on which one might hope to discover a definitive distinction between science and technology is that of aim, purpose, motivation. The aim of science, one might argue, is to explain the riddles of nature; the aim of technology, to solve the material problems of human life. The question about characteristic purpose will indeed furnish a criterion that separates the two concepts in a manner that is sharp, simple, and mutually exclusive. Unfortunately, however, it is valid only on the level of semantics. If we analyze actual historical events, we find that the motives behind actions are usually mixed and complex. Often scientific as well as technological purposes are pursued, simultaneously or at different times, by the same person, the same institution, and by the same methods and techniques.

It is becoming clear, then, that a practically usable criterion for making sharp and neat distinctions between science and technology simply does not exist. If this is true, it follows that technology and science are not mutually exclusive. Those models of their mutual relationship that treat the two as mutually exclusive opposites are therefore false.

Even the other models mentioned above that are subtler and more realistic are unsatisfactory: the metaphors of the spectrum or the two opposite poles are one-dimensional, whereas the science-technology relationship involves phenomena on many levels; mathematically, it could be compared with a function of a great many variables: a dynamic model that would do justice to all would be prohibitively complex.

The apparent complexity of the problem, however, is an illusion created by the attempt to give precise and rigorous definition to terms that in ordinary language are used only loosely. The words 'science' and 'technology' are useful precisely because they serve as vague umbrella terms that roughly and impressionistically suggest general areas of meaning without precisely defining their limits. Most successfully the two words are used in conjunction; 'science and technology' together refer to an entity that actually exists in our civilization but which is impossible to divide into two parts, 'science' and 'technology.' Within this whole the two words only set accents, referring to two sets of general styles and approaches that are contrasting as well as complementary.

We have thus tested two approaches by which historians try to deal with the science-technology relationship, one empirical-inductive, the other logical-deductive. We have concluded that both fail in solving the problem. But to say 'fail in solving the problem' only means 'fail in supplying the kind of solution we looked for.' What kind of solution have we dreamed of?

Perhaps we have an unreflected, tacit conception of an ideal solution that is unrealistic. Are we not trying to discover *the* relationship between science and technology? Are we not longing for a solution that has the form of a law expressing the relationship between science and technology in a positive and verifiable manner? Is our ideal an operational formula, possibly quantitative, that could be employed in practical government and in forecasting the future?

It is true that certain branches of social science are quite un-apologetically at work trying to describe social phenomena by quantitative laws. Such disciplines as operations research or technology assessment are indeed committed to the task of fore-casting future events. When they succeed, they do so by rigorously restricting the number of variables in a given poblem through simplifying assumptions of many kinds until the processes in question submit to mathematical description. Mathematical models may thus indeed offer valid representations of social phenomena, but their validity is limited to narrowly specified circumstances.

But history is not operations research or technology assessment. It is not concerned with prediction; the concept of the simplifying assumption is alien to it, and it has never established any laws — much less quantitative laws — that were not deeply controversial. If there is one thing that it cannot supply, either for the problem of the science-technology relationship or for any other, then it is a single grand solution expressed in terms of some universal law.

Are we then to abandon the problem of the science-technology relationship as unprofitable and inappropriate? Yes and no: the problem requires redefinition.

To begin with, the reflections above in no way affect the valid-ity of our interest in historical interactions and interchanges be-tween what can roughly be labelled 'theoretical' and 'practical' activities, that is, between man's investigations of the laws of nature and his actions and constructions aimed at solving life's material problems. Such inquiries, however, can only establish specific relationships such as, for example, those between the teachings of a certain school and the inventions of its pupils; be-tween a feature of some class of power machinery and the result-ing formulation of a thermodynamic law; between the discovery of a certain chemical synthesis and the subsequent rise of an in-dustry. But they cannot lead to a general formula for the 'science-technology relationship.' Indeed, such inquiries can be, and per-haps should be, conducted under complete avoidance of the terms 'science' and 'technology.'

By a slight change of viewpoint, the problem can be given a form which is realistic and meaningful and which for its solution requires precisely the characteristic skills of the historian.

The recommended change of viewpoint is this: so far we have defined 'science' and 'technology' in our own terms and have then tried to analyze their relationship through the course of history from our own vantage point.

Instead, we should recognize that the concepts of science and technology themselves are subject to historical change; that different epochs and cultures had different names for them, interpreted their relationship differently, and, as a result, took different practical actions.

Consider only the meaning of the words! We have seen how difficult it is to define what 'science' and 'technology' mean in contemporary English. And in English the two words are comparatively new. At earlier times, different words were used. Words like 'industrial arts' and 'natural philosophy,' however, were by no means identical in meaning with their modern correspondents, no more so than the phenomena that they connoted have remained constant.

The difficulties are compounded when we go into a different language. If we only compare English and German, two languages that are linguistically as well as culturally closely related, we find that 'science' is not quite the same as *Wissenschaft,* and 'technology' is not equivalent to either *Technik* or *Technologie.* What will happen if we deal with languages that are farther removed from our own, in terms of culture and chronology, is predictable; at best we may establish vague correspondences for our 'science' and 'technology.' It would be a mistake to think, however, that *techne* and *episteme* or *ars* and *scientia* meant to the ancient Greeks or Romans what 'science' and 'technology' mean to us.

Our words 'science' and 'technology,' then, may have equivalents in other languages, but what these terms have meant to their users in various cultures and epochs has depended on the given realities of the moment. This historical relativity of the concepts again underscores the futility of our attempts to construct permanent and universal definitions of the two concepts. But such strictly philological problems do not exhaust the question. In addition, we should consider the values associated with the words.

On our own scale of values, 'science' and 'technology' rank very differently. Consider how they have risen and fallen, both absolutely and relative to each other, in the last century alone, and how their estimation varies from country to country, even at the very present! The value ratings of the concepts 'science' and 'technology' are historical variables; this is reflected in the everyday language, in the public beliefs, and in the ideologies and philosophies of each historical era. Plato's opinion of the relative values of science and technology differed from the views of Saint Benedict of Nursia, Francis Bacon, Kant, Spencer, or Marx.

This leads us back to the original problem. Judging the relative value of the two activities and justifying such evaluations inevitably means postulating specific relationships between the two. This has been done all through history, by philosophers, teachers of religion, secular rulers, and others. In our own attempts to establish this relationship, we are thus simply following an ancient tradition. The results of such efforts, however, the relationships postulated at different times and places, have varied widely. They were expressions of cultural conditions in general and of the relative roles and reputations of science and technology in particular.

To analyze such past discussions of the science-technology relationship and to interpret them against their historical backgrounds would be a rewarding task for the historian. His task, in other words, would be not to discover what the science-technology relationship actually has been in history but what previous eras and cultures have thought it to be.

Such a redefinition of the problem has several advantages. To begin with, this new task is clearly defined, it can be attacked with the historian's proven tools, and it offers realistic prospects of useful results. Second, if it is true that all our quests for a definitive analytical solution to the problem of the actual science-technology relationship are in principle doomed to futility, then it must be instructive to learn about the struggles of previous generations with this problem. Third and most important, this new focus of inquiry might have some wholesome effects on our mental attitudes. It might make us conscious of how deeply our own views of the science-technology relationship as well as of its historical interpretations are determined by unreflected ideology. When we hear the views of Plato, Bacon or Marx on the issue, is not our first impulse to ask, who was right; while, if we were to respond as historians, we would ask, how did Plato (or Bacon, or Marx) come to his view? How much does it tell us about his historical situation and how much about his particular psychological make-up? How did his contemporaries judge such views?' Even such detailed analyses of the science-technology relationship as the Pentagon's 'Project Hindsight'[10] or 'TRACES,'[11] the National Science Foundation's answer to it, tell us more about contemporary ideological attitudes towards the relationship than about the relationship itself.

Some of the crucial debates in today's politics, debates in which we all take sides, have at their core conflicting views of the science-technology relationship. Is it coincidence that two of the stronger terms of professional condemnation among historians, 'Whig' and 'Marxist,' can be virtually defined as particular views of the relationship among science, technology, and social progress? If so, much political passion is linked with our issue; detachment cannot be taken for granted. It is precisely in this regard that the systematic study of earlier struggles with the science-technology relationship would pay off. By making us conscious of the ideological aspects of our own position and its historical roots, it might help us to achieve a measure of detachment from it.

REFERENCES

1.   Lynwood Bryant, 'The Silent Otto,' *Technology and Culture* 7 (1966): 184–200.
2.   James E. Brittain, 'The Introduction of the Loading Coil: George A. Campbell and Michael I. Pupin,' *Technology and Culture* 11 (1970): 36–57.
3.   Eugene S. Ferguson, 'John Ericsson and the Age of Caloric,' *Contributions from the Museum of History and Technology, U.S. National Museum Bulletin* 228 (Washington, D.C., 1961): 41–60.
4.   Robert P. Multhauf, 'Sal Ammoniac: A Case Study in Industrialization,' *Technology and Culture* 6 (1965): 569–86.
5.   Robert P. Multhauf, 'The French Crash Program for Saltpeter Production, 1776–1794,' *Technology and Culture* 12 (1971): 63–81.
6.   Otto Mayr, 'Victorian Physicists and Speed Regulation,' *Notes and Records of the Royal Society of London* 26 (1971): 205–28.
7.   Otto Mayr, 'Yankee Practice and Engineering Theory: Charles T. Porter and the Dynamics of the High Speed Steam Engine,' *Technology and Culture* 16 (1975): 570–602.
8.   *Dictionary of Scientific Biography* (New York, 1971). S.V. 'Edison, Thomas.'
9.   Ibid., S.V. 'Euclid.'
10.   C.W. Sherwin and R.S. Isenson, *First Interim Report on Project Hindsight (Summary)* (Washington, D.C., 1966).
11.   *Technology in Retrospect and Critical Events in Science* (prepared for the National Science Foundation by the Illinois Institute of Technology Research Institute, 1968).

9

Derek J. de S. Price

The parallel structures of science and technology

The evidence I should like to bring to bear upon our problem of transfer is that which has arisen in a series of researches whose aim has been simply to try to understand the nature of scientific work, historically and in the present. It is perhaps worth underlining the point that our work derives from scholarly curiosity rather than from the more usual objective in such investigations: an attempt to do some good, to cure some evil, or to find out how to organize something efficiently. I am therefore in a strong position to present evidence which connects reasonably well with what little we know about the history of science and technology, the sociology of science, documentation theory, or the like, but in a correspondingly weak position if one wants anything like a specific answer for practical means of organizing technology transfer.

DYNAMIC STRUCTURE OF SCIENCE

The body of research that I believe is most relevant derives from the quantitative investigations of all the things about science to which numbers can be attached. There has been a long tradition of such work. (See Price [1963] for a bibliography.) Counting papers and measuring manpower, expenditure of money, and several other things, but most recently a better set of quantitative data than ever before has become available. As a by-product of the commercial production of Garfield's *Science Citation Index*, we have available a corpus of several million papers, citations, and authors which are very readily susceptible to investigation. This very large population gives one a statistical certainty that was quite unreachable with the older and cruder head-counts, and we

164

are now in the position of being able to link together and cross-check several different sources of data and types of measurement.

Crucial to the whole investigation is that we can first derive reasonable proofs for the validity of these head-counts as a legitimate measure of scientific inputs and outputs. We have been able to show that in spite of the obvious lack of any control over the quality of the heads one counts, the same statistical laws are followed which one could get if all papers or persons were weighted by their magnitudes on any reasonable scale. The detailed results might be very different for any particular individual, but the distribution of the whole population would be unaffected.

Working from the population of papers, then, treating each as a sort of atom of knowledge, we have been able to derive something of a model to show how new papers are related to old ones. Since each paper carries an average of about a dozen citations back to past literature, these citations may be analyzed to find their pattern. Such analysis shows immediately that there exist two separate modes of citation (Price 1965a). The first, accounting for about half of all references back, may be called *archival*. It represents a raiding of the archive, almost completely independent of the age of the older paper being cited, and without structure. The absence of structure occurs as a random and patternless set of connections between the new papers and the entire body of old papers in that particular field. Some papers of course are more popular, more cited than others, but highly connected recent papers will probably contain quite different selections from the archive.

The second type of citation may be called *research front*. New papers use the other half of their references to connect back to the relatively small number of highly interconnected recent papers. In a particular field each recent paper is connected to all its neighbors by many lines of citation. A convenient image of the pattern is to be found in knitting. Each stitch is strongly attached to the previous row and to its neighbors. To extend the analogy, sometimes a stitch is dropped and the knitted strip then separates into different rows, each of them a new subfield descended from the first.

As the first fruits of this model we may derive some explanation of how it is that science has had, for several centuries, an uncannily constant rate of exponential growth of the literature. The number of scientific papers in all fields and for all countries, with only tiny perturbations for world wars, has been doubling every ten to fifteen years — depending on just what is measured or how it is measured. In our theory, each scientific paper ever published, lumping together good ones and bad ones, produces on the average one citation per year. Since it takes about twelve such citations to make a new paper we have a growth rate of about 8 per cent per annum, a doubling about every decade.

There are, therefore, grounds for believing that science grows as it does, so much faster than mere people, because old knowledge breeds new. Furthermore, the key process seems to be that in which very recent knowledge breeds new so much faster than it does when it later becomes packed down into the archive. Science grows very regularly, in a very structured way, and from its epidermis rather than from its body. In terms of our present problem this model may be taken as giving a mechanism for the 'transfer' from old science to new science.

The model may be further analyzed to give a little more of the nature of this, the simplest transfer that we have to consider. It has already been admitted that some papers are better than others, and equally truly some researchers are more powerful than others. An investigation of the distribution of quality in men and in papers shows immediately that there exists greater inequality than in any country's distribution of economic wealth and a bigger rat race than in any other competitive activity of mankind. The gap between a Nobel prize winner and a plain person is rather larger than that between an Olympic gold medalist and an ordinary mortal. Roughly speaking, if you have published n papers in your life, the chance of your reaching $2n$ happens to be about 1 in 4. As an inevitable consequence of a completely lopsided skewed distribution of this sort, it follows that a small number of the men who write scientific papers happen to be responsible in the aggregate for half the production and a good deal more than half of the value if one weights each paper in any reasonable way. This small number, a hard core, may be estimated at about the square root of the total number of writing scientists, and together they constitute the heart of what has become known as the Invisible College of all the good people who really count in that neck of the scientific woods.

Now the important thing for transfer studies is not so much that an Invisible College exists, but rather that with such a high rate of mortality, the large majority of people get lost or excreted while trying to make the grade to get into the core. This process is what we call Graduate Student Education. The mortality is so high not only because the work is difficult, but no matter how much the old science is packed down, the poor student has to get through the archive of centuries and then run much faster than the rate of progress of the research front in order to emerge at the growing edge of science. When he emerges with that particular training, that particular mode of approach, those little bits of research front material passed through, he will then (with luck) be able to make his new contribution at just that particular area of the research front. If he then lives with it for a while, always keeping up with his cohorts at this place on the wide front, he will participate in the knitting of new knowledge to the old which he and his friends have laid down.

Perhaps one should emphasize that this is the normal growth pattern of scientific research and most published work occurs thus. What does not usually happen is the *ab initio* growth of new knowledge coming from almost nowhere. Only in that convenient mythology of science that historians spend their lives trying to dispel, does it appear that the mountain-peak contributions of Newton, Copernicus, and Galileo arose from the native genius of isolated minds. Innovation in the sort of knowledge that is published in scientific papers arises when new facts, new experiments, new theories are added to the immediately preceding old ones in a very structured way. Transfer, from outside the knowledge of men is a rare thing, though it may be very important when it happens. I hazard a guess, however, that even when some outside force, like the invention of a cyclotron, makes knowledge growth burst out in new ways, it does so only in retrospect as we attribute a magnified power to the single great event and forget the less dramatic build-up of the host of related researches clustered around the clear main line.

THE PLACE OF PUBLICATIONS IN SCIENCE AND TECHNOLOGY

Armed with this sort of model for science, we may next turn our attention to technology. (Immediately there arises the hoary problem of defining science and technology. The usual course is to pick a definition from thin air — and then never use it.) With this model we have an interestingly new approach (Price 1965b). We have been talking about science in a way such that there is an implicit definition for it. Science, to fit this model, must consist of the scientific papers that are being cited, counted, and otherwise manipulated in such studies. I therefore propose, as a formal definition, to take as science that which is published in scientific papers. We may define these in turn, either crudely as articles in journals in the *World List of Scientific Periodicals,* or more artfully by making use of the knitted research-front structure. We may suppose quite reasonably (and the evidence bears it out) that the structure is what makes science different from the non-scientific scholarship that is published in other periodicals that are probably not in the *World List*. Along the same line of definition we shall define a scientist as a man who sometime in his life has helped in the writing of such a paper. If you complain that these definitions are too loose, I can tighten them as much as you like from the same theory. If we know the number of 'scientists' by my definition we can compute, let us say, the number of men who have published at least one paper a year over the last three years. If we know the total number of papers we may fairly easily compute the number that will ever be referred to in a subsequent paper by anybody except the original author.

It should be said at this point that this definition is useful because it can be used to tie together most of the known sets of statistics about pure science, and because it agrees very elegantly with the important sociological analysis of scientific private property, multiple discovery, and priority disputes. This analysis, due largely to the work of Merton (1962, pp. 447–85), is now a cornerstone in the historiography of science and fits well with the best integrated views of what happens in a scientific society. Because there is after all only one scientific world to discover, it is terribly important to lay claim to the discovery by publication. Beethoven's symphonies would never have existed without the man, but Planck's constant might just as well have been discovered by Professor Joe Bloggs. Indeed, scientific papers turn out to be such a good indication of science simply because the high motivation to publish exists because publishing is the only external sign that work has been done that can enter into the research front, make new knowledge, and hopefully (though more rarely) give the originator the immortality of his work in a perpetual archive.

With science defined in this way, what is to be the position of technology? Certainly not more than a part of technology could possibly be subsumed under such a definition. In the fields of electronics and chemical engineering there is, one can admit, considerable publication, but in many other well-known areas of technology there is no equivalent of a scientific paper. There are patents, of course, and these even carry citations back to previous patents. And in spite of the lack of technological papers that look like scientific papers there is an enormous technical literature – in fact, many more journals exist here than in the fields of science as it has been defined.

The difference between science and technology in this new sense is pointed to most clearly by the sociology of publication. The scientist, it was remarked, is heavily motivated to publish – this is the key to all the inner springs of his drive to do science.

In technology it is otherwise; the tradition, crudely speaking, is to conceal in order to have a new product or a process before others. It also happens that the scientist hears from his Invisible College about the new work on which he is going to build, and he therefore does not need to read the published journals very much, for by then they are old hat. However, in the other camp, the technologist is most eager, while revealing as little as possible himself, to pick up anything and everything that may be dropped his way by others. One may put the whole thing in an aphorism I have used before; the scientist wants to write but not read, the technologist wants to read but not write.

If technology is, so to speak, *papyrophobic,* we cannot use the measure of papers as we could with *papyrocentric* science to diagnose the features of a model for it. When one makes the attempt to analyze citations in most technical journals and patents, one finds no well-marked structure. Papers and patents at the best form short chains, each item taking off from a preceding one usually by the same author or company. It seems pretty clear that even though there exists a large mass of technical journals, the writing does not have the same function that it does for science. It seems to exist for a newspaper-like current awareness function, for boasting and heroics, and probably, above all, as a suitable burden to carry the principal content of advertisements which, together with catalogues of products, are the main repositories of the state of the art for each technology.

A point which tells particularly heavily for me, as a historian of technology, is this opposite polarity of science and technology in their attitudes to literature, which is precisely what makes it terribly difficult to write the history of technology in the same manner that one writes the history of science. The content of science is already embodied in papers, whereas that of technology first has to force itself into written form, and this leads to exactly that type of antiquarian writing which so often bogs us down. It is the preliminary stage to the writing of a history of technology, but it is not the history itself.

THE DYNAMIC STRUCTURE OF TECHNOLOGY

Since it seems obvious that technology is related in some way to science, and that it grows at much the same rate by any reasonable measure, it must follow (or lead) the cumulating structure of science. Now, the naive picture of technology as applied science simply will not fit all the facts. Inventions do not hang like fruits on a scientific tree. In those parts of the history of technology where one feels some confidence, it is quite apparent that most technological advances derive immediately from those that precede them. It may seem to happen from time to time that certain advances, particularly the spectacular and anomalous mountain peaks, as with science, derive from the injection of new science. In the main, however, old technology breeds new in just the same way as the scientific process already described. Because of this I now hypothesize that though it cannot be diagnosed by papers, technology has a structure that is formally identical with that of science. We shall define technology as that research where the main product is not a paper, but instead a machine, a drug, a product, or a process of some sort. In the same way as before there will be a highly competitive rat race, a similar exponential rate of growth, an archive of past technology, and a research front of the current state of the art.

One must enter the caveat that by choosing definitions that can do this work for me and by erecting such models, I have had to use words in a precise but not entirely ordinary way. Thus if a man is paid by a pharmaceutical company to develop a new drug but finds himself publishing a paper about the new chemical knowledge gained, he has been guilty of doing science on company time; but if a pure mathematician comes up with a new way of designing switchboards, it is technology, however pure be his mind.

PARALLEL STRUCTURES OF SCIENCE AND TECHNOLOGY

At any rate, with these definitions and a hypothesis we now have conjured up a pair of similar and parallel systems for science and technology. What makes them run in step, or in Toynbee's (1962, vol. 1, p. 3) apt image, behave like a pair of dancers to the same music, though it is imperceptible who is leading and who following? The Marxist answer would be to suppose that the music is the wishes and needs of society that make a certain type of science and a certain type of technology appear at any particular time. This, however, is quite inconsistent with the previous finding that it is the old knowledge rather than the wishes of society that is breeding new knowledge. One can, I suppose, create technology to order, just by wishing it. But ordinarily one is severely constrained by the old technology's having or not having the capacity to breed a particular desired thing. However much one is willing to pay for a cancer cure, one cannot develop it unless the state of the art, and perhaps the state of science, seems able to sustain it.

INTERACTION OF SCIENCE AND TECHNOLOGY

Rather than supposing that an outside force affects both dancers, it seems more reasonable to think that their action upon each other keeps them in step. Since we know so many cases in which science has passed into technology, and so many other cases in which the technology has made new science possible, it follows that we must have a complete interaction. The well-known lag between scientific and technological advance would seem to indicate that the dancers hold each other at arm's length instead of dancing cheek to cheek. To use the more precise language of the physicist, the relation between science and technology seems to be a weak rather than a strong interaction.

The mechanism for such a weak interaction is not hard to find. When a technologist is educated in the research front of the state of the art, he is necessarily subjected to some training in the ambient state of the science of his time. Similarly in his education, a scientist is taught the ambient state of technology. It follows then that men on the research fronts of science and technology will be able to use each other's ambient knowledge. It seems too that this will generally be the ambient knowledge that is on the average about one generation of students old — perhaps ten years.

We have shown already that 'transfer' from old science to new goes through the knitting process of cumulation. A similar process will therefore exist for the 'transfer' from old technology to new. It will occur to a certain extent as part of the archive through which each generation of students grows up. More important, it will occur through the knitting-together process in which the man making the new technology has just had contact, often through the Invisible College, with those who have just been laying down the immediately prior technology. Last, there is the most important type of 'transfer' that occurs from science to technology and vice versa. Again the medium of transmission is the person, and the method is that of the formal or informal education which a man gets in the current state of the art in science or in technology.

Here one must interject a remark of considerable importance to science policy: that the cumulations of science and of technology habitually interact, though weakly, in such a way that they may be said to exist in symbiosis, feeding from each other. There seems to be a body of pathological case histories in science and technology indicating that whenever society has tried to force science or technology away from such symbiosis, the results have been disastrous for both. Science without the by-play of technology becomes sterile, and in the several cases in which a materialistic

171

society decided it would pay for the technology which gave economic gain, and neglect the science which was just a chess game, the technology became moribund. If only there were strong interactions between science and technology one could point eagerly to the ultimate clear utility of pure science. With a weak interaction at present one can only have faith, though there has always been strong temptation to lie through one's teeth and claim direct benefit for pure science.

The mechanism which has been suggested to explain the interaction of science and technology has the important feature of noting that the lag between them is due to the way in which people are brought to the research front either in the science or in the technology, carrying with them the ambient technology or science of about a decade earlier. It has often been suggested by economists that this time lag was once very much greater than it is now, and that is has been decreasing markedly. As a historian of technology I cannot agree with this, and I feel that some error has been introduced by looking principally at those traumatic peaks of spectacular change, like penicillin and transistors, which are not very representative of normal interchange between science and technology. We can all agree, however, that there is a considerable lag in the weak interaction between science and technology, and that this is in contrast to the rapidity with which new technology arises directly in strong interaction from old technology and to the similar process by which old science generates new. The research front is basically contained in people, and the archive is somewhat distant from them.

It has often been supposed that some sort of communication difficulty holds back the optimum growth of technology. In terms of our new model this can be seen as something of a red herring. The papyrophobic character of technology appears to be a deep and long-standing tradition. Even in the Soviet Union, where industrial development is public property, the literature in technology seems to be vastly different from that in the scientific tradition. In the United States the fashion has been to try to force technology into the same literature cumulation as science by the device of the research report. Basically this represents a back-door method of publication, the output being forced as a fiscal device consequent upon the spending of public money, and the mechanism of report publication being used because there are no journals to take such material; not being commercial, it may not be published as a book. Evidently nobody really wants the material enough to buy it, but it has to be published just the same. There is simply no need for literature cumulation at the technological research front in the sense in which we use the term.

The effect of this unwanted research report literature in technology is probably to clog the information channels, and it should not, I feel, be encouraged. Publication should always be held as a privilege consequent upon having found something that somebody else would like to use. It is not a duty consequent upon the spending of public money. In any case, what communication difficulty there is seems due to the fact that though the scientists want to write and the technologists want to read, the scientists are writing for their colleagues in science, or sometimes for their imaginary archive; they are simply not writing the sort of material that the technologists want to read. This frustrates the technologists and makes them believe that somewhere in this pile of material, if only they could find it, there is the very valuable material they are looking for to make new products.

SUPPORT OF SCIENCE AND TECHNOLOGY

Next let us consider the role of the spending of public and private moneys in the purchase of research and the promotion of transfer. With science, it seems clear from the complete internationality of scientific knowledge and from the absolute steadiness of the growth curves of science that the breeding of new knowledge by old is hardly affected by the expenditures of any individual nation. If the United States put an embargo tomorrow on all pure research the world's growth might be abated slightly, but the growth rate would still continue at so near the present level that in very few years the United States would have no scientists and no teachers, and therefore no engineers who could monitor the new world position of the research front. Its technology would therefore dry up in the normal time lag — perhaps a decade.

Figure 1 provides an illustration of this theory from the data of Milton and Johnson showing the trends in the sources of support for a fair sample of scientific papers published in the United States 1920–1960. The growth in numbers of papers during this interval was completely regular apart from the local dip during the Second World War. The sponsorship, however, changes very suddenly in its trends; at the onset of the massive grant program at the end of the war there begins a swing that is just now tailing off because it cannot go much further. The government instead of the universities picked up the tab, but there was virtually no change in the normal rate of growth of scientific papers. Of course, one might argue that if the government had not picked up the tab the resources of the

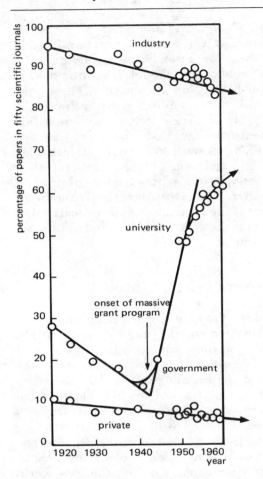

Figure 1 Trends in sponsorship of published papers in fifty large scientific journals (circa 1,000 papers in 1920; circa 10,000 in 1960). Source: Helen S. Milton and Ellis A. Johnson, Technical Paper ORO-TX-42, Operations Research Office, Johns Hopkins University, September 1961, and private communication.

universities would have been soon stretched to their limits, and the normal growth would not have continued. The United States would then have become an overdeveloped country much less rapidly then it has, and the concomitant growth of technology would have sloped off much more rapidly than has been the case.

The role of spending in technology seems to be rather different because each country has its own base in products and industries; there is not quite the strength of supranationality such that one can buy technological research and push it in any desired direction.

The space program is such a case in point. Again, however, one might take the daring suggestion that the chief function of NASA is that it picks up the tab for the necessary employment of a very large number of research technologists. Without them there would be a cut-off in technological growth similar to that which would have occurred in pure science without the National Science Foundation. With them one gets normal growth by means other than the direct buying of a product.

Seen in this light as the repository for a large section of the research front manpower in technology, NASA has a place in technology transfer with two distinct aspects. The first aspect is the transfer that occurs purely within the structure of technology, creating new results on the basis of old. The second aspect is the transfer that occurs in a more complicated way as part of the weak interaction that controls the relation between science and technology. It is crucial that on the basis of the model suggested, both types of transfer depend on the research front being contained in people and transferred by the movement of people. In the normal patterns of growth of both science and technology a considerable transfer occurs because people move from one university to another, from one industrial laboratory to another, from university to industry, and to the laboratories of federal agencies. Scientists and technologists are very much more mobile and migrate more, physically and intellectually, than the population at large; that they do so is probably due in large measure to the enormously competitive rat race of excellence which governs their careers. In a laboratory or institution that is growing only at the general approximately 7 per cent per annum growth rate of the overall structure of science and technology, it is usual to find that the increase comes about by something like a 17 per cent inflow of new staff and a 10 per cent outflow of old ones each year; it is this outflow that causes much of the human movement that creates transfer. With an agency or laboratory that is growing much more rapidly than this, very much less of the inflow is excreted again, and so the linkage is only at the beginning. With respect to NASA, if it does not fall on bad times that lead to massive departures from its research staff, there will be good linkage with the universities and schools that train its scientists and engineers, but there will be poor feedback of people to other industries. Perhaps one of the ways of engineering more massive transfer from NASA and defense technologies to private industry would be to encourage as far as practicable a larger continuous flow of people from government payrolls back to the training grounds, and most vitally, back to private industry.

A very similar suggestion has recently been made by Kapitza (1966) as a remedy for the falling scientific and technological productivity in Russia. He notes that there is far too much stability in the scientific establishment and that it would be generally helpful to fire people rather extensively so as to create more openings at the lower end. This means of course, not any decrease in the number of scientists and technologists employed, but a large increase in mobility and consequently a considerable increase in the communication that is effected by people acting as containers of the research front tradition and state of the art in science and technology. It seems indeed very likely that both in the US and in the USSR such a policy might come into effect whether one likes it or not for this particular purpose. Now both of the major scientific countries of the world are very rapidly filling with science and technology to the point where it is becoming increasingly difficult to muster more high-talent manpower and a bigger fraction of the GNP to research ends. Both countries are in this sense already overdeveloped, and the growth rates of science and technology in both countries seem to be dropping more and more rapidly behind the world total of growth. I would estimate that both countries are now about 10 per cent below the levels that they would have reached with uninterrupted exponential growth. As this gap widens between the leading countries and the rest of the world which must needs slowly overtake them, there will necessarily be widespread changes by those leading countries in the development of their fixed stock of precious resources for research. This should lead to a considerable movement of personnel from one area to another, and if this movement can be engineered with any sagacity it might well increase the transfer in science and technology to the point where the increase in pay-off compensates for the decreased possibility of absolute growth in man-power and money.

REFERENCES

Kapitza, P. L. (1966), *Theory, Experiment and Practice*, Moscow: Znanie, Series 9, part 5.

Merton, R. K. (1962), 'Priorities in scientific discovery: a chapter in the sociology of science', in B. Barber and W. Hirsch (eds.), *The Sociology of Science*, Free Press.

Price, D.J. de S. (1963), *Little Science, Big Science*, Columbia University Press.

Price, D.J. de S. (1965a), 'Networks of scientific papers', *Science*, vol. 149, pp 510–15.

Price, D.J. de S. (1965b), 'Is technology historically independent of science? A study in statistical historiography', *Technology and Culture*, vol. 6, no. 4, pp. 553–68.

Science Citation Index, published quarterly and annually by the Institute for Scientific Information, Philadelphia, Pa.

Toynbee, A.J. (1962), *Introduction: The Geneses of Civilization* (*A Study of History*, 12 vols.), Oxford University Press.

10

M. Gibbons and C. Johnson

Science, technology and the development of the transistor

The following passages from *Technological Innovation in Britain*[1] and the Swann Report[3] illustrate two views of the relationship between basic research and its commercial application.

> The justification for it [basic research] is that this constitutes the fount of all new knowledge, without which the opportunities for further technical progress must eventually become exhausted. Laboratories carrying out long-term basic researches which stretch techniques to the limit are at the same time a valuable forcing ground for new technical developments. The scale of applied research and advanced development must, however, be neccessarily determined by economic considerations, since these activities may culminate in a practical invention in something like three to seven years. A corresponding period may be required for the engineering development which transforms the invention into a full prototype ready for commercial development. Minor developments may take as little as one to two years; and newly installed production plant is normally brought into full use in about this same time. *(Technological Innovation in Britain.)*

> There is a general trend towards a marked shortening of the time between first discovery and its exploitation. This rate of change has been increasing in the past but need not necessarily continue to increase in the future. Two factors which can affect rate of change support this view. The first is the large scale of manufacture which produces economies of scale that are comparable with or can even outweigh the economies of a better process. The size of the resulting investment makes it less likely that these processes will be easily superseded since the new process, to be competitive, must be of comparable magnitude. The second factor which may inhibit further increase in rate of change is the need for continued and often protracted testing of performance of chemicals. This is particularly marked when their application impinges upon human health. (Swann Report.)

'A scientific discovery may take many years before it finds practical application. It follows, therefore, that the case for supporting long term scientific research cannot be argued on economic grounds.' In spite of these words of caution from a recent report, on *Technological Innovation in Britain*[1], there is still a widely held view, among scientists as well as economists, that fundamental discoveries in science provide the mainsprings of economic advance. This view conceals the importance of technological developments in innovation and may even suggest to some that technology is dependent for its development on discoveries in science. Current discussions of the decreasing time lag between scientific advance and commercial application betray the same bias by assuming that this time lag, whether increasing or decreasing, is a significant variable[2,3].

The relation between scientific discovery and technological development is complex, and only detailed case studies can give the appropriate insights into the precise nature of the interaction. One of the most frequently cited instances of a scientific discovery that led to technological advance and the setting up of a new industry is the transistor. The 'scientific discovery' of the transistor is presumed to exist potentially in a few papers on the quantum theory of semiconductors published by A. H. Wilson in 1932. This work is termed 'curiosity-oriented research' to signify that the author carried it out without suspecting that his work would have any practical application. For some, this type of unforeseen result seems to argue (albeit indirectly) for continuing support of all basic science because 'you never know what practical applications may result'. Although Wilson's principal contribution was to provide a theory of conduction in semiconductors, it is apparent from the earliest history of semiconduction that research was conditioned both by the development of associated technologies and by commercial and social stimuli, some of them military.

EARLY HISTORY OF SEMICONDUCTION

The observation of semiconductor phenomena dates from Faraday's discovery[4] in 1833 that silver sulphide possessed a negative temperature coefficient of resistance (its conductivity decreased with increasing temperature instead of increasing as for most metals). In 1839, A. E. Becquerel[5] noted the photoelectromotive effect of shining light on one electrode of a pair immersed in an electrolyte, and in 1873, Smith[6] discovered photoconductivity by observing the decrease in the resistance of illuminated selenium. A major advance was made in 1875 by Braun, who discovered that the junction between a semiconductor crystal and a metal point contact had asymmetric conduction properties and he systematically investigated this effect[7—10].

This 'point contact' rectifier was among the first semiconductor devices to have widespread applications, and observation of rectification phenomena soon led to increasingly sophisticated theories to explain them. In addition, the Hall effect (the transverse electromotive force developed when a magnetic field is applied to a current-carrying conductor) contributed to the understanding of semiconductor phenomena by providing a direct indication of the sign and number of the charges in the solid.

IMPACT OF RADIO TECHNOLOGY

During the early 1900s, a new technology was developing independently of heavy electrical engineering—radio. The possibilities of using a crystal rectifier as a detector of radio waves were investigated extensively[11—14], and the empirically developed 'cat's whisker' detectors were used until they were gradually displaced by thermionic valves (the triode was invented by Lee de Forest in the United States in 1906). The crystal detector was not displaced on a domestic level for some twenty years, during which time its characteristics and reliability were considerably improved. But a theoretical explanation of the phenomenon was still lacking. In 1930, when crystal technology was losing ground to a rapidly developing valve technology, three principal theories attempted to explain rectification: the electrolytic barrier[15]; local Joulean heating causing a Peltier voltage[16]; and cold emission across a gap[17]. These and other theories were advanced and refuted many times in the sixty years following Braun's discovery, but no definite reproducible experimental evidence could be obtained because of the lack of uniformity in the specimens used. Baedecker's remark in 1907 that many investigators used specimens of doubtful purity seems to have characterized the 'state of the art' at this period[18]. The extreme sensitivity of semiconductor phenomena to impurities, crystal structure, light and heat and mode of contact formation was not generally recognized. Even the best techniques of preparation and assay at the time seem to have been inadequate, and as a result there was little reliable experimental evidence to guide theoretical thought on the subject.

Meanwhile, other empirical discoveries continued to be made. Grondahl's discovery[19] of the copper oxide rectifier in the early 1920s stimulated further research with the material because of its commercially useful powerhandling capabilities, and revived interest in semiconductor rectification in general. The most satisfactory development of Braun's point contact rectifiers was made during the 1920s using natural lead sulphide (galena). Schleede and Buggisch pressed further and discovered that the sensitive parts of a lead sulphide crystal corresponded to regions of sulphur content in excess of stoichiometric composition[20]. It also became known that impurities could markedly alter the sign and number of charge carriers in semiconductors[21] and that the rectifying action took place at the semiconductor/semiconductor interfaces[22]. Although crystal detectors were receiving considerable attention, the lack of suitably accurate crystal growing techniques and an adequate theoretical understanding inhibited further development.

Electron valve technology continued to develop rapidly and extensive use was made of this, rather than crystal detector technology, during the subsequent explosive growth of the radio industry. The shift of commercial interest away from crystal detectors was an important factor in relegating the study of crystal structure to university laboratories, from which it did not emerge until the development of radar brought renewed interest in the subject.

NEW CONCEPTS FROM QUANTUM MECHANICS

On the theoretical front, no adequate explanation was available until the quantum theory of conduction, originally developed by Bloch[23], Sommerfeld[24] and others, culminated in A. H. Wilson's classic work[25,26] on conduction in semiconductors. He put forward a band theory of conduction, and explained the existence of both positive and negative highly mobile charge carriers (holes and electrons) as well as the effect of impurities on conductivity. It is not clear whether Wilson realized the commercial importance of his own work, but he was obviously aware that there were problems in explaining certain 'anomalous' effects in crystals, because this information would have formed part of the 'general knowledge' of physics at that time.

Wilson's work did not make the impact of a 'eureka' on crystal physics; rather, a process of diffusion followed and about ten years elapsed before his concepts became part of the stock-in-trade of semiconductor physicists. By 1935, no serious attempts had been made to explain rectification in terms of the band theory as anything other than an interface effect between different types of semiconductors. The theory of rectification developed slowly through key works by Schottky[27–29], Mott[30] and Davydov[31–33] which dealt with point contact, cuprous oxide and semiconductor/semiconductor rectifiers respectively. It is interesting that Schottky had put forward a theory which involved an asymmetric potential 'hump' across the contact boundary as early as 1923[34]. Attempts to use quantum tunnelling concepts in the early 1930s by Wilson[35] and Frenkel[36] were unsuccessful, chiefly because they predicted the wrong direction of easy conduction of electricity. Two other important contributions to the development of the theory of conduction of semiconductors, the significance of which was not recognized at the time, were from Tamm[37] on the existence of surface states and Frenkel[38] on photoconductivity, which introduced the idea of a potential difference arising from the different diffusion rates of majority and minority carriers in bulk semiconductors[39].

Wilson's work, which was among the first applications of quantum theoretical concepts to conduction phenomena, was done at a time when theoretical physicists were anxious to work out the implications of the quantum theory derived earlier by Schrödinger, Heisenberg and Dirac. The extension of this theoretical knowledge to various 'real life' situations was probably the principal motivation of theoretical physicists at this time. But Wilson's theory did not suggest the existence of the transistor effect; rather, it set forth the much needed model of conduction and rectification phenomena and provided experimentalists with concepts that could be applied to experimental situations.

DEVELOPMENT OF NEW TECHNIQUES

This diffusion of concepts proceeded at a leisurely pace until the Second World War when the necessity for radar detectors (rectifiers working at frequencies beyond the capability of valves) prompted much government sponsored work in the United States, principally at Purdue and Cornell Universities. This wartime programme led to the development of pure specimens of germanium and controlled doping techniques[40]. These specimens were subsequently utilized for the production of standard detectors, but more important, they made it possible for the first time to make p-n junctions with well defined properties. With accurate control of the impurity content of a semiconductor, the lifetime of the minority carriers could be increased. Hitherto, the significance of the minority carriers had largely been masked by the nature of specimens available.

At the end of the Second World War, many scientists were released from government laboratories and returned to industry and the universities. The Bell Telephone Laboratories decided that the current developments in solid state physics and technology were such that basic research in these fields was likely to yield a long term benefit. They accordingly assembled a multidisciplinary team to work on semiconductor phenomena. This team included the metallurgists Scaff, Ohl and Theuerer whose work at Purdue we have mentioned, and the physicists Bardeen and Brattain. The team was directed by Shockley.

In the introduction to his Nobel lecture, Shockley stated his strong interest in the practical applications of solid state physics, and there seems to be little doubt that this interest influenced the direction of his work. Based largely on Wilson's theory, Shockley had predicted the existence of the 'field effect' — the reduction of carrier states available in (and hence the alteration of the conductivity of) a thin layer of semiconductor when subject to the action of an electric field. Anomalous results, observed when this effect was sought experimentally, led to theoretical and experimental work proving the existence and statistics of surface states[40]. Stronger fields were subsequently used in an attempt to overcome this 'surface binding' and a low frequency field effect amplifier was produced.

INTERACTION OF SCIENCE AND TECHNOLOGY

In 1948, further work on the field effect amplifier produced a seemingly anomalous relation between the currents in the electrodes of the semiconductor block. This was immediately recognized by Shockley as being due to the injection of minority carriers, and thus the transistor effect was discovered[41,42]. Confirmation of this 'injection effect' was provided by Shive[43] in 1949, but Shockley anticipated these results and produced a junction transistor with a mode of operation firmly based on minority carrier injection, diffusion and capture as opposed to the earlier complex behaviour of the point contact devices.

From this brief description of the discovery of the transistor, it is evident that the language used to describe semiconductor phenomena had altered substantially by the late 1940s and that Shockley, Bardeen and Brattain had become well schooled in the quantum mechanical concepts of conduction originally put forward by Wilson and Frenkel.

But, as we have pointed out, Wilson's work by itself was not sufficient to produce a solid state amplifier. This development was also dependent on the improvement in metallurgical techniques of

crystal growing and doping. These essential techniques were following a development independent of the theoretical development
set in motion by Wilson in 1932. Without these improvements in
the doping and purity of crystals the transistor effect could not
have been observed. The spur to the development of these techniques came from the massive research and development effort in
search of more efficient detectors for the radar equipment needed
by the armed forces.

In spite of considerable enthusiasm at all levels (there was even
a slump in valve sales[44] caused by public fear of their obsolescence),
the production of reliable transistors had to await the development of the associated technologies. In 1952, a review article in
the *Proceedings of the Institute of Radio Engineers*[45] could still
sum up transistors as being difficult to design from theory, tedious
to manufacture from design and unreliable once produced. But
in applications where speed of operation, light weight, low power
consumption and mechanical reliability were essential there was
a great incentive to rapid development.

SCIENCE AND TECHNOLOGY IN COMPLEX RELATIONSHIP

The early history of the transistor suggests that the transistor
could be more accurately described as a technological development than a scientific one. That is to say, the transistor arose
primarily as a result of technology building on technology — in
this case, rectifier technology. The role of Wilson's theoretical
model of the conduction process in semiconductors was undoubtedly of great importance but, as we have tried to show, it was just
one of many factors contributing to the final result.

This case study suggests that the relation between science and
technology is symbiotic rather than that technology is applied
science, which seems to be the prevalent view. The important
problem is not concerned with choosing, rather academically, a
suitable starting point from which the transistor can be seen as
the logical development, but rather the mechanism whereby new
concepts like those suggested by Wilson and Frenkel diffuse into
existing practice and stimulate not only technology but science itself.

It is perhaps unwise to try to draw general conclusions from
one case study. But, at a time when there is renewed interest on
the part of economists and others in measuring the contributions
of science to general economic well-being, the history of the
transistor shows that the relation between science and technology

is complex. It deserves to be investigated more fully before any realistic assessment can be made about the economic and social benefits that derive from scientific research.

REFERENCES

1. *Technological Innovation in Britain.* Report of the Central Advisory Council for Science and Technology (HMSO, London, 1968).
2. Baker, in *Technology and Social Change* (Columbia University Press, 1964).
3. Newth, F. H., *The Flow into Employment of Scientists, Engineers and Technologists,* Cmnd 3760, 103 (HMSO, London, 1968).
4. Faraday, M., *Exp. Res. in Electricity,* 1, 122 (Bernard Quaritch, London, 1839).
5. Becquerel, A. E., *CR Acad. Sci.,* 9, 711 (1839).
6. Smith, W., *J. Soc. Telegraph Eng.,* 2, No. 1, 31 (1872).
7. Braun, F., *Pogg Ann.,* 153, 556 (1874).
8. Braun, F., *Wied. Ann.,* 1, 95 (1877).
9. Braun, F., *Wied. Ann.,* 4, 476 (1878).
10. Braun, F., *Wied. Ann.,* 19, 340 (1883).
11. Adams, W., and Day, R., *Proc. Roy. Soc.,* 25, 113 (1876).
12. Bose, J., *US Patent No.* 755840 (1904).
13. Dunwoody, H., *US Patent No.* 837616 (1906).
14. Austin, L., *Phys. Rev.,* 24, 508 (1907).
15. Huizinga, M., *Phys. Z.,* 21, 91 (1920).
16. Jolley, L.D.W., *A.C. Rectifiers,* 436 (Chapman and Hall, London, 1928).
17. Hoffman, G., *Phys. Z.,* 4, 463 (1921).
18. Baedecker, K., *Ann. Phys.,* 22, 749 (1907).
19. Grondahl, L., *US Patent No.* 164335 (1925), also *Phys. Rev.,* 27, 813 (1926).
20. Schleede, H., and Buggisch, H., *Phys. Z.,* 28, 174 (1927).
21. Heaps, C., *Phil. Mag.,* 6, 1283 (1928).
22. Guntherschulze, A., *Electronic Rectifiers and Valves* (trans. by de Bruyne, A.) (Springer, Berlin, 1926).
23. Bloch, F., *Z. Phys.,* 52, 555 (1928).
24. Sommerfeld, A., *Z. Phys.,* 47, 1 (1928).
25. Wilson, A.H., *Proc. Roy. Soc.,* 133, 458 (1931).
26. Wilson, A.H., *Proc. Roy. Soc.,* 134, 277 (1931).
27. Schottky, W., *Phys. Z.,* 31, 916 (1930).
28. Schottky, W., *Z. Phys.,* 113, 367 (1939).
29. Schottky, W., *Z. Phys.,* 118, 539 (1942).
30. Mott, N.F., *Proc. Roy. Soc.,* A.171, 27, 144 (1939).
31. Davydov, B., *J. Tech. Phys. USSR,* 5, 87 (1938).
32. Davydov, B., *J. Phys. USSR,* 1, 167 (1939).
33. Davydov, B., *J. Phys. USSR,* 4, 335 (1941).
34. Schottky, W., *Z. Phys.,* 14, 63 (1923).
35. Wilson, A.H., *Proc. Roy. Soc.,* 136, 451 (1932).
36. Frenkel, J., *Phys. Rev.,* 36, 587 (1930).
37. Tamm, I., *Phys. Z. Sowjetunion,* 1, No. 6, 733 (1932).
38. Frenkel, J., *Phys. Z. Sowjetunion,* 8, 185 (1935).
39. Torrey, H.C., and Whitmer, C.A., *Crystal Rectifiers* (McGraw-Hill, 1948).
40. Brattain, W., *Nobel Lecture* (1956) (Elsevier, Amsterdam, 1964).
41. Bardeen, J., *Phys. Rev.,* 71, 717 (1947).

42.   Bardeen, J., and Brattain, W., *Phys. Rev.,* 74, 230 (1948).
43.   Shive, J., *Phys. Rev.,* 76, 575 (1949).
44.   Kraus, J., *The Electron Tube Industry in Britain* (1935—1962), Technology and Culture, 9, 544 (1968).
45.   Morton, J., *Proc. Inst. Radio Eng.,* 40, 1314 (1952).

PART FOUR

The interaction of science and society

There was no way in which we could satisfactorily represent here the full range of material which relates science to the wider context. We had to rest content with indicating some of the more important available perspectives in our Bibliographical Notes; and we limited our immediate concern to that aspect of the subject which bears most directly upon our general objectives. Accordingly, this section only considers the relationship between the *culture* of science and that of the wider society. Given this, our title may seem a little strange. It is more usual to speak of 'the social uses of science' or 'the social implications of scientific knowledge'. But to use either of these formulations is to acquiesce in the widely accepted *a priori* assumption that scientific knowledge is quite unaffected by so-called 'external' factors; whereas one of our primary concerns is with this as an *empirical* question.

The basic form of the relationship between the culture of science and the culture of the wider society is simply that which appertains between any two associated forms of culture. As such, it is formally equivalent to the relationship between science and technology, already discussed in some detail in the previous section: without exception, every *formal* point made there about the science/technology relationship applies also to the relationship between science and the wider society. Much exposition can be saved here simply by referring back to the Table in the introduction to Part Three, and inviting the reader to substitute a reference to the general culture for every reference to the culture of technology (T). The hierarchical model in the Table then corresponds to the relation of science and society conceived of as the 'social use of science'. And the symmetrical model provides a more satisfactory representation of the general form of the relationship — one which makes no *a priori* assumption of one way determination, and one which recognizes the status of all forms of culture as cognitive *resources* rather than as entities with cognitive *implications*.

187

If we are guided by the symmetrical model, then we will not presume anything in advance about the role of 'external' factors in the cognitive development of science. We will accept instead that the knowledge and culture of science is maintained and developed by specific people, in specific institutions, in specific contexts, and that how far these people keep their esoteric culture distinct from and unconditioned by what is going on in the wider social context is a wholly contingent matter, needing separate ascertainment in every individual case (see also Barnes, 1974, Chapter 5; 1982, Chapter 5.3).

This means that the general question of how far and in what way science is affected by the surrounding culture could well lack a worthwhile answer. The best answer it can hope to elicit is a statement of what has proved *typical* as indicated by available empirical studies. As far as the basic concerns of the social sciences are concerned the question is a *small* one, and that may indeed be the most important thing to understand about it. A basic flaw on both sides of many of the current debates about the 'social conditioning' of scientific knowledge is that they transform this small question into a central issue, and grossly exaggerate its significance.

Part of the reason for this is historical. There was a time when scholars, including most sociologists, had no perception of science as a form of culture, and characterized scientific knowledge as a special set of representations validated entirely by 'reason' and observation. It followed that 'social' factors in the development of science could only be external intrusions disturbing the proper operation of 'reason'. And arguments about the presence or absence of 'social' factors concerned the fundamental question of how far science was 'socially conditioned'. Although today interest is firmly focussed upon scientific knowledge as a form of culture, and hence as constitutively 'social', a residue of the older perspective remains: sociological analysis of scientific knowledge is still sometimes equated with the search for external disturbances (Laudan, 1977).

The second, more fundamental, reason is of course that science counts for many people as a crucial emblematic symbol; its separation and sacralization, or conversely its pollution and defilement, are regarded as intensely meaningful activities (Bloor, 1976). And, traditionally, to display the relevance of an 'external' factor in scientific inference or judgment is to pollute the sacred symbol. As a result the question of the level of external determination of science has assumed a disproportionate importance, even in the field of the sociology of scientific knowledge itself, where most workers well understand its basic insignificance. Practitioners in

this field have stressed the matter of fact, non-evaluative orientation of their work *ad nauseam* over the last decade, and have
insistently rejected any connection between the invocation of
'external influences' and the debunking of the knowledge in
question. But they have found that those aspects of their work
which arouse general interest are precisely those which are relevant
to the tasks of sacralization and defilement, and not unnaturally
they have been led to attend to these aspects at the expense of
others. If one asks whether or not there was an ideological dimension to Newton's *Principia,* or whether or not the scientific response to Darwin's *Origin* was made on purely technical grounds,
or to what extent if any the reception of an acausal quantum
mechanics was a matter of expediency, one finds oneself addres-
sing questions with appeal. Thus, issues important to the defence
or the debunking of science, although explicitly rejected by sociologists as irrelevant to their concerns, have tended to creep back
into the field and have continued to exist as clandestine sources,
imparting structure and significance to research.

Accordingly, we must briefly attend to the question of how far
and in what way scientific knowledge has been affected by the
wider context. Is there any worthwhile general account which can
be given of this? One widely held view is that there is: the findings
of empirical, historical research are said amply to confirm that
scientists have been well insulated from the surrounding culture
by their institutional arrangements. According to this view it is
perfectly legitimate to assume the continuing autonomy of scientific inference and scientific judgment: the assumption amounts
to an inductively well-grounded trust in the continuing efficacy of
those institutional arrangements.

This viewpoint can however be called into question, precisely
in terms of its empirical basis. Consider first the way that historical
research has concentrated upon selected strands of scientific
development: mechanics and astronomy, electricity and magnetism,
atomic theory, and other paradigm cases of the most basic forms
of science. Compare this with the present distribution of scientific
research encompassing fields such as agricultural research, computer science, ecology, electronics, engineering sciences, forestry,
all the numerous sciences relating to medicine, meteorology,
metallurgy, oceanography, parasitology, toxicology, and many
more. When one looks at what scientific research is actually
involved with today, an extraordinarily large proportion proves
not just to be oriented to specific human ends and objectives, but
to be carried out in fields which are explicitly ordered around
such ends and objectives. The standard view of these fields is

that the rate and direction of their research is indeed externally determined, but that their cognitive processes remain untouched, and that the generation of knowledge proceeds entirely in terms of its own 'internal logic'. This is simply false.

Krohn and Schäfer offer a case study of how theory actually does develop in this kind of context. They show how the theories of agricultural chemistry are constructed to apply to a man-created domain, and how its concepts are developed to match that domain rather than inherent features of 'nature'. More profoundly, human goals and objectives become incorporated into the structure of theory, so that it takes on an irredeemably *teleological* form; and this form becomes part of the preconditions of work in the field: 'Agricultural chemistry does not provide insights into *why* agriculture functions but rather constitutes a design of *how it must* function' (200). Finally, the teleological character of the theory and the goal orientations expressed in it are relevant to its evaluation as a 'good' theory: its acceptability as a paradigm for research is directly dependent upon these factors.

A lack of comparable studies makes the evaluation of Krohn's and Schäfer's claims difficult. However their suggestion that connections between theory and goal-orientation are characteristic of the applied sciences receives support from a quite different perspective in the work of Robbins and Johnston (1976). They use the case of occupational toxicology to illustrate how cognitive differentiation within science can be stimulated by the relationship of scientists to clients with practical needs. Although this study does not undertake a systematic analysis of the relationship between the theory of toxic thresholds and the practical goal-orientations to which toxicologists are obliged to be sensitive, it compensates through the breadth of its discussion. Above all, it brings home as no other account does the profound *importance* of the relationship between theory and goal-orientation, and the difficult questions it raises concerning the scope of applied-scientific knowledge and the form of its validity.

One final point must be made about the evaluation of Krohn's and Schäfer's contribution. Its specific merits as a case study need to be assessed separately from the merits of a very general theory of scientific development to which its authors relate it. This is the highly controversial 'finalization' thesis (Böhme, van den Daele and Krohn, 1976; Schäfer, 1979). Although much of the criticism directed at this thesis is intelligible in terms of the normative commitments of its supporters and its detractors (Pfetsch, 1979), it is certainly more open to question than most of Krohn's and Schäfer's more specific claims.

The 'finalization' thesis distinguishes two kinds of sciences: basic sciences address the structure of reality itself, conceived as independent of human interests, whereas the theories of goal-oriented sciences construct and organize nature teleologically, in relation to specific goals. And these two kinds of science are equated with two separate stages in scientific development. In the first stage a science operates much as Kuhn describes: the paradigm of the basic science is developed and articulated by autonomous scholars, whose work is unaffected by 'external' considerations. Eventually, however, this kind of work brings a science to 'theoretical maturity', when most of the fundamental problems are solved, and when their solution accordingly ceases to be the overwhelming concern of the field. At this point the science has the potential to enter a second stage, when appropriate goal-oriented specializations and developments of its basic general theory can be initiated, to suit whichever external goals and objectives are decided to be the most important foci for further work. (Something of the intensity of the backlash against the finalization thesis can be understood by considering this point: it is easily read as a suggestion that the direction of modern physics be set in the hands of politicians and science policy sciolists.)

Since practically every tenet in this general account is controversial and unsatisfactorily supported by evidence, there is no choice but to treat the whole with considerable scepticism. It is interesting to note how heavily the overall scheme relies upon a realist account of basic science. A realist perspective points up the contrast between the two kinds of science; it rationalizes the exclusion of external influence from the paradigm development stage of basic science; and it makes intelligible the otherwise highly suspect notion of 'theoretical maturity', which presumably occurs when all relevant real universals (fundamental particles, kinds of atom, kinds of force, etc.) have been characterized. On an instrumentalist view of basic science, in contrast, the distinction between the two kinds of science must be made in terms of levels and kinds of goal-orientation; there is no *a priori* argument for the absence of external influence from basic science; and the notion of 'theoretical maturity' cannot be given a meaning.

For all the controversy which it has engendered, the finalization thesis is not entirely incompatible with accepted wisdom upon external determinants; indeed on the key matter of the relative autonomy of basic science it shares identical presuppositions. Far more controversial would be claims that images and representations, particularly images of social and political relationships, have been drawn from the surrounding context and incorporated into the culture of science; or claims that the external context has engendered variation in scientific judgments of matters of fact; or similar claims concerning the evaluation of scientific theories and research programmes. (For briefly presented, readily intelligible, concrete examples of these three processes see Miller, 1972; Provine, 1973; and Ben-David, 1960; respectively.)

Processes of just this kind have been documented by Shapin in two extended interpretative review essays (1980, 1982). He argues that they feature far more frequently in the history of science than is commonly supposed. Part of his case rests upon the findings of recent empirical work which has effectively combined the methods and perspectives of the social sciences with those of social and intellectual history (see, for example, Barnes and Shapin, 1979; MacKenzie, 1981). But for the most part he relies upon the standard repository of historical materials — that is, upon the very archive which provides the foundation of the viewpoint he opposes. Shapin argues that the presupposition of the autonomous 'non-social' character of science has been so strong in the past that it has resulted in contra-indications being overlooked or misinterpreted: once the presupposition is set aside the mainstream tradition of the history of science, encompassing such central figures as Boyle, Newton, Priestley, Pasteur, Darwin and Mendel, gives ample indication of the (cognitive) relevance of the wider context. Shapin is able to cite a remarkably large number of historical case studies wherein scientific judgments can be seen to be related to considerations which lie well beyond the immediate technical context, and are indeed in most cases reasonably described as political or ideological. And effects upon practically every aspect of scientific culture are considered, from general models and images, through abstract theoretical structures and statements of fact, to pictorial representations and the structuring of perceptions (in which last categories Shapin's own empirical work [1979a] is probably the best available model for sociological research).

Shapin's reviews offer what comes close to an alternative history of science to that provided by the traditional autonomist account, and it will be interesting to see how historians respond to them. But they are not designed only, or even mainly, for this purely historical end: they use historical materials to demonstrate the value of sociological approaches to scientific knowledge generally, and lay no particular stress on 'external' influences on knowledge. Anyone in the social sciences who uses them as bibliographic resources in considering the problem of the autonomy of science will also find more important sociological issues discussed and illustrated. For example, they bring sociological and anthropological perspectives to bear upon the problems surrounding the social use of science (see also Shapin, 1979b).

The standard anthropological approach to a culture is not to transform it into a formal system of concepts and statements, and then to interpret it as a disembodied set of ideas. Verbal culture is always studied as it is manifested in a *context of use*. One does not think first of the verbal culture and its significance, and then secondly of how that culture is used: one understands the culture as something which has evolved through use, which is accepted because of its utility, and which indeed is meaningful only as it is used and changes in meaning as its use changes. There is no alternative route to understanding meaning other than by attending to usage, and no alternative basis for meaning other than in usage.

Accordingly, a routine anthropological study of the social use of science would presume a feedback from that usage to the meaning of the terms involved, and since the same terms are, by definition, used in scientific contexts also, it would presume a feedback into the context of science. The presumption would only be discarded if special mechanisms could be identified which kept scientific and everyday usage completely separate, so that identical terms (think, for example, of 'struggle for existence' or 'natural selection' or 'race' or 'heritability') actually came to stand for different concepts, a 'scientific' and an 'everyday' variant in each case. But the study of the social use of science has actually proceeded under the opposite assumption, that use has no effect upon meaning, and the existence of mechanisms of separation effective at the semantic level has simply been assumed. The social use of science has hardly ever been studied in what is commonly accepted to be the proper way.

The interest of Brian Wynne's contribution is precisely that it studies late Victorian, Cambridge physics in what, anthropologically, is the proper way. Wynne shows that important aspects of the verbal culture of the Cambridge physicists (natural philosophers) were developed through usage in two contexts, one narrowly technical, one explicitly political. As a result the specific, Cambridge conception of the luminiferous ether and the associated sameness-relation 'ethereal phenomenon' were constructed with regard to the requirements of use in both contexts. Goals and objectives from the narrow 'scientific' context and the wider 'social-political' context were relevant to the development of the ether concept of these natural philosophers. Accordingly, it is pointless and question-begging here to speak of a 'scientific' concept being used outside science, or for that matter of an

193

'ideological' concept being used inside. The concept of the ether developed by usage in two contexts, neither of which can be meaningfully set above the other. And continuing usage in either context built upon previous usage in both, so that, in so far as there were two separate forms of verbal culture present, a symmetrical interaction occurred between them.

We shall conclude with some very brief remarks upon how work such as Wynne's relates to the evaluation of knowledge. This, of course, is entirely a normative question, of no sociological interest, but a number of points can be made which bear upon it, and given its endless fascination it is perhaps worthwhile to make them. The significance attributed to Wynne's findings is likely to vary according to the view that is taken of the 'internal' culture of science. The more strongly judgment and evaluation in science is idealized, the more clearly will Wynne's findings stand as documents of distortion in that judgment and evaluation, and the more negative will be their epistemological significance for the associated knowledge. On the other hand, the findings need not be given any such significance at all: this is unnecessary if, for example, the instrumentalist point of view adumbrated in Part Two is accepted. There, the verbal culture of science was characterized as a conventional system, and its development was related to interests and objectives situated in the esoteric context of science. Pickering offered a concrete study illustrating this: his treatment of theory choice was, among other things, an account of the micro-political structure and the micro-politics of physics. In the case documented by Wynne (as in many others in the *embarras de richesses* cited by Shapin), the difference is merely that the development of the conventional structure is related also to goals and objectives sustained in the exoteric context: the reference is to macro-political factors, as well as micro. There is no fundamental difference between the results of the two studies: both relate scientific judgment to contingent contextual factors.

It might be said that there are contingencies and contingencies. Those contingent goals and interests established within the sub-culture of science inspire the generation of genuine technically-applicable knowledge. Those established in the wider context all too often inspire the generation of ideology. Clearly, therefore, it is desirable that science should not become oriented to external goals and interests at the expense of internal ones, and hence it is entirely proper to think of external contingencies pejoratively, as intrusions and biases.

This sentiment may well be admirable; but the argument which seeks to justify it is not. It is not possible to equate different kinds of goal or interest with different kinds of knowledge. Goals and interests inspire not the creation of knowledge *ab initio,* but the manner of using and developing existing knowledge: Pickering's and Wynne's studies both illustrate this clearly. Goals and interests help us to understand specific acts of cognition or concept application, and through this to understand not culture and knowledge, but cultural change and the growth of knowledge — conceived of as the development and articulation of existing culture and knowledge in an immense historical sequence of separately determined acts of concept application. If we consider a body of knowledge developing in this way in the context of a diverse range of interests, it emerges that competition between the interests is in no way a necessity. The adaptation of a body of knowledge to the requirements of one set of interests does not have to be at the expense of its adaptation to another set: a body of knowledge can evolve to be multifunctional (Barnes, 1981, 1982). Thus, from an instrumentalist perspective it is remarkably difficult to formulate a convincing account of any special inadequacy in the research of Wynne's natural philosophers, or in the physical knowledge which they produced.

11

Wolfgang Krohn and Wolf Schäfer

Agricultural chemistry:
a goal-oriented science

The fundamental concept of agricultural chemistry is that of the *reproduction cycles of organic processes*. As Liebig wrote:

> Our present research in natural history proceeds from the conviction that laws of interaction exist not only among two or three but rather among all phenomena which in the realm of minerals, plants and animals condition life on the surface of the earth. Thus none of them is separate but at all times joined to one or several others, all of them linked together without beginning and without end. The sequence of these phenomena, their origins and their departures can be compared to the tidal movement within a cycle.[1]

It can be supposed that Liebig did not derive the idea of cyclical organization only from agronomy and biology (where it is suggested phenomenally by the ontogenetic scheme and by the seasonal cycle) but also from the political economy of Adam Smith and John Stuart Mill, to whom he frequently refers.[2] Liebig criticized the agronomists for failing to make the distinction between efficient practices which reduce natural working capital and those which raise returns and interests. For the cycle to be maintained in equilibrium the capital turn-over must exclude diminishing returns. Otherwise extended reproduction would lead sooner or later to a breakdown:

> Agricultural production does not differ fundamentally from the regular industrial enterprise. Factory-owner and manufacturer know well that their investment and working capital must not continuously diminish if their business is not to come to an end. In the same way, a judicious agricultural enterprise requires the farmer who wants higher yields to increase the number of active components in his soil which are conducive to production.[3]

196

Presumably the concept of 'cycle' as an analogy to capital circulation is of consequence precisely because it makes the transition from natural cycles to cycles based on cultural norms more easily conceivable. Treating the difference between capital turn-over (= natural cycle) and capital returns (= agricultural earnings) by analogy requires 'agriculture to dispose of a reliable yardstick to measure the value of its experiences' regarding investments.[4] Does an increase in year x represent returns or is it concealed loss of capital? Only science can give prognostic answers. For example, does loosening the soil by mechanical means bring higher returns or does it increase its exhaustion?

The notion of the cycle has various connotations — e.g., the revolution of the stars in astronomy; the circulation of the blood in medicine; the circulation of money in economics; the notion of feedback cycles in technology; supply-demand mechanisms in capitalistic economics.[5] Two independent factors can be found underlying this variety: firstly, the assumption of a starting position, which will be re-established by a chain of metamorphoses; secondly, the notion of an equilibrium which is maintained by means of control mechanisms. In the tradition of political economy these aspects were initially separated from each other. The physiocrats, and Quesnay in particular, concentrated on the production-consumption cycle and thus on the reproduction of expended substances. The English school, and Smith in particular, had as its focus the mechanisms required to maintain a labile equilibrium. With Marx, the two aspects became combined.

Though Liebig's concept of cycles bears a strong resemblance to its equivalent in political economy, the concept's specific meaning grew out of chemical considerations: what explains the fact that although the nutrition of plants and animals constantly consumes certain resources of the atmosphere and of the soil, these resources remain constant over time? There are three possible solutions. Either these resources are infinite, or they are spontaneously generated, or they are regularly reproduced. The first and second solutions can be falsified by chemistry. If the quantity of oxygen consumed by animals were calculable and the atmospheric mass known, the exhaustion of the oxygen constituent in the atmosphere could be estimated. According to Liebig:

> In 800,000 years the atmosphere would no longer contain any trace of oxygen. It would, however, have become entirely unsuitable for respiration and combustion processes much earlier for already a reduction of its oxygen content to 8 per cent . . . would have produced a lethal effect on animals.[6]

The second solution is ruled out by a successful basic hypothesis

197

of the chemical paradigm: chemical elements cannot be transformed into each other. Therefore, the concept of the cycle is the
only possible scheme whereby the chemical conditions of life
processes can be conceptualized. Nature *must* follow cycles if
what is empirically observed is to be made intelligible.

Theoretically the concept of the cycle presupposes the concept
of an 'ought-value', namely a norm which determines that the
cycle has come to its completion: the initial conditions in respect
of which all later events are interpreted and analyzed. Yet if the
initial conditions are marked out as the 'ought', to be achieved by
future events, the degree of deviation — that is, the actual value —
and possible corrections must be detectible. Thus Liebig defines
circulation 'on the surface of the earth' as 'continuous change,
constant disturbance and recovery of the balance'.[7]

Liebig's conception of circulation contained both factors of reproduction and self-control. It was, however, impossible for Liebig
to offer a teleological explanation for these processes. Thus, the
principle of circulation made new demands on causal analysis in
chemistry. In the complex interplay of the processes of composition
and decomposition in the atmosphere, in the soil and in plant
physiology (including physically determined osmoses and diffu-
sions), a guiding thread had to be found to enable scientists to
follow all these chemical transformations.

A preliminary form of the concept of cycle was the concept of
metamorphosis developed in organic chemistry. This relied on a
concept of balance. In economics, the keeping of a balance sheet
presupposes that inputs and outputs of a qualitatively different
nature can be traced as transformations of a given value. This was
precisely the problem in chemistry and physiology. Which element
and which product (the plant as product of the soil and the air;
the animal as product of the fodder and the air) has what equivalent among the production factors? The discovery of these modifications of matter (called metamorphoses by chemists and physiologists at the time) was among the pioneering activities of the
physiologists of the eighteenth century.

From a methodological point of view, the principle of cycles of
reproduction involves two experimental essentials. In the first
place, experimentation in agricultural chemistry demands the
observance of *in vivo* conditions. Such experiments must intend
to identify *all* and *only* the necessary conditions of physiological
processes.[8] In his *Organische Chemie* Liebig accused the plant
physiologists of conducting experiments which were 'valueless for
the decision of any question' (p. 42) because they had no theory
of procedures to control their experiments:

> They conduct experiments, without being acquainted with the circum-
> stances necessary for the continuance of life — with the qualities and
> proper nutriment of the plant on which they operate — or with the
> nature and chemical constitution of its organs. These experiments are
> considered by them as convincing proofs, whilst they are fit only to
> awaken pity. (p. 41)

Similarly, Liebig reproached the animal physiologists, to whose
field the *in vivo* postulate also applied. His criticism referred
particularly to a question of practical consequence, i.e. whether
carbohydrates could be converted into fats or not:

> Without being acquainted with the conditions or even asking whether
> such conditions exist they first of all exclude everything which would
> make it possible to answer the question. The animals are put into a
> state of artificially induced disease, deprived of all nourishment; with
> the greatest care they exclude all those matters which play a part in
> blood formation and in the sustenance of vital functions acting on fat
> formation. They then believe that these miserable and cruel experiments
> have furnished proof that sugar, a nitrogenous substance, cannot be
> converted into fat, another nitrogenous substance.[9]

To answer a question by experiment requires knowledge of the
conditions under which the question is answerable. Liebig introduced this theory-dependent relation of theory and experiment into agricultural chemistry by means of the concept of 'chemical
metamorphosis'.

The second methodological aspect in experiments relates to the
time horizon in which agricultural experiments are framed. When
experimenting in agriculture, 'it takes a year or even longer before
one has all the results . . .'[10] Besides the fact that experiments can
find and explain innumerable details of physiological processes, a
final answer to the question of whether the chemical experiment
has made transparent one of the segments of the actual cycle will
be known only after at least one period has passed; and indeed,
because of a minimal accumulation of marginal changes, it is often
known only after several periods.[11]

THE CONSTRUCTIVE VARIATION OF CYCLES

The principal of reproduction cycles and its implications, however
important, form only one part of the paradigm of agricultural
chemistry. So far nothing has been said about the 'culture' of agricultural chemistry. This principle refers to the 'conditions necessary
for the life of all vegetables . . .'[12] But agricultural chemistry is not
merely the science of natural cycles and their stability, but also

the science of their variation in keeping with human needs and purposes. In what manner these purposes affect the content of the paradigm, will be briefly outlined.

While the immanent finality of nature, in a manner of speaking, is rooted in the enduring stability of natural reproduction, the farmers' immanent interest is:

> to cut as many shoes as possible out of the inexhaustible stock of leather in the soil and the best teacher will be the man who has made the most of such intensive cultivation of the soil . . .[13]

Nonetheless, once it has become evident that over-exploitation of the soil has catastrophic consequences, the farmer's interest and that of society no longer coincide. Society's interest must be in 'achieving big and ever increasing harvests for ever and ever'.[14] To solve this problem is, according to Liebig, the purpose of agricultural chemistry. Its object then is no longer confined to the discovery of the laws of reproduction governing vegetable life, but to construct natural reproduction cycles which will benefit man. The purposes that have to be achieved in such constructions with the means of nature are not purposes of nature. Thus, agricultural chemistry becomes a social science operating with the methods of natural science. At first sight this seems to be the usual connection between science and technology: once discovered, natural laws enable technical constructions within nature. But in agricultural chemistry the social purpose is not the *result* of scientific knowledge but the *condition* of scientific inquiry. Agricultural chemistry does not provide insights into *why* agriculture functions but rather constitutes a design of *how it must* function.

The inter-relation between goals and theory construction in agricultural chemistry becomes evident as soon as one apprehends that knowledge of agricultural chemistry is directed towards goals which are not natural goals. The general notion of 'achieving big and ever increasing harvests for ever and ever' finds specification in a whole set of norms (or 'oughts'), when the concept of 'increase' is operationalized. 'Increase' as signifying an additional yield is a concept which is not derived from nature:

> In a free and uncultivated state the development of every part of a plant depends on the amount and nature of the food afforded to it by the spot in which it grows. A plant is developed in the most sterile and unfruitful soil as well as in the most luxuriant and fertile, the only difference which can be observed being in its height and size, in the number of its twigs, branches, leaves, blossom and fruit.[15]

It makes no sense to say that a dwarf-pine is less perfect than a pine grown in the forest only because their individual organs differ; a dwarf-pine 'increased' in its height and size would not be capable of existence. In this sense, if we take agricultural chemistry to be a science of the continuous removal of cultural utilities from nature, it is a science which treats laws of social nature. These laws make reference to norms which are superimposed upon natural qualities: in part on the qualities of plants, in part on those of the soil, in part on those of the reproduction cycle.

With respect to plants, such norms are, according to Liebig: to obtain an abnormal development of certain parts of plants; and to obtain an abnormal development of certain vegetable substances. Or, to put it more concretely, fine pliable straw, giving strength and solidity, or more corn from the same plant, or a maximum of nitrogen in the seeds, and so on.[16] The norms for the soils are in particular: different degrees of fertility (depending on the sum of nutrients which can be absorbed), and the duration of fertility (depending on the sum of nutrients present in the soil).[17] Norms which relate to the reproduction cycles are derived from the selection of the crops and their succession.[18]

Now it is obvious that by whatever method one tries to order these norms systematically, they will still not be coincident with a scale of natural states of things. In this respect agricultural chemistry differs from classical medicine, which is concerned not with diverse forms of health but with eliminating disease. Deviations from an 'ought' state of affairs which is thought to be naturally given, can, in principle, be ordered on a scale. However, when such deviations are defined as the goals of various maximization strategies (something which in medicine would be interpreted as a theory of alternative forms of possible good health and the means of improving health) it will be the goal-orientations themselves which determine, or more precisely, which have a determining influence on the descriptive pattern and the methods of the theory.

The teleological structure of 'finalized' theory in agricultural chemistry signifies that the construction of suitable soils, manures and plants (and ultimately of a suitable ecological system) is not just an application of scientific results, but the goal of obtaining knowledge. This is because the concept of suitability is a theoretical concept which cannot be derived from the state of things as found in nature but itself structures possible states nature can adopt. Agricultural chemistry is explicitly a cultural conceptualization of nature. This statement should not lead to misunderstanding: it is not intended to say that the actual results of agricultural chemistry (causal explanations and laws) are not objective natural laws. They are laws of nature in the sense of being theoretically derivable from general laws formulated by chemistry, physics and physiology, and in the sense that the resulting experimental and technical practice entails objective experimental rules and procedures. Therefore, if we speak of finality in this connection this does not refer to the form of these laws but to the possibility of their being conceptualized.

Thus the hard core of agricultural chemistry is not the concept of reproduction cycles but its social goal-orientation, achieving a continuous increase of crops with perpetual continuance. The goal so defined has a complex structure which seems at first sight inconsistent: if on the one side there is posited the connection between all processes of animal, plant, soil and atmospheric chemistry as constituted by a system of circulation, and if on the other side it is stipulated that an increasing mass of substances will be removed from circulation, how can stability of circulation be ensured? Liebig's paradigm cleared the way to change the various cycles, according to social goals, without disturbing the natural equilibrium given. The removal of any substance from a system of circulation necessarily implies the integration of that same substance into other chemical-physical processes. In these processes, each 'disappearance' of a substance has to be linked to a reorganization of cycles. This reorganization must, however, safeguard the reproduction of the initial conditions of the original cyclical process. Only then are 'ever increasing yields' no longer an idle dream but a realizable goal.

THE 'PROGRESSIVE PROBLEMSHIFT' IN AGRICULTURAL CHEMISTRY

For the institutionalization of agricultural chemistry it was decisive that the paradigm formulated by Liebig possessed an important theoretical impact and was empirically generative. This can be demonstrated by six points.

1.   The concept of the cycle — that is to say the first part of the paradigm — led Liebig to the discovery of fundamental chemical processes such as the nitrogen cycle, the hydrogen cycle and the interdependent oxygen-carbon cycle. This made possible the explanation of the constant composition of the media (atmosphere and soil) and it led to inferences about the decrease of carbon in the air prior to the emergence of the animal world and the equivalence of the mass thus lost to natural carbon deposits.

The heuristic relevance of the category of the cycle can scarcely be overestimated. The concept affects animal physiology, links

animal and plant physiology and can be used practically in the process of 'recycling'. The concept connects interdependence with causality and structures a systematic concept for the reproduction of physical-chemical-biological processes of the earth.
2.   The concept of metamorphosis gained from organic chemistry performed a key function in agricultural chemistry. Berzelius, referring to it, wrote:

> In our study of organic bodies our attention was primarily centered on the particular product . . . while we almost neglected the by-products. Gradually we began to notice some, finally all of them. Actually, it was Liebig and Wöhler who were the first to take up these studies . . . and who shed light on an entirely new field in the study of organic chemistry . . . that of the chemical metamorphosis of *organic oxides*. The excellent work they did on the transformation of uric acid came just at the moment when it could best serve us as a guide in these complex questions . . . The time has come now to pay the closest attention to these metamorphoses.[19]

In agricultural chemistry the concept of chemical metamorphosis guided the analyses of the processes of fermentation, decay and putrefaction as well as of nutrition and respiration. Here knowledge of all the products formed in such transformations was fundamental for an understanding of the chemical unity underlying the reproduction of animal and plant life. The same held true for knowledge of the effects of substances artificially introduced into nature.[20] The chemistry of organic decomposition exhibited a systematic connection with agricultural chemistry and thereby acquired a consistent pattern itself.
3.   The experiments performed by rational agriculturists became susceptible to scientific interpretation only within the frame of agricultural chemistry:

> What Thaer found good and proper in his fields at Moeglin came to be held as good and proper for all German soils and what Lawes discovered on the small strip of land at Rothamstead came to be held as axioms applying to all English soils.[21]

Leibig did not intend to advocate inductionism. On the contrary, he realized that there must be a theoretical standard explaining the range of techniques, for only on this basis could 'trial' become 'experiment':

> If the agriculturist, having no proper scientific principle to guide him, carries out experiments . . . then there will be small chance of success. Thousands of agriculturists carry out similar experiments in various ways, the result of which in the end will be an agglomeration of practical experiences, comprising a method of culture, whereby the goal sought will be obtained for a certain area. However, the next door

neighbour will fail using this method. Imagine the amount of capital and effort lost in *these* experiments! How totally different, how much safer, the path followed by science; once on it, science does not subject us to the danger of failure and bestows upon us all guarantees for success.[22]

In agricultural chemistry a series of new experimental devices were generated, permitting laboratory experiments with artificial nutrient fluids, analysis of physical transformations and serial research on comparative physiology.[23]

4.   It took some twenty years before artificial manure was obtained; but the early failures were not owing to theoretical errors as much as to insufficient knowledge. Apart from this, agricultural chemistry proved successful in predicting the effects of manure of different kinds upon plants. In the history of artificial manure, the difficulties encountered were of two kinds.[24] Firstly, the almost exclusively theoretical derivation of the chemical composition of manure proved incompatible with some of the empirical conditions. Liebig had expected rain-water to wash off the manure if it consisted of water-soluble substances and had not allowed for the fact that the physical properties of the soil prevented the manure from being absorbed by the plant. It took him some sixteen years to recognize why this occurred. Secondly, Liebig had underestimated the influence of nitrogen. He believed that plants could provide for themselves sufficiently with the nitrogen they absorbed from atmospheric ammonia.[25] It was only with considerable hesitation that Liebig conceded this point, because to recognize that manures needed to contain nitrogen clouded the clear differences between the traditional humus theory and Liebig's own theory of minerals. This did not seem desirable from either a strategic or an institutional point of view.

5.   The construction of artificial manure was based on two fundamental insights which, under the label of the 'law of the minimum', came to form a kind of technical natural law: a) the growth of plants is determined not by the organic but by the mineral substances of the soil (known as the 'mineral theory of plant nutrition'); and b) the substances contained in the nutrient salts are not intersubstitutable.

The theory that the growth of a plant depends on the mineral substance which is exhausted first had been discussed by Sprengel.[26] Liebig did not think highly of Sprengel's contribution. Sprengel could present clear reasons for the importance of mineral substances to the growth of plants by the analysis of ashes.[27] However, he did not distinguish between the actual constituents of plants and the required quantities of each constituent (or their possible substitutes) which, according to theory, were indispensable. Plants absorb all soluble salts from the soil. As Liebig wrote:

> Because of this capacity no conclusion can be deduced for the necessity of the presence of certain salts in plants or of their presence being merely accidental. All inorganic constituents present in the ashes could not be considered necessary without further investigation.[28]

This distinction between *actual* constituents present in plants and *necessary* ones makes the 'law of the minimum' non-trivial; it cannot be established by analytic observation only, but on the basis of theory and experiment for particular plant species. As an experimental law, it had strategic and technical significance: it rendered possible the discovery of absent substances and thereby their addition or substitution.

6.   As Liebig's theory could not be invalidated by various faulty suppositions — furthermore could falsify these very suppositions — its stability as a *theory* became established. The initial failure to construct artificial manure, which prevented academic acceptance of agricultural chemistry, did not undermine Liebig's theory. Faulty suppositions (for instance, the view that plant cells did not need to respire oxygen or Liebig's views about the irrelevance of micro-organic processes) did not destroy the structure of agricultural chemistry. Rather, once corrected, they served to extend and develop it.

The 'progressive problemshift' which occurred in the paradigm of agricultural chemistry indicated that agricultural chemistry had become more than just the successful application of accumulated chemical knowledge. Nonetheless, there exists an important difference between a discipline of this type and, say, classical mechanics or inorganic chemistry. Agricultural chemistry had nothing as fundamental to offer as, for example, axioms of mechanics or the chemical laws of simple and multiple proportions. Yet it is one of the characteristics (although not a necessary one) of a 'finalized theory' that its foundation does not rest on the relevance of its propositions within the construction of the 'natural' reality, but is structured by the urgency of social problems. To win acceptance, the theories which can take these purposes into account and find the respective problem-solutions need no longer evince the same fascination as do fundamental theories. On the other hand, they must possess paradigmatic stability for evaluative standards of progress to emerge and to define the next layer of problems to be considered. If such paradigmatic autonomy does not exist, the respective field will, as a rule, become the assembly-point of findings from a great variety of research fields. In this sense, even if there had not been Liebig's paradigm, there would have developed an agricultural science eclectically assimilating the results of physiology, botany, physics, and so on. But such a science would have lacked the characteristics of a progressive problemshift based upon paradigmatic structures. Finalized disciplines take their rise in this middle-ground between an inter-disciplinary eclecticism, which is preformed *only* by social needs, and a scientific interest generated solely by theory-guided problem constellations.

THE INSTITUTIONALIZATION OF AGRICULTURAL CHEMISTRY

So far our analysis has focused on the historical development of the cognitive structure of agricultural chemistry. The question now to be asked is whether that structure is associated with any determinate type of institutionalization. This seems likely, for if a discipline is based strategically on goals external to science it may be expected that its institutionalization will be strategically pursued.

For Liebig, the decision to shift the focus of his concern from organic chemistry to agricultural chemistry was not motivated by scientific reasons alone. His 'insuperable disgust and aversion against chemistry's activities at the present time' was based on the belief that chemistry was *asocial* and that 'useful applications cannot be obtained therefrom, neither for medicine, nor for physiology, nor for industry'.[29] He wished to prove 'that if physiologists and agronomists do not use the culture of chemistry, no permanent and valuable progress can be expected in physiology or in agriculture'.[30]

Liebig's view was that chemistry must turn to agriculture to prevent catastrophes of huge dimensions. Such a motivation rules out any 'ivory-tower' mentality of the 'pure' scholar. In order to institutionalize his concepts Liebig extensively used tactical and strategic methods. His behaviour was frequently characterized (and criticized) as over-ambitious. There is no need to deny these allegations since a science which comes into being for external reasons, and which operates on the basis of external goals, requires the active introduction of its findings and methodology into society. In his writings Liebig frequently touched on the problems of the establishment of new doctrines. His explicit opposition to the humus theorists and physiologists was grounded not only upon his mineral theory but also upon his understanding of the principles underlying effective opposition:

Glancing at the history of science reveals that if a new doctrine takes the place of a prevalent one, the new doctrine will not be a further development but the very opposite of the old one.

For a long time I believed that in agriculture it would suffice to teach the truth, in order to disseminate it, as is common in science, and to disregard errors; finally, however, I have realized that this was wrong and that the altars of the lie have to be shattered, if truth is to be based on a firm foundation.[31]

Liebig acted accordingly.

In the years following 1840, Liebig was so convinced of his theory that almost nothing shook his belief in the effectiveness of artificial manure, the formulas of which were mainly derived theoretically. Together with the manufacturer of his manure, he waged a publicity campaign in the English press which they hoped would promote the rapid diffusion of the manure. Widespread introduction of the new manure would have considerably increased the demand for agricultural chemists in rural areas, in industry, and at universities. However, the manure's inefficacy soon became obvious. As far as the institutionalization of agricultural chemistry was concerned the manure programme proved a complete failure. As one historian has commented, 'Liebig's teachings seemed so much discredited that the *Journal of the English Agricultural Society* refused to publish Liebig's answer to reports by Lawes and Gilbert.'[32] It took Liebig fifteen years to explain the failure of the manure. His battle against the 'practical men' of rational agriculture over acceptance of his claims took equally long. This underscores the significance of institutional success for a 'finalized' science. Within chemistry, the success of the reproduction cycle theory might have been sufficient for acceptance.

For Liebig, the training of students was of paramount importance for the institutionalization of his theory. As he was one of the first to found a laboratory in which students received regular training in the art of experimentation, he could easily promote agricultural chemistry as a special field of training. Liebig's purpose was to tackle the complex problems of agricultural chemistry by an effective division of labour and organization of research. But behind this organization of a system of division of labour there was yet another problem to which no solution had been found until Liebig's time: the standardization of experimental conditions and the practicability of speedy simultaneous analyses. Liebig's purpose in his activities as designer of experimental apparatus was to simplify the apparatus even at the cost of precision.

The regular training of students presupposed the existence of job opportunities. In several reports Liebig attacked his contemporary academic colleagues for their incompetence, hoping to push his students into their chairs.[33] Reflecting on his success in Austria, he wrote: 'For many years no candidate who had not received training in Giessen or had not completed his studies there could get a teaching post'.[34] In Hessen, Saxony and Bavaria these efforts proved successful; later, they were successful in Prussia, too. For this policy of 'public relations' Liebig did not need to found a special journal. He held a dominant position in the *Annalen der Chemie und der Pharmacie* (the leading German chemistry journal and, at the same time, the leading international journal of organic chemistry), and thus he could give agricultural chemistry the backing of organic chemistry which was developing with extraordinary success. To this end Liebig in 1840 modified the traditional classification of the *Annalen* (in which investigations in agricultural chemistry would have appeared only under the item 'Mixed Notes') and had all relevant publications, including his own works and polemics with other scholars, appear in the *Annalen*. In addition, Liebig ensured that agricultural chemistry was given a prominent place in many of the handbooks and textbooks published.

In Munich in the last period of his life (1859–1873) the institutionalization of agricultural chemistry was completed: in 1862 the first professorial chair for agricultural chemistry was established in Halle, followed by Leipzig in 1869, Giessen in 1871, Göttingen in 1872, Munich in 1874, Königsberg and Kiel in 1876, Breslau and Berlin in 1880. Accordingly, the academies of agriculture which had been the castles of the oppositional rational agriculture were dissolved: Regenwalde in 1859, Möglin in 1862, Waldau in 1868, Tharandt in 1869, Hofgeismar in 1871, Eldena in 1877, and Proskau in 1880. Others were taken by Liebig's students.

CONCLUSION

The theory of agriculture to which Liebig first gave a relatively complete formulation marks the beginning of a new type of theory-building in the natural sciences. We have called this type 'finalized science', a term chosen to indicate that human needs and interests explicitly take part in forming the subject field of a science. As teleological causes they are the cognitive impetus of theory-building. Thus the science of agriculture is a theory about the most rational organizing of nature to satisfy human needs and at the same time a science about the rationality of human needs *vis-à-vis* nature.

Obviously such a science differs from the classical disciplines of physics, chemistry and biology. In the latter the order of nature — the structure of reality conceived as independent of human interests — constitutes the subject of research, although of course applicable results follow from research and the patterns of scientific thought are not isolated from the prevailing cultural environment. Newton's classical mechanics, Einstein's theories of relativity or Watson's double helix are in this sense knowledge that is valid, independent of human needs and interests. On the other hand, no clear distinction can be made between agricultural chemistry and applied or engineering sciences or medicine. In these fields specific segments of nature are conceptualized under the perspective of satisfying human needs.

However, our concern is not to distinguish agricultural chemistry (as finalized science) from applied sciences, but rather to give the concept of applied science a precise meaning. The term 'applied science' is misleading in suggesting that it is nothing but the *application* of a given science, and not a contribution to science itself, resulting in the notion of the superiority of 'pure' over 'applied' science. In certain instances an applied science or technology may not require theoretical work but only need to establish certain strategies to gain new knowledge. As a rule, however, the production of applicable science will of necessity entail a process of theory-construction evolving independently from the mother disciplines. Such additional theoretical work is necessary not only because the respective application is a specialization, the refinements of which have no major significance for the mother disciplines; but also because these refinements cover *a limine* different types of objectivity: on the one hand, inquiry into nature unrelated to any specified purpose; on the other, nature constructed purposefully in terms of specific goals.

Historically agricultural chemistry is one of the first examples of successful goal-oriented theory. Goal-oriented sciences have since modern times been theoretical in the sense that they have formulated models and analogues for the phenomena observed. The eighteenth century offers many examples of the closer relationship between science and technology in textiles, agriculture and transport. But it was only in the latter half of the nineteenth century that Liebig and — a little later — the pioneers of mechanical engineering (Franz Reuleaux) and refrigeration and heating (Carl von Linde) for example, formulated theories which were deductive, comprehensive and heuristic and thus fulfilled the function of a paradigm.

Traditional philosophy of science has tended to neglect this type of applied sciences. Their theoretical status and their relationship to pure sciences have not been adequately clarified. Presumably, two causes may be held responsible for this neglect. First, object areas and problems limited to the study of disease, machines, manure and the like do not possess the dignity natural philosophy was ready to accord to universal properties of nature (e.g., forces, chemical compounds, and so forth). Secondly, inquiry into the artificial, the purposefully constructed, reality is held to possess less dignity than the 'natural state' of nature.

It is impossible to discuss on the mere basis of a case study the importance of the role of pure science in the progress of applied and technological sciences. We intended our case to show in what manner applied sciences are theory-developing research strategies. Finalized sciences, on the one side, presuppose mature disciplines in which entire areas of reality have been subject to scientification (e.g., fields of physics, chemistry). On the other hand, they presuppose advanced technologies, which furnish new processes, but encounter limitations, which cannot be overcome by mere empiricism. Special developments occur when specific technical or social goals are introduced into a basic science already matured. The philosophy of science reveals but one aspect of modern science — its reductionist programme. Another important aspect of modern science has been discussed here in terms of our concept of finalized science: that is, the construction of reality according to scientific methods and theories, directed by the changing goals of man in his natural environment.

REFERENCES

1. J. Liebig, *Die organische Chemie in ihrer Anwendung auf Agricultur und Physiologie* (Braunschweig: Vieweg, 1840), I. intro., 87.
2. *Ibid.*, 134ff.
3. *Ibid.*, 147ff.
4. J. Liebig, *Chemische Briefe* (Leipzig/Heidelberg: Winter, 1878) x.
5. O. Mayr, 'Adam Smith and the Concept of the Feedback System', *Technology and Culture* 12 (1971) 1-22.
6. J.Liebig, *Die Chemie in ihrer Anwendung auf Agricultur and Physiologie*, 2 vols (Braunschweig: Vieweg, 1862) I. intro., 18. As there exists no authorized translation of the first volume, we were obliged to translate the quotations ourselves.
7. J. Liebig, *Chemische Briefe* (Heidelberg: Winter, 1844) 178.
8. Liebig, *Die organische Chemie*, 36ff.
9. J. Liebig, *Bemerkungen über das Verhältnis der Thierchemie zur Thier-Physiologie* (Heidelberg: Winter, 1844) 40.
10. Liebig, *Die Chemie*, I. xiv.
11. *Ibid.*, I. 290.
12. *Ibid.*, I. 137.
13. *Ibid.*, I. intro., 5.

14.   *Ibid.,* I. intro., 12.
15.   *Ibid.,* I. 157.
16.   *Ibid.,* I. 159.
17.   *Ibid.,* I. 252 f.
18.   *Ibid.,* I. 257.
19.   J.J. Berzelius, 'Über einige Fragen des Tages in der organischen Chemie', *Annalen der Pharmacie* 31 (1839) 20f.
20.   Liebig, *Die organische Chemie,* 199ff.
21.   Liebig, *Die Chemie,* I. xiii.
22.   *Ibid.,* I. 180.
23.   See, respectively, J. Liebig, 'Die Wechselwirtschaft', *Annalen der Chemie und Pharmacie* 46 (1843) 73, and O. Kellner/H. Immendorf, 'Beziehungen der Chemie zum Ackerbau', *Die Kultur der Gegenwart,* 3, Teil, 3. Abteilung, 2. Band (Leipzig/Berlin: Teubner, 1913) 415, 464.
24.   Cf. L. Schmitt/H. Ertel, *100 Jahre erfolgreiche Düngerwirtschaft* (Frankfurt: J.D. Sauerländer, 1958).
25.   Liebig, *Die organische Chemie,* 84.
26.   C. Sprengel, *Die Bodenkunde oder die Lehre vom Boden* (Leipzig, 1837).
27.   C.A. Browne, *A Source Book of Agricultural Chemistry* (Waltham, Mass: Chronica Botanica Company, 1944) 233.
28.   J. Liebig, *Die organische Chemie in ihren Beziehungen zu den Herren Dr. Gruber und Sprengel* (Heidelberg: Winter, 1841) 36.
29.   J. Carriere, *Berzelius und Liebig. Ihre Briefe von 1831-1845* (Munich/Leipzig: J.F. Lehmann, 1893) 210-11, letter to Berzelius dated 26 April 1840.
30.   *Ibid.,* 215, letter to Berzelius dated 3 September 1840.
31.   J. Liebig, *Die Chemie,* I. intro., 13, and I. xv.
32.   Volhard, *Justus von Liebig,* 2 vols (Leipzig: J.A. Barth, 1909) II. 37.
33.   J. Liebig, *Über das Studium der Naturwissenschaften und über den Zustand der Chemie in Preussen* (Braunschweig: Vieweg, 1840); J. Liebig, 'Der Zustand der Chemie in Österreich', *Annalen der Pharmacie* 25 (1838). 339-347; J. Liebig, *Herr Dr. Emil Wolff in Hohenheim und die Agricultur-Chemie* (Braunschweig: Vieweg, 1855).
34.   Volhard, *Justus von Liebig,* I. 369.

12

Brian Wynne

Natural knowledge and social
context: Cambridge physicists
and the luminiferous ether

I am afraid that this 'nature' is herself nothing but a first custom, just
as custom is a 'second nature.' — Pascal, *Pensées*

The suggestion that moral or social concerns may be expressed in
the idiom of natural laws no longer seems extraordinary, even
though it may be treated as a threat by some. It is taken for granted in this chapter that ideas about nature even in their most
institutionalized form—that is, in scientific thought—may come to
reflect contextual factors in one way or another. However, this
hard-gained vantage point only heralds the beginning of challenging problems confronting the historian and sociologist of science.
The important questions concern the ways in which different
sorts of social factors interact with other factors, including empirical, technical, and intellectual ones, to produce different forms of
scientific knowledge.

One of the most important problems in this area concerns the
multiple uses of scientific concepts, especially as between scientific
and political or moral contexts. One argument holds that this
'social use of scientific concepts' is simply a case of knowledge
being drawn from science to underline a point, without any
reciprocal influence upon the context of use within science.
However, it is increasingly clear that this may not always, or even
typically, be the case (Forman, 1971; Young, 1973). It is now
argued that one context of use can never simply be assumed to be
prior to or independent of another, even if it is the context of
natural science, and that interaction between contexts of use
should always be examined symmetrically with causal connections
in both directions being sought (cf. Shapin, 1979). Much will
depend upon the specific social nature of and relationship between
those contexts.

One useful way of approaching an elucidation of the relationships between cultural context and natural knowledge is via the postulate that different social groups articulate and develop their concerns and social interrelations through forms of knowledge which contain tacit meanings embedded in them, and which are in a process of continual renegotiation, defence, and development. This is not an uncommon position in anthropology and sociology; it is stated particularly clearly by Mary Douglas in her *Implicit Meanings* (1975). She (1975:281) asserts the connection between society and natural knowledge in uncompromising and completely general terms:

> Apprehending a general pattern of what is right and necessary in social relations is the basis of society: this apprehension generates whatever *a priori* or set of necessary causes is going to be found in nature.

I would like to discuss, in an unashamedly exploratory way, the extent to which natural scientific knowledge can be justly and usefully regarded as a form of displaced, tacit moral discourse. I shall discuss one example—that of late Victorian physics. Although the exploration will thereby be narrowly focussed, I offer in compensation the analysis of a concrete case, from a well-documented historical situation.

THE SCIENTIFIC CONTEXT: LATE VICTORIAN PHYSICS

The physics of the last quarter of the nineteenth century has attracted the attention of a number of historians of science (Sviedrys, 1970; Swenson, 1972; Brush, 1967; Wilson, 1971; Goldberg, 1970), but as yet probably not in the degree which it deserves. During this period, as in so many others, British physics (or 'natural philosophy') was dominated by Cambridge and recent émigrés from Cambridge to the provinces. At Cambridge during this period were the men at the forefront of late Victorian physics—Lord Rayleigh (the third Baron Strutt), J. J. Thomson, G. G. Stokes, Joseph Larmor, James Clerk Maxwell. Close behind in eminence came Balfour Stewart, who went from Trinity to the Chair at Owens College, Manchester (where he taught Thomson and sent him to Cambridge); P. G. Tait, who went from Peterhouse to the Chair of Natural Philosophy at Edinburgh; William Barrett, who returned from Trinity, Cambridge to the Trinity, Dublin Chair of Natural Philosophy; and G. F. Fitzgerald, Fellow of Trinity College with Rayleigh and Thomson, and their colleague in the Cavendish Laboratory. Sir Oliver Lodge was a close friend of both Fitzgerald and Thomson and, although not a Cambridge don, worked with them at the Cavendish for short periods. W. M. Hicks was a Fellow of St. John's, friend of his co-Fellow Larmor, and later Professor of Natural Philosophy at Sheffield.

Although these physicists spanned a quarter century or more of scientific activity, and differed in their detailed interests and attitudes, nevertheless there were certain fundamental tenets which they held in common, against various competing trends and movements in scientific thought. It was they who bore the standard of British Victorian physics well into the twentieth century and who continued steadfastly to articulate their paradigm in the face of eventually irresistible pressure for change (Goldberg, 1970).

One way of conveying something of the distinctive style and flavour of the work of this school of physicists is via a discussion of one of their most important concepts, that of the ether. This, of course, has been a widely employed concept throughout the history of natural philosophy. But it must be stressed that it possessed a quite distinctive significance and mode of use in the present context; that this was the result of an active transformation of their received conception of the ether on the part of the Cambridge physicists; and that this transformed conception is not readily intelligible as being 'required by the state of experiment and observation.' In this period, which followed Maxwell's development of the electromagnetic field equations, many conceptions of the character and status of the ether appeared plausible. Indeed Maxwell's own ambivalent attitude to this, and similar scientific concepts, neatly symbolizes the point (Kargon, 1969).[1]

Perhaps the most fundamental shifts from the received view were the unification of ether and ponderable matter, and the transformation from material theories of the ether to exactly the reverse, namely ethereal theories of matter. J. J. Thomson (1908) himself observed that this was the nature of the transformation in which he had taken a leading part between 1885 and 1900. Doran (1975) has documented Larmor's central role in this transformation to a thoroughgoing electromagnetic conception of nature rooted in and unified by ether. Lodge (1889) expressed this view cogently in his *Modern Views of Electricity,* as did Hicks (1895). Michelson (1899:75) expressed the prevailing consensus thus:

> Suppose that an ether strain corresponds to an electric charge, and ether displacement to the electric current, these ether vortices to the atoms—if we continue these suppositions we arrive at what may be one of the grandest generalisations of modern science—of which we are tempted to say it ought to be true even if it is not—namely that all the phenomena of the physical universe are only different manifestations of the various modes of motion of one all-pervading substance—the ether.

A central characteristic of the late Victorian ether was its *supremacy* over matter and its non-material nature. It was 'matter of a higher order' with 'a rank in the hierarchy of created things which places it above the materials we can see and touch' (Fleming, 1902:191). According to Stokes (1883:17):

> We must beware of applying to the mysterious ether the gross notions which we get from the study of ponderable matter. The ether is a substance, if substance it may be called, respecting the very existence of which our senses give us no direct information; it is only through the intellect . . . that we become convinced that there is such a thing.

Larmor (1900:vi) was even more categorical when he explained that:

> Matter may be and likely is a structure in the ether, but certainly ether is not a structure made of matter. This introduction of a suprasensual ethereal medium, which is not the same as matter, may, of course, be described as leaving reality behind us; and so in fact may every result of thought be described which is more than a record or comparison of sensations.

Matter, electricity, indeed all physical phenomena were to be made intelligible as properties of a 'suprasensual' ether. Thus, the ether took on a transcendent, *unifying* role within science. It was simplicity lying behind diversity, coherence behind disorder. It established continuity and connection between disparate particular events. Larmor (1908:43) put it as follows:

> Our conviction of an orderly connexion between things constitutes the conception of a *cosmos*. We have placed the foundation of this in the existence of a uniform medium, the ether, the physical groundwork of interstellar space. . . . The only ground for postulating the presence of this medium is the extreme simplicity and uniformity of the constitution which suffices for its functions. Needless to say, there remain many unresolved features, some still obscure, but hardly contradictory. But should it ever prove to be necessary to assign to the aether as complex a structure as matter is known to possess, then it might as well be abolished from our scheme of thought altogether. We would then fall back on simple phenomenalism, proximate relations would be traced, but we need not any longer oppress our thoughts by any regard for a common setting for them; the various branches of physical science would cultivate with empirical success independent modes of explanation of their own, checked only by mutual conservation of the available energy, while the springs of their orderly connexion would be out of reach.

And Lodge (1913:15, 17) considered the ether to be:

> the universal connecting medium which binds the universe together, and makes it a coherent whole instead of a chaotic collection of independent isolated fragments.

> The ether of space is at least the great engine of Continuity. It may be much more, for without it there could hardly be a material universe at all.

The physicists were willing explicitly to acknowledge their *faith* in the all-unifying medium. Here they were reacting against positivistic and naturalistic competitors who sought to do away entirely with entities whose existence could not be empirically observed. Fitzgerald (1896:441) rebuked one such competitor thus:

> Professor Ostwald ignores such theories as that of vortex atoms, which postulate only a continuous liquid in motion; but, it may be, this is omitted because it is merely a way of explaining the atoms. He also ignores metaphysical questions, such as whether motion be not only the objective aspect of thought, and also whether an intuitively necessary explanation of the laws as distinct from the origin and consequent arrangement of phenomena is not postulated by the fact that the Universe must be intelligible. Consequently, his attempt to deal with nature in a purely inductive spirit is unphilosophical as well as unscientific. The view of science which he puts forward—a sort of well arranged catalogue of facts without any hypotheses—is worthy of a German who plods by habit and instinct. A Briton wants emotion—something to raise enthusiasm, something with a human interest. He is not content with dry catalogues, he must have a *theory* of gravitation, a *hypothesis* of natural selection.

Thus did Fitzgerald unashamedly defend a 'metaphysical' science, basing itself in 'necessary' *a priori* principles of reality such as the all-unifying ether as much as in empirical observables. Indeed, antipathy to empiricism and positivism extended right down to matters of detailed scientific method. The 'Cambridge' style vigorously criticized overemphasis upon precision and decried the tendency in other approaches in physics to deny existence to anything which did not yield to highly accurate measurement. Nor was this a later *ad hoc* defence of ether against scientific progress, for even Rayleigh, perhaps the most utilitarian of this group, was emphasizing this point as early as 1884, against the naturalistic complacency of the previous decades which assumed that 'all the really great discoveries had been made and that nothing remained but to carry them to the next decimal place' (d'Albe, 1923:256). Lodge (1913:9) as usual was most fluent in his encapsulation of the point:

> The simple laws on which we used to be working were thus simple and discoverable because the full complexity of existence was tempered to our ken by the roughness of our means of observation. . . . Kepler's laws are not accurately true, and if he had had before him all the data now available he would never have discovered them.

The Cambridge group championed, instead of precision, the

imagination as paramount in going beyond the 'cowardly security' of empirical sense-data. And they urged cultivation of 'the neglected borderlands between the branches of knowledge' (Rayleigh, 1884) which naturalism was fragmenting, with its denial of any common setting or *plenum* in nature for developing specialist disciplines. Probably the most fruitful research programme of the era, Thomson's, on the conduction of electricity through gases of greater and greater rarefaction (which led *inter alia* to discovery of the electron), was the outcome of such an approach, presaged by Crookes' (1879) famous investigation of 'matter in a fourth state,' or 'radiant matter' where:

> We have actually touched the borderland where matter and force seem to merge into one another, the shadowy realm between known and unknown which for me has always had peculiar temptations. I venture to think that the greatest scientific problems of the future will find their solution in this borderland and even beyond, here it seems to me, lie ultimate realities, subtle, far reaching, wonderful.

These ultimate realities in the shadowy unknown would have been regarded by scientific naturalists as so much metaphysical baggage. But they were central to the Cambridge approach, which indeed increasingly, if paradoxically, felt the need for a scientific demonstration of their presence. This approach insisted upon the ethereal constitution of matter and its continuity with radiation, ontological realism, and an essentially transcendent continuity in nature which could be mirrored in a systematically connected natural philosophy. And all these themes were, of course, intimately bound to each other and mutually supporting. It remains to be asked why late Victorian physics should have manifested such a particular and distinctive pattern of concepts and principles. So far these concepts and principles have been discussed only in the context of use provided by the esoteric scientific forum of the time. When one examines the social context more generally, however, it is immediately apparent that the same concepts and principles enjoyed a useful life in other areas. And this raises the possibility that by broadening the scope of enquiry some progress towards answering this intriguing question might be made.

THE SOCIAL CONTEXT. LATE VICTORIAN CAMBRIDGE

The history of the nineteenth century in Britain is dominated by the episode of industrialization and its effects. By the end of the century the lion's share of political power and influence had passed to the main agents and beneficiaries of industrialization, the bourgeois middle class, at the expense of the traditional landed interest. Even the last bastions of traditional control—the Services, the Church, and the universities—were exposed to the 'vulgar,' meritocratic invasions of liberalism, the political manifestation of industrialism (Haines, 1969; Rothblatt, 1968). An earthy, antimetaphysical materialism, and a fundamental individualism, found proponents even within the confines of Oxford and Cambridge.

One way in which this trend was manifested was, needless to say, in new conceptions of science and scientific education. There was a growing stress upon the vital role which a properly organized science could play in an industrial state, and demands for the incorporation of science into a thoroughgoing utilitarian frame of reference (MacLeod, 1972). Such demands were pressed increasingly by an emerging group of practising scientists, who advocated an explicit policy of scientific professionalization, and a consequent radical reorganization of the old universities. For E.R. Lankester (1910), a typical 'professionalizer,' Oxford and Cambridge purveyed 'an empty sham of ancient culture rather than . . . the real and inspiring scientific culture of the modern renaissance' (1910: 7). They were nothing more than 'a couple of huge boarding schools, which are shut for six months of the year' (p. 8), and it was 'monstrous' and 'disastrous' that they should so dominate the country's learning and values with a profoundly anti-utilitarian, anti-industrial ethos of knowledge.

Not only did the professionalizers, in the name of industrial and scientific progress, demand such material changes in the institutions of learning as to threaten the power base of the traditional upper classes, but they also posed a corresponding threat in the realm of ideology. Part of the social programme of professionalization under the utilitarian ethos required that a public image of neutrality and objectivity be cultivated. This in turn required that any traces of metaphysics be expunged from science, at least in its public image. The social correlates of this implied a power struggle, initially between the established Church and the professional vision of science, which imperialized no matter what realms of experience with its secular naturalist ethos.

Scientific naturalism was the ideology of the advocates of the freedom and domination of professionalized science. It was also, thereby, the cosmological backcloth of industrialization. The social and moral connotations of naturalism were that, since (mechanist-materialist) science could comprehend all there was to experience, then society must be persuaded to look towards rational, scientific, and secular ideas to solve its ultimate problems and interpret experience, rather than towards Christian or otherwise metaphysical or 'anti-progressive' modes of thought (Turner, 1974b). Scientific men and institutions must replace their ecclesiastical counterparts as the leaders and educators of culture and political virtue. The professionalizers talked of 'The Scientific Basis of Morals' (Clifford, 1875) and referred to 'the scientific knowledge of life as the one sure guide and determining factor of civilisation' (Lankester, 1910:13). Destiny was to be placed firmly in human, scientific hands, not in divine ones (Turner, 1974a; Brown, 1973). And, whereas the divine hand had previously been ideological maiden to the landed aristocratic master, the newly raised hand of utilitarian and professionalized scientific materialism was maiden to the industrial bourgeoisie.

In opposition to this threat from the aggressively ascendant industrial middle class, there arose a world-view which attempted to destroy or at least mitigate that threat. A sense of general social crisis was expressed which was attributed to the debilitating consequences of industrialism and its materialist credo (Rothblatt, 1968; Maurice, 1882:674; 1892:459). And there was a stress upon metaphysical unity to give coherence to a moral universe and a society which was seen as falling apart under the atomistic nihilism of the scientific naturalists. The professionalizers and their cosmology were regarded as the prologue to utter social fragmentation and chaos. With their concern to decouple scientific knowledge from metaphysical entities, to leave each scientific discipline free to wander its own, increasingly specialized and fragmented, utilitarian pathway, and to deny any transcendent reality suffusing the vulgar material world, the professionalizers were advancing a model of the social role of knowledge which was reprehensible. It abandoned causes and meanings for mere descriptive laws and formulae. It claimed no wider role than technical usefulness, or, when it did claim a wider 'moral' role, it was on the basis of comprehensive social planning rooted in middle-class materialism.

That Cambridge dons should be foremost in formulating this reaction to naturalism is hardly surprising given Oxbridge's central social position in the traditional pantheon, and given that its own institutions had now fallen under direct threat from the aggressive materialist tide. Rothblatt (1968) has described the 'Revolution of the Dons' of the Cambridge colleges against the encroachments of the vulgar middle-class industrialists who had gained a power-base within the University itself. (The Cavendish physics laboratory had originally been set up under the University's auspices as a utilitarian, professionalized scientific training centre to counter the colleges' dominating ethos of culturally and spiritually inclined natural knowledge [Shaw, 1926; Sviedrys, 1970; 'Cavendish', 1910]. However, the colleges fought back successfully and managed to 'infiltrate' the Cavendish with their own champion, J.J. Thomson, who was perhaps the leading figure of late Victorian physics.)

The Cambridge intellectuals, Maitland, Seeley, Sidgwick, Maurice, and others (including their Oxford counterpart, Matthew Arnold) warned of the need for a new, unifying intellectual universe to underpin a revamped moral and political universe of unity and harmony. As Rothblatt put it (1968:119), they were out to find 'a new sovereign authority which would end anarchy, discredit each man doing as he likes, put down the working classes and heal the social divisions.' Believing that England was disintegrating under the vacuum of instrumental amorality embodied in the new scientific professionalism and industrial materialism, these intellectuals,

> The England of the aristocracy and the Church behind them, . . . examined every available intellectual theory in the hope of finding a guide for their actions. Every new or old idea that came to hand, scientific, historical, theological, logical, or aesthetic, was anxiously inspected for its possible prophetic content. [Rothblatt, 1968:244]

Above all they searched for a scheme which evoked a sense of organic unity and harmony (Maurice, 1882, 1886; Backstrom, 1974). Against the social managerial hubris of scientific naturalism, they emphasized the absolute importance of organic *system,* of factors which would heal the wounds of atomism and which lay beyond the manipulatory powers of fragmented and reductionist bodies of knowledge. Against the professionalizers' interest only in narrow instrumental capability, they stressed the need for unified conceptions of nature, to the extent of stressing the unseen, 'spiritual' aspects of nature and experience. This was entirely consistent with a wider moral role for natural knowledge. Indeed, they explicitly advocated (e.g. Maurice, 1882; Seeley, 1882) the equation of science and religion in a new form of 'higher law' administered by a scientific clerisy. Maurice (1882: referred to the need that 'for the good order of society and for the sake of avoiding possible perils hereafter, certain forms of expression must be kept up which seem to imply that the infinite can be brought within the range of our cognisance,' and he explicitly included 'the man of science, just so far as he testifies to right, order, truth, in the government of men and the operation of nature.'

Clearly, the concepts of the organic unity of knowledge, metaphysical realism, and the unseen world articulated in general opposition to the ideology of naturalism are strikingly reminiscent of the themes and principles articulated by the late Victorian physicists in esoteric scientific contexts. The hypothesis of systematic, significant relationship must be very plausible. But what real evidence exists of the relationship, and what kind of relationship was it? These are difficult questions, but one thing they suggest is the need to examine how the physicists interacted with other aspects of Cambridge culture. This immediately leads to some striking support for the initial hypothesis.

One venture in which many Cambridge intellectuals were involved, indeed which was dominated by them, reflects many of the important undercurrents in the cultivation of natural knowledge at this time, and links more or less directly with physics. This was the field of psychical research. Part of the incipient disintegration which the upper classes felt (or felt it necessary to articulate) was fuelled by the 'underground explosion' (Howe, 1972:61) of occultism and antirationalism which occurred from about 1870 onwards. It is no coincidence that this anarchism in the realm of knowledge was strongly linked to political anarchism. Knowledge appeared to be so bound up with society that to break down the intellectual order seemed to be identical to breaking the social order.

One such current of decentralist antirationalism was spiritualism, which flourished among the 'lower orders' of society as a *mélange* of grass-roots politics, religion, and alternative science (Podmore, 1902; Crookes, 1874, 1926; Nelson, 1969; Hill, 1932; Metcalf, n.d.). At times patronizing towards this grass-roots celebration of the supernatural, and at other times more aggressively antagonistic,[2] the Society for Psychical Research (SPR) was formed in 1882 by Henry Sidgwick, Edmund Gurney, Frederick Myers, and William Barrett the physicist (all fellows of Trinity College, Cambridge) to investigate the phenomena of spiritualism and related matters (Gauld, 1968; Nicol, 1972). It championed 'the unseen world' explicitly as a weapon against the limitations of the materialist cosmology, and used an elaborate scientific approach to establish the scientific reality of events and agencies beyond the ken of that cosmology and its adherents. But it also sought 'scientifically' to control, from its own upper-class social base, all excursions into the unseen world beyond current boundaries of ordered experience, so as to discredit the 'morally degenerate' interpretations being derived from such excursions (Wynne, 1977).

221

The activities and intellectual approach of the SPR incorporated science as part of the weaponry of social negotiation of legitimate meanings and social relations. Ironically, psychical research turned out to be a naturalization of the supernatural—a form of *scientific* supernaturalism which attempted to trump the naturalism of the professionalizers with a more comprehensive, 'transcendent' naturalism. It wanted scientific authority for its own metaphysical cravings and social ambitions. As Henry Sidgwick himself put it, in thoroughly utilitarian terms (James, 1970:7):

> What we aimed at from a social point of view was a complete revision of human relations, political, moral and economic in the light of the science directed by comprehensive and impartial sympathy and an unsparing reform of whatever, in the judgment of science, was pronounced to be not conducive to the general happiness. . . . [The] time had come for the scientific treatment of political and moral problems.

The difference from the utilitarianism of the industrialists and professionalizers was only that the social base of the reform, and the specific content and form of its ideological medium, science, was that of the upper classes. They too, under threat, were forced to abandon the authority of pure tradition for the authority of alleged empirical demonstration and natural law. No matter that the science concerned contained no meaningful social directives, except a vague concept of universal unity and 'spiritual' direction of matter; it was the authority to socially interpret reality for society at large which was crucial.

It is worth outlining the intimate social connections between the upper-class Cambridge intellectuals, the leading members of the SPR, and the physicists who constituted the orthodoxy of the late Victorian period, to illustrate how difficult it would be for those physicists to be unaware of, or uninterested in, the wider conflicts and concerns of their social context.

William Barrett, physicist and founder member of the SPR with his fellow Trinity dons Sidgwick, Gurney, and Myers, was also a friend of Balfour Stewart, who was a Fellow of Trinity and later Professor of Physics at Owens College, Manchester. Stewart and P.G. Tait, Fellow of Peterhouse, and later Professor of Physics at Edinburgh University, wrote the anonymous book, *The Unseen Universe,* in 1874, and followed it up with *Paradoxical Philosophy* (1878). F.D. Maurice was Master of Trinity, Sidgwick's tutor and friend.[3] Sidgwick's wife was the sister of A.J. Balfour, Tory Prime Minister, ex-Trinity undergraduate, and member of the SPR. Balfour was made President of the B.A.A.S. in 1904 and in his Presidential Address made an eloquent philosophical plea on behalf of the ether. Sidgwick, Balfour, and Mrs. Sidgwick were all Presidents of the SPR in its first thirty years. Lord Rayleigh, the eminent physicist and inheritor of huge landed estates, married another sister of Balfour; was an undergraduate at Trinity with Balfour; became Master of Trinity College (after Sidgwick) and Cavendish Professor of Physics at Cambridge; and was also President of the SPR after Sidgwick. Mrs. Sidgwick acted as a research assistant for Rayleigh in the Cavendish laboratory. J.J. Thomson came to Cambridge from Owens College where he had been taught by Balfour Stewart, became a Fellow of Trinity, succeeded Rayleigh as Cavendish Professor of Physics, and also as Master of Trinity College. He became a close friend of Rayleigh and tutored his son, the 4th Baron Strutt. Thomson, although never President of the SPR, was a lifelong member of its Council. Significantly enough, he chose to devote two chapters of his autobiography (1936) to psychical and related research, defending it against its 'professionalist' detractors. A.J. Balfour was an accomplished philosopher and cultivated an élite social-intellectual circle which included Sir Oliver Lodge (Jolly, 1974:115). Lodge was an ardent psychical researcher and an eminent scientist, being F.R.S. and Professor of Physics at Liverpool University. He was a friend of Sidgwick, Myers, Stewart, Barrett, and Gurney, and a close friend of J.J. Thomson, Balfour, and Fitzgerald. Another leading physicist-mathematician of this period was Joseph Larmor, Fellow of St. John's College, Cambridge, whose affairs he administered together with the biologist William Bateson. Larmor was a friend and close colleague of J.J. Thomson and Lodge.

It is clear therefore that the dominating élite of British physics at this time was actively involved in psychical research, which was itself so embroiled in controversy that they must have been aware of the wider implications of such involvement. It was not self-evidently a 'natural' move to make. Furthermore, the physicists were intimately connected, socially and intellectually, with the élite of conservative politics and of moral and political philosophy. The two contexts, of social and political debate, and of scientific research, have been pulled closer together and rendered less clearly distinguishable. But still more than this can be established. Many of the writings of the physicists themselves were not addressed purely to esoteric audiences, and some involved obvious political statements. Moreover, in such writings the concept of the ether itself was employed in legitimating a particular vision of the social order.

In *The Unseen Universe* (1874), Stewart and Tait laid out, with all the authority of physics, a rigorous scientific model of the non-material world and its hierarchical connections with the different levels of the material world. It was a deliberate attempt to justify immortality and the superiority of Spirit. It cut into the defensive parallelism between God and nature implied in natural theology, by attempting to demonstrate that Divine providence operated through new and higher natural laws. In this work the ether appears as 'not merely a bridge between one portion of the visible universe and another,' but 'a bridge between one order of things and another' (1874:158).

Stewart, an SPR member, also wrote a paper which appeared as an appendix to Barrett's book *On The Threshold of the Unseen* (1895) in which he sought a 'higher law' than the materialists could see in nature. It was a cardinal tenet that this 'higher law'—of manifest moral significance—would make itself known in experimental science, which was why the SPR's activities were 'of unusual importance.' Spiritual realities would be incorporated in a higher natural law, and thus be demonstrable through science. This was the intellectual counterpart of the institutional fusion of science and clericalism in the clerisy. One can extend Heimann's (1972:73) remark that 'Though the *Unseen Universe* can be regarded as a popularization of science for an ideological purpose, it was intended as a contribution to the philosophy of nature,' by noting that it was both—the latter being the *means* of the former. Analogously, in Oliver Lodge's *Ether and Reality* (1907:179) we find that ether 'is the primary instrument of Mind, the vehicle of Soul, the habitation of Spirit. Truly it may be called the living garment of God.' The 'higher laws' situated in the ether belonged to 'a different order of being—an order which dominates the material, while immersed or immanent in it.' Nature 'must be guided and controlled by some Thought and Purpose, immanent in everything, but revealed only to those with sufficiently awakened perceptions . . . [S]uch reasoned control, by indwelling mind, may be undetectable and inconceivable to a low order of intelligence being totally masked by the material garment.' Lodge went on to note the pedagogical and homiletic possibilities of the ether concept:

> The Ether connexion is simpler, more direct, more informing, less dependent on code, more immediately intelligible than anything connected with language. Pictures appeal to children before words. Pictures made appeal to very early humanity before language. Visible things were apprehended before words. [1907:178]

On the basis of these, and less explicit but nonetheless suggestive statements in Larmor, Rayleigh and indeed several others amongst the Cambridge physicists (cf. Wynne, 1977), it is evident that their science had recognized social value for this group of men. It was a medium of moral demonstration, a symbolic universe of classic conservative reaction (Mannheim, 1938), rooted in the wider intellectual community of the Oxbridge colleges and with natural knowledge centrally embedded within it. To these scientists and intellectuals, science's true role was a broader moral one, in contrast to the utilitarian, antimetaphysical conception of science held by such 'professionalizers' as Huxley, Tyndall, Lankester, Carpenter, and Clifford.

The 'Cambridge School' inextricably linked physical theory to uses beyond those of the esoteric technical-empirical world, particularly in its integral relationship with psychical research, and more broadly with aspirations to the restoration of a spiritually-based social unity to mend the divisive and fragmentary movements in thought and politics associated by the conservatives with the rise of the middle classes. There is a striking analogy between the ineffable, unseen ethereal basis of matter and the articulation of an ineffable spiritual and transcendent basis of social reality—the conservative ideological response to bourgeois individualism. The systemic relational essences underlying material nature symbolized the enduring organic ties of society and acted as the 'natural law' witness to a higher spiritual or political law, to be revealed to the ignorant masses by an upper-class intellectual élite. Thus the Cambridge physicists championed the imaginative leap to the 'common setting' of thought and the cosmos (Larmor, 1908), and held an *a priori* sense of the basis of all matter and energy in the superior spiritual realm of ether which ignorant and vulgar materialists could not apprehend.

SCIENTIFIC CONTEXT AND SOCIAL CONTEXT

How to relate the two contexts of use of physical theories and principles is a problem as important as it is difficult. Not the least of the questions it raises is whether two distinct separate contexts can be adequately identified in the first place. But, assuming that they can, the manner of their interaction must be specified as well as its implications for the culture which they shared. On balance, the present case does not appear to be well described as an example of the social exploitation of scientific concepts and theories. Indeed, any presentation which implies one-way traffic from the scientific to the more general context seems inappropriate here. Much more plausible are hypotheses which treat the two contexts as in a symmetrical interaction, or even those which give priority to the general context insofar as the themes and principles discussed in this chapter are concerned. Such hypotheses would imply that features of the general context influenced the cognitive content of late Victorian physics in important and systematic ways. Since this is not the kind of claim which is routinely acceded to at the present time, it is worth reviewing the considerable evidence and powerful arguments which support it.

First, it must be emphasized how radically the late Victorian physicists had transformed their received view of the ether. Doran (1975) has documented the gradual shift from an earlier elastic-solid conception through the late Victorian period, culminating in Larmor's fully electromagnetic, ether field theory of nature. The 'ether' remained throughout, but its scientific meaning changed in absolutely fundamental respects. It is no exaggeration to say that 'the ether' remained at the centre of a full 180° swing in the conception of matter and nature. Whereas previously matter was the baseline from which ether was a conceptual extension, latterly exactly the reverse was true. At the same time the philosophical status of the ether became increasingly well defined: it was supra-sensual but real; ontologically prior to matter, the underlying basis of, and explanation of coherence in, observable phenomena. Thus, for all that they had a received concept of ether, the ether of the Cambridge physicists was essentially their own construction. And it was a construction which other physicists did not deem to be required by the technical state of their discipline.

Secondly, although the constructed ether and associated metaphysics of the 'Cambridge School' are difficult to understand entirely in relation to the technical concerns of the esoteric scientific context, they are very readily intelligible in the more general context of use. It is abundantly clear how they functioned in moral discourse, and provided the response to naturalism with an alternative representation of the physical world which appropriately legitimated an alternative social ideal. Moreover, having examined the participation of the physicists in Cambridge intellectual life and particularly in the SPR, there is no difficulty in understanding how the modification of culture in one context could feed back into or interact with usage in the other. Even without the evidence of the more popularizing and polemical writings of the physicists, there is much to show that their natural philosophy was conditioned by broadly-based social factors.[4]

Thirdly and finally, if the reality of this connection is accepted, there can be little doubt of its importance as a systematic and continuing factor in the technical context. At the level of theory-construction it is noteworthy that the basic assumption of the ethereal, continuous electromagnetic, nature of matter on the part of the late Victorian physicists also led them to assert the continuous radiation nature of the new emanations, x-rays, gamma-rays, alpha-rays, and beta-rays, which were being found at the turn of the century. This was in sharp contrast to the inheritors of the professionalist mantle, Rutherford and Bragg, who adopted an atomist, particulate paradigm for the full catalogue of emanations (Wynne, 1977; Stuewer, 1971).

Analogously, organic continuity, ontological realism, and the ethereal theory of matter, were put forward as standards of judgment, as the 'Cambridge School' expressed their adverse evaluations of the theories of the naturalists and energeticists. One of many good examples is the attack upon the principle of the Conservation of Energy mounted in Stewart and Tait (1874). They argued that the unseen, non-material world of ether, continuous with the world of matter (indeed the constitutive basis of it) and the seat of all energy, must exhibit energy transfers to and from the material realm. Thus, measured energy transformations in the material realm alone would inevitably show discontinuity and non-conservation. Apparent discontinuities in material nature were thus 'in reality so many partially concealed avenues leading up to the unseen' ethereal realm (1874:192). Conservation of Energy would not hold, therefore, in the material world as observed and defined by scientific naturalists such as Huxley, Tyndall, Clifford, and the German energeticists. Against the assertions of these groups that conservation did always apply, the late Victorians argued that the realm of applicability of energy conservation went beyond the confines of the materialist cosmos and incorporated the realm of ether and matter together. Stewart referred to the Conservation of Energy principle as 'a weapon against visionaries' who might, like him, be seeking that unseen spiritual realm (Barrett, 1895: 309). Scepticism with respect to the Conservation of Energy continued to mark the scientific thought of the late Victorians, as illustrated by Fitzgerald (1896) and Lodge (1913). The physical conception of this matter and ether integration involved a hierarchy of connected levels, with ponderable matter low and vulgar, and ether the repository of a 'higher law' in nature. The matter and ether integration, and the ethereal conception of material 'strain centres' in the ether, allowed these physicists to put forward the possibility that material particles might dissolve into ether, and vice versa, under certain conditions. This integrated approach to the Conservation of Energy, to the structure of matter, and to matter-radiation interactions, was entirely at odds with the non-ether, 'corpuscular' approach of the naturalists.

In summary then, the present case appears to be one where the concepts and principles of a science were developed and sustained not only (or perhaps not even) for their technical value, but very much also for their social value. Scientific thought developed in particular ways related to its possible functioning in the general social context rather than the esoteric scientific context. How far general insights can be drawn from this particular case it is difficult to say. Certainly late Victorian science was both less differentiated internally and less divorced from the wider cultural context than science is today. Indeed, the rigidity of these boundaries was one of the issues dividing the 'Cambridge' physicists and their antagonists. Apart from the particular interest of these materials, we can suggest that our findings have a bearing on other episodes where scientific concepts are found in employment beyond the esoteric scientific context and where scientific development is impossible to account for on purely 'internal' criteria. In such cases there is every reason to look to the general social context to explain the particular course of scientific growth. And in these cases the burden of proof lies firmly on those who reject the social context of use as a formative influence on scientific knowledge.

NOTES

1.   As Kargon (1969) observes, Maxwell's own epistemological stance was more conventionalist or pragmatic with respect to the ontological status of scientific concepts than most of his Cambridge peers could accept.

2.   Lodge's patronizing attitude and his fear that spiritualists were vulnerable to subversive persuasions is evident from his correspondence, e.g., to J.A. Hill (1932:25).

3.   Maurice's attitude towards spiritualist phenomena is indicated in his biography in a letter dated 1870:

> [it] is truth and not fiction, the deliverance from dreams, not the indulgence of them, to hold fast the faith that the veil of flesh has been rent asunder, that for all and for each the invisible world has been opened; that we *must* have converse with it and its inhabitants whether we desire the converse or shrink from it. [Maurice, 1892:625]

Maurice appears in many ways to have been more ambivalent regarding the ultimate naturalistic basis of progressive social evolution than his friend, Henry Sidgwick, who, whilst loathing the restricted, 'vulgar' naturalism of the materialists, nevertheless craved for a naturalistic licence for his own form of 'spiritual' upper-class paternalist 'progressivism.' In this, Sidgwick went much further than true conservatives such as Larmor and Bateson were prepared to go, even though their responses were much the same. The similarities of outlook between Larmor and Bateson are discussed by Wynne (1977), in particular their common abhorrence of reductionism and its social managerial implications. Larmor clearly carried this over into his evaluations as a physicist.

4.   Some contradiction is evident between late Victorian intellectuals' profession of a spiritual, organic nature and the utilitarian leanings of some of them, such as Rayleigh and Lodge. However, the structural elements of proper science and its social role which the late Victorian 'metaphysical' school articulated can be regarded as stereotypes articulated in *pure,* extreme form in demarcation against the threat from an alternative social group and its ideology. It is probably generally true that a necessary (but not sufficient) condition of the authority of such symbolic universes is that they should be pure and unambiguous, in order to eradicate doubt. In the course of symbolic action between competing social groups, ideologies will inevitably tend to be articulated in forms which are more extreme than the social reality which they are articulated to underpin. Furthermore, to regard apparent contradictions between ideologies, or symbolic universes, and actual social practices as problematic is to imply that the ideologies contain an intrinsic, ideal logic which leads inevitably to the related social practices. Although this is widely held to be so, inside and outside science, it is more valid to see them as weapons in an authority struggle with the ideological justifications of threatening social groups (be they anarchists or industrialists). Thus the social patterns and the symbolic universes so articulated may be genuinely influential upon the nature of natural knowledge developed by each group, without being intrinsic logics of that science, nor the sole determinants thereof. The main elements may be determined less by such supposed logics than by the contingencies of (a) what cultural symbols or parts thereof are available, (b) what would *appear* to be authoritative (i.e., related to established symbols of order, authority, and so on), and (c) what elements would demarcate the groups' ideals more clearly in opposition to the existing symbolic universe of a threatening social group. The relative influence of such ideological determinants and of empirical validity upon the detailed theories of science is something which can only be worked out in specific cases.

REFERENCES

Backstrom, P. (1974) *Christian Socialism and Cooperation in Victorian England.* London: Croom Helm.

Barrett, W.F. (1895) *On the Threshold of the Unseen.* London: Routledge & Kegan Paul (reprinted, 1907).

Brown, A.W. (1973) *The Metaphysical Society.* New York: Octagon.

Brush, S. (1967) 'Thermodynamics and history.' *Graduate J.* 7:477-565.

Cavendish (1910) *A History of the Cavendish Laboratory,* 1871-1910. London: Longmans Green.

Clifford, W.K. (1875) 'The scientific basis of morals.' *Contemporary Rev.* 26:650-662.

Crookes, W. (1926) *Researches into the Phenomena of Spiritualism.* London: Psychic Bookshops.

(1879) 'Radiant matter.' *BAAS Reports*: 167.

(1874) 'An inquiry into the phenomena called spiritualism.' *Quart. J. of Sci.* 5:3-19.

d'Albe, E.F. (1923) *The Life and Work of Sir William Crookes.* London: Fisher Unwin.

Doran, B. (1975) 'Origins and consolidation of field theory in 19th-century Britain: from mechanical to electromagnetic views of nature.' *Historical Studies in the Physical Sciences* 6:133-260.

Douglas, M. (1975) *Implicit Meanings.* London: Routledge & Kegan Paul.

Fitzgerald, G.F. (1896) 'Ostwald's energetics.' *Nature* 53:441-442.

Fleming, J.A. (1902) *Waves and Ripples in Water, Air and Aether.* London: S.P.C.K.

Forman, P. (1971) 'Weimar culture, causality and quantum theory, 1918-1927.' *Historical Studies in the Physical Sciences* 3:1-115.

Gauld, A. (1968) *The Founders of the Society for Psychical Research.* London: Routledge & Kegan Paul.

Goldberg, S. (1970) 'In defence of ether.' *Historical Studies in the Physical Sciences* 2:89-126.

Haines, G. (1969) *Essays on German Influence upon English Education and Science, 1850-1919.* Connecticut College: Archon Books.

Heimann, P.M. (1972) *'The Unseen Universe:* physics and philosophy of nature in Victorian Britain.' *British J. for the History of Sci.* 6:73-79.

Hicks, W.M. (1895) 'Theories of the aether.' *BAAS Reports*: 595-606.

Hill, J.A. (1932) *Letters from Sir Oliver Lodge, Chiefly on Spiritualism.* London: Cassell.

Howe, E. (1972) *The Magicians of the Golden Dawn.* London: Routledge & Kegan Paul.

James, G. (1970) *Henry Sidgwick.* London: Oxford Univ. Press.

Jolly, W.P. (1974) *Sir Oliver Lodge.* London: Constable.

Kargon, R. (1969) 'Model and analogy in Victorian science.' *J. of the History of Ideas* 30:423-436.

Lankester, E.R. (1910) *Science from an Easy Chair.* London: Methuen.

Larmor, J. (1908) 'On the physical aspect of the atomic theory.' *Manchester Memoirs in the Physical Sciences* 51:1-54.

(1900) *Aether and Matter.* Cambridge: Cambridge Univ. Press.

Lodge, O.J. (1913) 'Continuity,' *BAAS Reports:* 1—47.

(1907) *Ether and Reality.* London: Hodder & Stoughton.

(1889) *Modern Views of Electricity.* London: Macmillan.

MacLeod, R.M. (1972) 'Resources of science in Victorian England,' pp. 172-196 in P. Mathias (ed.) *Science and Society, 1600-1900.* London: Cambridge Univ. Press.

Mannheim, K. (1938) 'Conservative thought,' pp. 67-131 in *Essays on Sociology and Social Psychology.* London: Routledge & Kegan Paul.

Maurice, F.D. (1892) *The Life of F.D. Maurice, Told Chiefly from His Letters.* London: Macmillan.

(1886) *Social Morality.* London: Macmillan.

(1882) *Moral and Metaphysical Philosophy.* London: Macmillan.

Metcalf, H. (n.d.) *The Evolution of Spiritualism.* London: Hutchinson.

Michelson, A.A. (1899) 'A plea for light waves.' *Proceedings of the American Association for the Advancement of Science* 37:67-78.

Nelson, G.K. (1969) *Spiritualism and Society.* London: Routledge & Kegan Paul.

Nicol, F. (1972) Essay Review of A. Gauld's *The Founders of the Society for Psychical Research* (1968). *Proceedings of the Society for Psychical Research* 55:341-367.

Podmore, F. (1902) *Modern Spiritualism.* London: Methuen.

Rayleigh, Lord (1884) 'Presidential Address to the BAAS.' *BAAS Reports*: 1-32.

Rothblatt, S. (1968) *Revolution of the Dons.* London: Cambridge Univ. Press.

Seeley, J.R. (1882) *Natural Religion.* London.

Shapin, S. (1979) 'Homo phrenologicus: anthropological perspectives on an historical problem', pp 41-71 in B. Barnes and S. Shapin (eds.) *Natural Order,* London, Sage.

Shaw, N. (1926) 'The Cavendish as a factor in a counter-revolution.' *Nature* 118:885.

Stewart, B. and Tait, P.G. (1878) *Paradoxical Philosophy.* London: Macmillan.

(1874) *The Unseen Universe.* London: Macmillan.

Stokes, G.G. (1883) *On Light.* London: Macmillan.

Stuewer, R.H. (1971) 'William H. Bragg's corpuscular theory of x-rays and gamma-rays.' *British J. for the History of Sci.* 5:258-281.

Sviedrys, R. (1970) 'The rise of physical science at Victorian Cambridge.' *Historical Studies in the Physical Sciences* 2:127-154.

Swenson, L.S. (1972) *The Ethereal Aether.* London: Univ. of Texas Press.

Thomson, J.J. (1936) *Recollections and Reflections.* London: Bell & Sons.

(1908) 'The late Lord Kelvin.' *Cambridge Rev.* (16 January):5.

Turner, F.M. (1974a) *Between Science and Religion.* London: Yale Univ. Press.

(1974b) 'Rainfall, plagues, and the Prince of Wales: a chapter in the conflict of religion and science.' *J. of British Studies* 13:47-65.

Wilson, D. (1971) 'The thought of late Victorian physicists.' *Victorian Studies* 23:29-48.

Wynne, B. (1977) 'C.G. Barkla and the J phenomenon: a case study in the sociology of physics.' M. Phil. thesis, University of Edinburgh.

Young, R.M. (1973) 'The historiographic and ideological contexts of the nineteenth-century debate about man's place in nature,' pp. 344-438 in M. Teich and R.M. Young (eds.) *Changing Perspectives in the History of Science.* London: Heinemann.

PART FIVE

Science as expertise

Processes of social differentiation are at the same time processes of cognitive differentiation: the division of intellectual labour is part and parcel of the division of labour as a whole. Hence modern, highly differentiated societies increasingly rely upon specialized knowledge and competences, and the equally specialized roles and institutions associated with them. In such societies the problem of credibility arises in a particularly acute form, since those who implement policy are almost invariably deficient in the relevant expertise, and have to depend upon the submissions of trusted advisers, both as the basis and the legitimation of their decisions.

It is hard to exaggerate the practical importance of this problem. It pervades whole areas of policy: military technology and strategy; nuclear power and energy planning; pollution control; health and transport policies — the list is endless. Experts in these areas are locked in a perpetual dialogue with politicians and professional decisionmakers; and, in the West in recent years, they have also found themselves increasingly involved, both as consultants and adversaries, with the general public and 'grass-roots' pressure groups. The West has also seen the rise of anti-establishment professional experts: Ralph Nader and consumer group advisers; nuclear physicists in Friends of the Earth; practitioners of 'Public Interest Science' (BN).

Unfortunately, we know very little about the basis of credibility: the importance of the problem is matched only by its complexity and its comparative neglect; very little current work has any serious bearing on it. There are, of course, some superficial accounts, but these typically offer reassuring mythologies, designed to legitimate rather than to analyze the current distribution of cognitive authority. For example, faith in scientific expertise is sometimes justified by massively idealized accounts of the nature of rationality, or by references to the remorseless operation of 'scientific method'.

And there are even those who imagine that science faces no real problem of credibility at all, because it is a repository of 'truth': they conceive of truth as in itself a cause of belief, an intrinsic property of a knowledge-claim which presses the mind to its acceptance. (The Greeks knew better: *their* mythology vividly dramatizes, in the figure of Cassandra, the sad disjunction of truth and credibility.)

A good way of grasping the nature of the problem is to investigate what forms of logical argument and reasoned justification are available to the expert to maintain credibility, if it is challenged in a specific case. To examine the most that can be expected of logical argument in ideal circumstances is to examine the simplest form that a credibility problem can possibly take.

By definition, an expert is the representative of a trusted institution, and/or the bearer of trusted knowledge and competence. In defending his pronouncements with reasoned arguments, the expert must seek to exploit this trust. Thus he will argue that his pronouncements are 'what science says', the 'best' assessments in terms of accepted scientific knowledge. There is no way in which the expert can go further and, *ab initio,* give a reasoned justification of science itself. Reasoning is simply incapable of functioning without a starting point, and therefore cannot provide a justification for expertise at this most basic level. If science itself is called into question, then the scientific expert can only retire gracefully; his very premises for reasoned argument have been removed. (Recall that, in Part 2, the conventional character of scientific knowledge was displayed and emphasized. This reinforces the point: even with full awareness of the nature of scientific knowledge, an irreducible element of commitment remains.)

How, then, might an expert, in a specific case, defend his pronouncements as 'what science says'? Two possibilities come to mind. One is to display the pronouncements simply as deductions derived from an accepted theory — that is, as the logical implications of that theory. But we have already noted the impossibility of this in Part 2. Whether a particular case falls under a theory, whether the theory applies to it, can never be determined by deduction. Any alleged deduction from the theory to the case presupposes that the theory applies to the case: it involves the *judgment* that the case is the kind of phenomenon or situation to which the theory applies.

The second possibility involves induction. A recommendation about particular case Y may be said to be justified because Y is the same as Y', a phenomenon well known to scientists, about which much accepted scientific knowledge exists. The recommendation is justified because it is accepted knowledge of Y' applied to Y. This is an entirely acceptable form of justification, and is no doubt in many cases also a good empirical description of how an expert actually arrives at his recommendations. Unfortunately, however, it is not an incorrigible justification. It involves the *judgment* that Y and Y' are 'the same', and this particular judgment can always be challenged without calling scientific knowledge in general into question (see Part 2).

Precisely because scientific knowledge has an inductive character, its applicability can always be challenged in every particular case. The scientific expert may find his authority accepted in general, but his specific recommendations nonetheless rejected as illegitimate extensions of what is already known. He may find the immense prestige of science, and the respect accorded to him as its representative, curiously inefficacious.

Scientific experts often experience this kind of difficulty. The use of animals in carcinogenicity tests affords an interesting example: can animals and humans be assumed to be 'the same' in their responses to particular chemical insults, or can they not? Or again, in the heated disputes over the fluoridation of public water supplies, most chemists maintain that, at the very low concentrations involved, the 'same' state of affairs is produced by the presence of both the (natural) calcium fluoride (CaF_2) and the (artificially added) sodium fluoride (NaF), since both are completely dissociated, to yield 'identical' fluoride ions (F^-). They therefore regard evidence that naturally fluoridated water is harmless as confirmation of the hypothesis that artificially fluoridated water would be, too; and they can recommend that the F^- concentration be everywhere increased to 1 ppm by the addition of NaF, so that the benefits in dental health known to occur where this concentration exists naturally can be more widely available. But, to anti-fluoridationists, NaF is a poisonous industrial by-product; they see a crucial difference between it and 'natural' CaF_2, and deny that inferences about the former can be made from what is known about the latter. It only requires one competent authority to postulate the (natural) existence of *complex* calcium/fluorine ions for this counter judgment to appear legitimate.

The diuternal vulnerability of the expert is neatly, if inadvertently, demonstrated in the short paper by Oteri, Weinberg and Pinales. The authors are lawyers specializing in the defence of drug cases, and their paper deals with the cross-examination of forensic scientists. The situation of the forensic scientist in the court of law serves as a marvellous allegory of the expert in society: and the multiplicity of strategies available to defence counsel to cast doubt upon sameness relations, and to call technical conclusions into question, graphically brings home the predicament of any expert who forfeits trust.

No expert, or body of experts, or institution embodying expertise, can use logical methods to establish a firm connection between accepted knowledge and their specific recommendations. In this sense, they cannot command credibility on a strictly logical basis. Let us now ask whether an audience can ever *accord* credibility on such a basis. Clearly, if what we have already said is well conceived, there is no way in which an audience can *deduce* that a source of expertise deserves credibility. An audience may, however, arrive at such a conclusion inductively. People may, for example, agree that the expert proffered correct, or profitable, advice in previous cases A, B, C, and accordingly consider it reasonable to trust him in case D. Or people may ascertain that the expert competently deploys a technique that has generated correct, or profitable, recommendations in the past, and hence consider it reasonable to trust him to the extent that he uses that technique in the future. There is no reason to doubt that chains of inference of this kind constitute very important pathways to credibility, nor that they could become very much more important than they presently are.

On the other hand, such inferences are not oases of reasonable behaviour in a desert of irrationality. They are inductive inferences, and hence, as we have now emphasized many times, they involve judgments of sameness. And the form these judgments take will depend upon the context in which they are taken: inductive inference is not a logical *rather than* a social process, but is both at once (Bloor, 1976; Barnes, 1982).

Take the simple pattern of inference set out above. Effective advice has been proffered in cases A, B, C; therefore the source of the advice should be trusted in case D. In some circumstances, such an inference may indeed be found plausible and attractive: but in other circumstances it may be resisted. Apparent success with A, B and C may then be attributed merely to good fortune. Or case D may be perceived as significantly different to A, B and C, and thus situated beyond the expert's competence. Indeed, our own experts commonly respond in just this way, both in evaluating each other and in assessing the claims of such putative experts as acupuncturists, scientologists, water diviners, osteopaths and faith healers, whose apparent effectiveness has manifestly not resulted in any automatic acquisition of credibility (see also Mulkay 1979b).

It is true that such responses are particularly common when a putative source of expertise is viewed with hostility, or where its recommendations in a particular case are unwelcome; but the temptation to label the responses as 'irrational' has nonetheless to be resisted. An emphasis on differences in order to resist analogies and block inductions is no more inherently unreasonable than an emphasis on similarities in order to stress analogies and legitimate inductions.

Modulators of Credibility

The main thrust of our discussion so far has been designed to show that the credibility of expertise cannot be satisfactorily established by strictly logical arguments, or by the kinds of procedure which rationalists admire and accept. This is the tragedy of the expert, and of the society which relies on him: it is therefore our tragedy. It is of immense proportions, and it seems that all the obvious escape routes eventually turn out only to have led deeper into the mire. Perhaps fortunately, we are not called upon to consider its normative aspects here, and can confine ourselves to credibility as a sociological problem. From this perspective, the most important point to grasp is that judgments of credibility and evaluations of expertise are invariably and essentially modulated by the contingencies of the settings in which they are made. What little relevant research has been done to date serves only to show how many such contingencies there may be, and hint at the complexities of their operation. We need concentrated empirical study, in case after case, to build up an understanding, not only of specific judgments and evaluations, but of the basis of credibility generally.

The possible contingencies, and their relationships, could hardly be listed and described — let alone adequately categorized, discussed and illustrated — in the space at our disposal. We have therefore chosen to close this set of readings with one of the most impressive and detailed case studies published so far: the study by Gillespie, Eva and Johnston of the divergent assessments of the carcinogenicity of a persistent pesticide in Britain and the USA. The relevant contingencies which influence the selection of experts, the scientific judgments they reach, and their credibility and reception, here include a wide range of institutional, historical, political, legal, economic and cultural factors. A more striking illustration of the complexity of the problem would be hard to find. We will not try to spell out how the aspects we have previously emphasized feature in this particular case: rather, leaving that exploration to the reader, we will attempt here the more modest task of introducing a few of the main points which are emerging in the area.

In discussing what he called the 'political resources of American science', Ezrahi (1971) listed four of the major modulators of scientific credibility — or, as he put it, four categories of the 'political visibility' of science:

> 1 The relation and relevance of scientific pictures of reality or images of nature to prevailing social, political and religious beliefs.
> 2 The relation of technologies generated by different fields of science to prevailing social values and concerns.
> 3 The degree of accessibility of a given science to the public.
> 4 The degree of peer consensus among the scientists of any given field. (Ezrahi, 1971: 120)

These, as Ezrahi says, 'refer to the publicly perceived features of science which affect lay attitudes towards science', attitudes which can be either positive or negative: they could therefore define both political resources and political liabilities — indeed, the 'sign' of any one category can change rapidly.

> The political skill of scientists should therefore consist largely of the ability to exploit relevant social beliefs and attitudes in order to manage the public images of science so as to improve its political visibility and its capacity to evoke public support. (Ibid.)

Mary Douglas' paper indicates the full extent of the problem raised by this superficially straightforward 'engineering of trust'; the vicissitudes of the Environmental Movement, and the Race/IQ and Sociobiology debates (BN), could all illustrate her case. She asks how ecologists, the images of nature they purvey, and the social implications that appear to flow from those images, acquire credibility, and, in pressing this question to its logical conclusion, she sensitizes us to some of the issues hidden in Ezrahi's first category. Trust is no minor matter: it pervades the whole cognitive domain. In Douglas' view, set as it is within the Durkheimian tradition in social anthropology, the cosmos itself seems as it is because of the trust we place in our institutions, and in our *moral* order: weaken that trust, and certainty about the cosmos goes. The point is well recognized by political agents. The business of maintaining the social order involves making out some experts as being consistently right (and others wrong): this is particularly so in modern society, where consequential political decisions are seen to flow from technical expertise. Frequent U-turns on, for instance, the desirability of nuclear power (or, for that matter, the truth of monetarism) cannot be tolerated: social and political commitment is inextricably intertwined with beliefs about the world; and those beliefs are legitimated by expertise. Where expertise is challenged, institutions falter, and flux and chaos threaten.

Expert Disagreement

Ezrahi's fourth category ('peer consensus') is clearly a major modulator of credibility. Where experts publicly disagree, their authority and influence are reduced. Dorothy Nelkin has documented a range of disputes in which this element appears. In an early paper, she discussed two cases (the proposed siting of a nuclear power plant on Cayuga Lake, NY, and the extension of Logan Airport, Boston), and suggested six 'propositions which may be generalizable to other controversies involving conflicting technical expertise':

> First, *developers seek expertise to legitimize their plans and they use their command of technical knowledge to justify their autonomy.* They assume that special technical competence is a reason to preclude outside public (or 'democratic') control.
>
> Second, *while expert advice can help to clarify technical constraints it also is likely to increase conflict,* especially when expertise is available to those communities affected by a plan. Citizens' groups are increasingly seeking their own expertise to neutralize the impact of data provided by project developers. Most issues that have become politically controversial (environmental problems, fluoridation, DDT) contain basic technical as well as political uncertainties, and evidence can easily be mustered to support or oppose a given proposal.
>
> Third, *the extent to which technical advice is accepted depends less on its validity and the competence of the expert, than on the extent to which it reinforces existing positions.* Our two cases suggest that factors such as trust in authority, the economic or employment context in which a controversy takes place, and the intensity of local concern will matter more than the quality or character of technical advice.
>
> Fourth, *those opposing a decision need not muster equal evidence.* It is sufficient to raise questions that will undermine the expertise of a developer whose power and legitimacy rests on his monopoly of knowledge or claims of special competence.
>
> Fifth, *conflict among experts reduces their political impact.* The influence of experts is based on public trust in the infallibility of expertise. Ironically, the increasing participation of scientists in political life may reduce their effectiveness, for the conflict among scientists that invariably follows from their participation in controversial policies highlights their fallibility, demystifies their special expertise and calls attention to non-technical and political assumptions that influence technical advice.
>
> Finally, *the role of experts appears to be similar regardless of whether they are 'hard' or 'soft' scientists.* The two [cases] involved scientists,

engineers, economists and lawyers as experts. The similarities suggest that the technical complexity of the controversial issues does not greatly influence the political nature of a dispute.

In sum, the way in which clients (either developers or citizens' groups) direct and use the work of experts embodies their subjective construction of reality — their judgments, for example, about public priorities or about the level of acceptable risk or discomfort. When there is conflict in such judgments, it is bound to be reflected in a biased use of technical knowledge, in which the value of scientific work depends less on its merits than on its utility. (Nelkin, 1975: 51-54)

Here, technical expertise is seen as a flexible resource, to which all parties in a dispute can have access, and which they can 'use'; essential uncertainties and ambiguities in the content of scientific advice can always be exploited in a conflict; but expertise then loses political credibility. This simple picture has since been elaborated in further empirical studies; Nelkin has edited a most useful collection of these in *Controversy* (1979). Her own analysis has been enriched by this material; we reproduce here an extract from her introduction to the book, which succinctly summarizes some of the issues now raised.

Disputes among experts can also expose the uncertain nature of scientific knowledge-claims, and lead experts to re-enact in the public arena the kind of strategies we have previously seen them adopting in their laboratories and esoteric journals. Allan Mazur (1981: Chapter 2) draws on case material from his studies of the low-level radiation and fluoridation disputes to illustrate how actors can attempt to discredit opponents' claims. He clearly demonstrates the limitations of any 'rational' model of expertise and its evaluation; and, as in Nelkin's work, he shows the extent to which experts can depart from stereotyped 'academic norms'. As to the effects of such disputes on the public, Mazur amply endorses an earlier analysis of the fluoridation debate:-

The voters do not have to reject completely the expertise of public health officials for fluoridation to be defeated; they need only to perceive the controversy as a dispute among conflicting experts and to be concerned about the costs of making a wrong choice . . . (Sapolsky, 1968: 431)

Licensing and Autonomy

The fluoridation debate also raises the issue of the definition of 'relevant expertise', and hence the problem of accrediting experts:

The health officials believe that fluoridation's safety and efficacy are technical public health problems, for which they can present the only

legitimate scientific position. They do not perceive the antifluoridationists to be experts in public health problems and on this basis easily
dismiss their arguments. The opponents, however, number among
themselves persons with scientific or professional credentials. In most
communities there is at least one doctor, dentist, university science
professor, or research scientist who will speak out against fluoridation.
Several of these persons have gained national reputations through
their fight against fluoridation, and are active in referendum campaigns
throughout the country. While the public health officials may consider
these professional antifluoridationists to be marginal at best in their
scientific professions, and to be speaking outside their particular fields
of competence, the voters might well consider them to be qualified to
discuss the technical issues and to be granted the status of fluoridation
experts. (Sapolsky, 1968: 431)

As Sapolsky notes, one tactic adopted by American antifluoridationists has been to form, or gain support from, organizations
with grand, if spurious, titles. The average voter cannot distinguish
these from the 'proper' professional organizations: much of the
activity of the latter, therefore, focusses on establishing their claim
to 'genuine' expertise — to procedures of 'licensing' of experts,
and of 'boundary maintenance'. In this, the fluoridation debate
may be typical. Participation in the AntiBallistic Missile (ABM)
debate, for instance, forced the Operations Research Society
of America to review their accreditation, professional ethics and
autonomy (BN); similar concerns were forced on ecologists by
environmental disputes and legislation (Nelkin, 1977c). These
professional strategies also attempt to establish the *disinterestedness* of experts. Opponents can easily discredit experts by revealing
hidden interests: given the origin of NaF, for instance, the dis-
closure of an industrial connection can severely embarrass a
profluoridation technical witness.

Such controversies put a discipline's professional body in a
dilemma: how far can it go in determining who is a qualified
expert, and what are the boundaries of certified knowledge,
without arbitrarily excluding deviant or innovative schools of
thought that may constitute the orthodoxy of the future? A
discipline may feel obliged to try to contain conflict, especially
where its external authority may suffer if it does not: even internal
disputes may have implications for disciplinary credibility if
they are sufficiently colourful and advertized; the Race/IQ and
Sociobiology debates are cases in point here.

Another case which illustrates the political problems faced by
scientists in attempting to maintain disciplinary control is the
dispute over Recombinant-DNA (r-DNA) research (BN). Here,
the researchers themselves expressed concern at the possible

public health hazards of r-DNA techniques, imposed a voluntary moratorium, and initiated studies to determine how the hazards might be best assessed and controlled. Confident that their 'responsible' actions would only reinforce the public trust and esteem which they felt the potential medical benefits of their research had already earned them, they publicized their concerns. But some politicians, and citizens' and environmentalist groups, alarmed at the nature of the hazards, linked the research with 'Frankenstein' images, and with fears of authoritarian manipulation: they pointedly questioned whether those with an interest in advancing the research were the best people to decide how it should be monitored and controlled. Even the link with medical technology backfired: left-wing critics questioned whether so much public money should be devoted to research from which drug companies would benefit, and argued that the potential benefits did not justify the risks. Faced with unanticipated public hostility, proposed bureaucratic restraints, divided opinion within their ranks, and a situation slipping beyond their control, many r-DNA researchers attempted to re-assert disciplinary autonomy and the traditional values of academic science, arguing (as several had all along) that any attempt at 'democratic' control of scientific research was improper.

In the long run, as the r-DNA hazards are discounted, this conservative strategy of boundary maintenance may succeed. But these scientists did, for a time, put their autonomy voluntarily at risk, and unwittingly allowed experimental tests of the force of Ezrahi's third category ('public accessibility'). At a number of places, lay panels informed themselves about r-DNA research in order to advise on local safety policy. In general, their reports were sympathetic to the research continuing, and proposed safety guidelines not essentially dissimilar from those which were then accepted as reasonable by r-DNA researchers. The scientists, delighted, decreed these experiments 'successful'. But the apparently mutual goodwill of the r-DNA case may be unusual. As it is, committed opponents of r-DNA research have a different view: they grumble that these panels have been 'taken in by the experts'. And, had the panels not made proposals to their liking, the scientists could easily have dismissed their findings, arguing that 'laymen cannot properly understand the technical issues', that 'only scientists can know how best to conduct research'. In short, credibility and mutual trust were both contingent and negotiable: we need not conclude that lay accessibility and involvement, on this model, necessarily reduces conflict, or enhances expert credibility.

Facts and Values

One response to increasingly frequent clashes of expert opinion, problems of credibility, and the political confusion they engender, has been the 'Science Court' proposal (Mazur, 1977, 1981). In this proposal, technical issues crucial to political decisions would be argued out in front of scientific judges, whose conclusions could be taken, by decisionmakers, to have 'presumptive validity'. The proposal has been fiercely criticized (BN): two of the main criticisms expose interesting views of both science and the political process, and highlight intractable problems associated with the political role of expertise.

The first concerns the 'fact/value' distinction (Mazur, 1981: Chapter 3; Nelkin, 1977d). The Science Court proposal rests on the possibility of separating out clearly defined factual issues, the resolution of which will facilitate open discussion, and the establishment of consensus, on related value issues. To Mazur, this separation 'seems quite feasible' (Mazur, 1981: 125): but Nelkin differs. Scientists can often fail to agree on a 'value-free' specification of the problem. The values of the scientists themselves, and consequent effects on conceptualizations of problems, can lead them to 'talk past each other', and prevent technical consensus being reached: Robbins and Johnston (1976) emphasize this point, and it reappears in our final Reading. And, in any case, technical consensus may be unrelated to the *resolution* of value differences. For instance, it was finally agreed that the Cayuga Lake nuclear plant might increase the lake's surface temperature by no more than $0.5°C$, with a slight increase in eutrophication: supporters of the plant took this change to be trivial; but opponents argued that it was now *agreed* that an *irreversible* degradation of our natural heritage *would* take place – and no such wilful change, however small, was ethically justified. Indeed, Nelkin suggests the Science Court proposal is caught in a 'Catch 22': if a technical consensus *can* be reached, the point will be marginal to the main issues; and if the technical point *is* of central importance to the political debate, conflicting values will so permeate the experts' discussion as to thwart any technical consensus (see also Martin, 1979).

The second criticism concerns the importance of a *prior* agreement, among all parties, as to the Science Court agenda, composition and procedures. As in public enquiries in general (e.g. the UK Windscale Inquiry: BN), power resides with those who can decide the timing and topic of the hearings and the kind of expertise which will be held relevant to its proceedings, select the 'judges', and command the resources to summon expert witnesses.

However 'open' the proceedings may appear to be, disputants can justifiably see the inquiry as a mere ritual validation, with 'the cards stacked against them'. The point is neatly made in the brief case study by Barry Casper and Paul Wellstone, of a dispute concerning the siting of a power line in Minnesota. Here, the utilities tried to define the issues in the technical terms of health and safety, but the farmers objected. No Science Court proved possible, and the modified arrangement proposed by the farmers, with explicit inclusion of value issues, was not strictly a 'Science Court' at all.

Scientism

The Minnesota utilities' behaviour exemplifies a point already made in Nelkin's Reading, and in Sapolsky's discussion of fluoridation — namely the tendency to 'convert' value issues into technical discussions. Antifluoridationists often adopt their stance as a moral protest against 'mass medication': but they are faced with public health officials justifying fluoridation in technical terms, so they are forced to try to destroy the credibility of that kind of case. These tendencies are all part of what Habermas (1971), in a classic (if somewhat abstract) paper, calls 'the scientization of politics' — the 'scientism' by which the infiltration of technical expertise determines the *conceptualization* of political problems, the *language* in which they are expressed, and the *institutional forms* by which decisions are reached. It is this process which tends to be overlooked by any analysis which stresses the role of technical expertise as an ambiguous and flexible 'resource', potentially available to all parties engaged in adversarial dispute (cf. Benveniste, 1972). Technical expertise is an *institutionalized* 'resource', and the political problem it poses involves more than ensuring that all parties have the financial resources to deploy it. Who can frame it is at least as important as who has access to it. To control expertise is to control the *terms* in which disputes are conducted: indeed, that reconceptualization may be essential to attaining any 'control' at all.

Donald Schon's Reading illustrates this point neatly. Schon describes how modern commercial organizations are forced to reconceptualize essential 'uncertainty' in terms of quantifiable 'risk': he explains how institutional structures conspire to consolidate this tendency, and lead to an essential conservatism in 'rational', planned innovation (cf. the notion of 'disjointed incrementalism': Lindblom, 1968). Habermas (1971) extends this scientism into national and international politics. (Elaboration of the processes leading to the social institutionalization of technology can be found in McDermott [1969], Dickson [1974] and Winner [1977]; for an excellent historical discussion of related issues, see Noble [1977]; and for consideration of the sources of 'inflexibility' in the rational planning of technical innovation, see Collingridge [1981].)

Essentially, scientism involves an attempt both to extend the *scope* of accepted expertise, and to establish its *scientific* pretensions (Cameron and Edge, 1979). How far can our accepted experts press their claims to be competent? Can a medical expert, with his command of ECGs and his knowledge of the 'brain stem', redefine 'death' — or, for that matter, with amniotic samples and chromosome counts, 'life'? We are faced again with a 'sameness relation' problem: what is 'the same' as the matters our accredited experts know about, and what is 'different' — and hence beyond their competence and authority? No unproblematic answers are possible. As Cameron and Edge put it:

> Scientism can be seen as an attempt to 'colonize' territory where scientific language, techniques, approaches, models and metaphors (unproblematically established elsewhere) have previously been thought inapplicable [or not been considered at all]. But it has been argued that *science itself* also advances in this kind of way. To those engaged in developing econometrics, for instance, the task is that of 'making economics more scientific', and their efforts are widely held to be legitimate, however scientistic they may seem to their critics; the extension into biological disciplines of techniques [and associated reconceptualizations] pioneered by physicists has raised similar controversies (and also won Nobel Prizes) . . . [It] is worth considering whether the same social and cultural mechanisms may not operate both *within* and *outside* the scientific community: the appropriateness of the new deployment of established scientific resources is then a matter of (differing) judgment by the several parties affected and involved . . . The categories of 'scientific' and 'scientistic' can then both be seen as socially grounded and evaluatively deployed: the distinction between them (if any) is a matter of dispute, with a very broad no-man's-land of debatable instances. (Cameron and Edge, 1979: 6, 8)

Risk and Consensus

The exigencies of modern decision making have spawned a succession of 'scientific' techniques: 'cost benefit analysis', 'policy sciences', 'systems analysis', 'risk assessment (or accounting)', 'technology assessment', 'environmental impact statements', 'technological forecasting', 'futurology', and so on. These techniques, and their associated experts, have explicitly scientistic pretensions: 'The thoroughness and objectivity which characterize technology assessment make it a logical extension of the scientific method' (see Wynne, 1975b: 149). Credibility can be sought by an obsession with quantitative methods, and other aspects of 'scientific form' (such as academic societies and journals), or (as above) by appeal to notions culled from the philosophy of science. The strategies adopted, within the scientific community, by those anxious to establish the scientific credentials of 'worm running', parapsychology and creationism are instructive in this regard (Travis, 1981; Collins and Pinch, 1979; Moore, 1974). Sometimes the aping of scientific practice merely mimics the ritual performances of accepted experts (Wallis, 1976); sometimes a scientistic rationalization serves, after the event, to give a spurious legitimacy to decisions determined by expediency and characterized by muddle (Sapolsky, 1972).

Recently, attempts to use 'risk accounting' to demonstrate the relative safety of nuclear power have led to controversy (BN). Much of this criticism has focussed on the presence of 'hidden' values embedded in quantification: in assessing the effects of low-level radiation, for instance, what weight should be given to the likely genetic consequences in remote future generations? But what blatantly separates the disputing parties, at least rhetorically, is another 'sameness' judgment: to supporters, nuclear risks are of the same kind as other everyday hazards, and can therefore be compared with them on a single scale; but opponents see nuclear risks as *essentially different,* and declare such comparisons invalid. As Hannes Alfvèn, the physicist and Nobel Laureate expressed it, in a widely-quoted statement:

> Fission energy advocates say that in all other technologies one accepts certain risks, and there is no way of completely eliminating all hazards. They claim to have taken all reasonable precautions to eliminate risks. What more can one do? This may very well be true, but it is irrelevant because we are facing risks the nature of which we have never before experienced. The consequences of nuclear catastrophes are so terrible that risks which usually are considered to be normal are unacceptable in this field. (Alfvèn, 1972: 6)

Surrey and Huggett have attempted to 'spell out' what these essential differences are:

> In our view, the distinctive feature of the perception of nuclear-specific risks arises from the combination of the collective nature of the decisions which impose risks upon large sections of the population, including all those who live within a certain radius of the plant; the large scale of a potential accident, even though the risk is extremely small; the possibility of 'long-term' deaths from cancer and possible genetic effects

upon future generations; the association by some people of nuclear power and nuclear weapons; and the fact that, because of long-lived radioactive byproducts, the commitment to nuclear technologies cannot be phased out in 30—50 years' time if more satisfactory and safer technologies are then available. (Surrey and Huggett, 1976: 288)

But such lists are not notably illuminating, and are certainly not exhaustive: better, perhaps, to acknowledge that the parties differ on fundamental values, and then to see this 'assertion of difference' as a 'rational' political strategy. Wynne (1978) has discussed the differences between radiation standards in the US and UK in such a perspective. He sees the intrusion of 'risk accounting' procedures as an attempt to maintain the authority of *social* institutions, by allowing a form of conceptualization mediated by technical experts to dominate political decisions:-

It is important to note that there is relatively little disagreement between the [US Environmental Protection Agency (EPA)] and the UK authorities on the estimation of risks from given exposures. The differences arise when interpreting the significance of the risks. The EPA takes the view that exposures are planned and avoidable, so should be assessed in terms of the increase of risk entailed by a given discharge over the existing cancer risk from natural radioactivity and other sources. In stark contrast, the British authorities spend a great deal of time stressing the insignificance of particular doses when compared to natural background, or to other risks in society such as car-driving, rock-climbing or wine-drinking. While comparing risks may be illustrative, to use it as a justification of discharges raises some very awkward sociological and moral issues which tend to be ignored. Despite this, the comparative approach looks set to become the new method by which the 'irrationality' of the public's attitudes to radiation risks is to be 'scientifically' defined, and thus the social question of acceptability again subsumed under the control of the scientific-administrative establishment. (Wynne, 1978: 210)

In a long and important paper, Wynne (1975b) criticizes 'technology assessment' for the way in which it, and its practitioners, *assume* a consensus model of society, and systematically *devalue* social conflict. Attempts to impose such conceptualizations and expert language on the political process can be seen as attempts to exploit the general trust in, and authority of, technical expertise to 'engineer consensus' in society, by redefining 'rationality' (and 'irrationality'). However, as we have seen, exposure of such expertise to public scrutiny in adversarial proceedings can easily reduce its authority, and hence its ability to carry out this task: if conflict is to be contained, therefore, institutional procedures in which technical expertise and public participation both play a part require careful design; for the focus of credibility and trust then shifts from actors to *social institutions* (Wynne, 1980; Surrey and Huggett, 1976). As Wynne argues, 'all public decisions involve strong elements of ritual' (Wynne, 1980: 197), so every public inquiry necessarily has its purely ritual function (see also Nelkin and Pollak, 1979). The UK Windscale Inquiry is a striking example (Wynne, 1982). The political role of expertise, and its authority and credibility, is contained within a broader complex of social institutions which maintain social order, and serve powerful interests. Our final reading illustrates many of these issues: it also points to developments which may be required if credibility and social order are to be maintained.

A View from the Top

Such is the structure of interests, and the relationships of trust and power, in any modern industrialized society, that expertise has tended predominantly to be linked with 'the higher levels' of government and industry. Indeed, expertise is trained and hired to satisfy the demands of powerful interests. In recent years, the consolidating effect of expertise on power structures has come to be widely realized. When their role in the system as a whole is considered, individual experts, acting entirely in good faith, may become mere legitimators: as Nelkin noted, in her 'local' case studies, experts are cited when they confirm prejudices, overlooked when they do not. The fates of the UK Wootton Report on Cannabis, and of the US Presidential Commission Report on Pornography, illustrate the essential powerlessness of experts who step out of line.

On the other hand, those at the 'higher levels' do share many of the interests of the rest of us, and no doubt recognize that some decisions can be taken only after the most careful assessment of expert advice upon their longterm consequences. It is sad that so little is known, particularly in the British context, about how such decisions are arrived at. But here is the former UK Prime Minister, James Callaghan, describing, early in 1978 in a hitherto unpublished speech, how one crucial decision came to be made. It is a striking illustration of what we earlier described as 'the tragedy of the expert.'

> One of the most difficult decisions recently taken by the Government in an area of rapidly advancing technology has been the choice of nuclear reactor systems. We have adopted a flexible strategy, namely that our thermal reactor programme for the present should not depend upon an exclusive commitment to one reactor system. The period of intensive preparation that went into this apparently simple decision was

248

an interesting illustration of the various factors that a Government must take into account in reconciling different objectives.

To begin with, the Department of Energy and the electricity supply industry needs efficient, economical, reliable plant to meet consumers' needs.

The Department of Industry is concerned that the nuclear industry itself should feel that it has a healthy confident future.

The public and the Department of the Environment have an important concern that their safety and the environment both be safeguarded.

The Treasury is concerned with the cost of building nuclear power stations and the Department of Trade with the balance of payments, implications for imports and exports.

We drew on many sources of advice — the Atomic Energy Authority as R and D organisation; the National Nuclear Corporation as builder; the electricity supply industry as customer; the Nuclear Installations Inspectorate as independent safety adviser; the nuclear industry itself; and the experience of other countries.

I have said on a previous occasion that in my experience the Government has never been so well informed before reaching a decision. I was faintly surprised at the certitude shown by some who offered their opinions — but alas they were not all of the same mind. So in the end, after all the welter of scientific, technical and industrial advice had been proffered, we had to synthesise it and reach as laymen a decision that was based on our own judgment.

In the light of the previous history of Government decisions on nuclear reactor policy, I would be very rash to assert with utter confidence that we have got it right. Only time will tell. But what I can say is that I don't believe we could have done any better in the present state of knowledge and given all the factors we had to reconcile. (Callaghan, 1978: 3)

13

J.S. Oteri, M.G. Weinberg

and M.S. Pinales

Cross-examination of chemists
in drugs cases

In a drug case several lines of defense are available. The first line is the facts. Do they support a motion to suppress? Entrapment? Alibi? etc. The second is the Constitution. Is the statute constitutional on its face? As applied? The third line is the one most often ignored: an attack on the analysis of a drug. Is the substance seized from the defendants and analyzed by the official chemists in fact the drug which is specified in the indictment? If it is not, the Government case fails. Despite the assumption of nonscientific people that chemists are infallible, recent cases have demonstrated the possibility that the overworked state laboratories have more than once made significant analytical errors. The defense attorney can either expose these errors, or create reasonable doubt as to the validity of the chemist's opinions, remembering always that the chemist's testimony is merely his opinion and not the gospel.

The Massachusetts statutes require that the Department of Public Health 'shall make, free of charge, a chemical analysis of any narcotic drug . . . when submitted to it by police authorities. . . .'[1] The results of the analysis shall be furnished, upon request, to police officers in a signed certificate and this certificate 'shall be prima facie evidence' that a chemical analysis has been made and that the substance is what the chemical concludes it to be.[2] Nevertheless, the certificate is *only* prima facie evidence and the right of a defendant to cross-examine the chemist is not denied by these statutes.[3] In other states, e.g., Ohio, no prima facie evidence statute exists, and the chemist must testify unless there is a stipulation as to his report. Prosecutors seldom subpoena chemists for a trial; they expect a stipulation. A motion to dismiss should then be made for want of prosecution, although it will be overruled; the next trial date will usually be peremptory. Occasionally the Government will attempt to admit the report as being in the 'ordinary course of business.'[4] This, however, is objectionable as a denial of due process and the right to confront one's accusors.[5]

If defense counsel are to pursue this line of defense, the two critical questions are: (1) Can the chemist be successfully attacked? (2) How can it be done? It is the premise of this article that many chemists are not well prepared to testify and that in the past they have remained relatively invulnerable only because defense attorneys have not adequately prepared themselves for cross-examination.

I.   ATTACKING THE EDUCATION AND QUALIFICATIONS OF THE CHEMIST

Basic tools for cross-examination

The first and most vulnerable area of attack is the education of the expert. A series of inquiries ought to be directed toward the limited formal training of the chemist in the detection of the drugs in question:

Q   Sir, while attending high school how many chemistry courses did you take?

Q   How many hours per week did these courses require?

Q   What percentage of your total time in high school was spent in the study of chemistry?

Q   How much of the time in high school which you spent studying chemistry, was spent on the study of (the substance in question e.g. marijuana, heroin)?

Q   How many times did you, while studying chemistry in high school, analyze (the substance in question)?

The chemist may have to answer 'None' to the last two questions since the possession of marijuana or heroin is criminal and no high school teacher could ever have been trained in this area even for educational purposes. If he answers affirmatively he can be pressed for details and his credibility can be completely destroyed.

Similar questions should be used to examine the expert's training in college and in graduate school. The paucity of his experiences in analyzing the substance in question should be compared to the entire curriculum which dominated his studies.

Finally, the on-the-job training of the chemist should be scrutinized. When did the chemist first analyze the type of substance which the defendant is accused of possessing or selling? Under whose auspices? What was his formal training? What sort of training aids did he have? Any formal courses? Indicate to the jury that

251

if the expert's teacher was doing the test wrong so was the expert. Also examine him on how many books or articles he has written. Few chemists are also authors.

II.   ATTACKING THE FOUNDATION OF THE ANALYSIS

A.   Authentication

After completing the first line questioning the examiner ought to inspect the authenticity of the samples used in the analysis. Each and every link in the chain of possession should be examined with care.

The normal procedure is for the arresting or searching police officer to confiscate the suspected contraband. The police officer then delivers it to the chemical laboratory where it is registered in a log book, assigned a number, and given an identifying card by the clerk. After a sample of the seized material has been analyzed, it is returned to the officer in a sealed, plastic bag which he retains in custody until trial. The following areas of examination of both chemist and police officer may, as examples, yield valuable testimony:

> 1.   Was there a time lag between the date the substance was seized and the date it was analyzed?
> 2.   If the answer to (1) is yes, is the drug stable? Is it subject to chemical change? Could the analyzed drug be the decomposed result of a time lag?
> 3.   Was the officer who seized the drug the one who delivered it to the laboratory? In the interim, where was it kept? Who had access to it? Was it sealed when delivered to the chemist? Was it sealed at all times before that? If not, there is the possibility that other substances were added to it.
> 4.   How many other samples did the delivering officer possess? How fastidious was he in keeping the drugs separate?

Any chemist will say that he analyzed the material delivered to the laboratory. He can be challenged on the care given at the laboratory to the samples, but he cannot testify to what happened before he received the samples. Was there a mix-up? His only response is 'I don't know.' Unless the history of the sample from seizure to analysis is carefully authenticated, the validity of the analysis is irrelevant.

252

B.   Carelessness at the laboratory

The examiner should spend some time pointing out the number of chemists at the laboratory, the number of samples examined at the same time as the sample in question, and the opportunity for inadvertent errors because of the volume of the laboratory's work. This line of inquiry should be expanded to include the possibility that traces of other substances found, for example, as residues in the chemicals or instruments used to perform the tests in question, destroyed the validity of the analysis.

C.   An inference of bias

The examiner should induce the witness to disclose that he is either a Government employee or that he is paid by the prosecution, and that he has testified x number of times, always for the Government.

III.   ATTACKING THE TESTS USED TO ANALYZE THE DRUG

The examiner should learn all the possible tests that could be used for the analysis of any particular drug. He should then ask the chemist if he applied each one. The following lines of attack often produce testimony that suggests that the tests were not reliably performed:[6]

1.   Demonstrate that he did not perform all the possible tests. Leave the jury with the impression that the chemist did not perform all the tests either because he was careless, or because he had something to hide, or because he just failed to exhaust the steps necessary for proof beyond a reasonable doubt.

2.   Ask the witness if he took notes while performing the tests. Then ask if he looked at these notes to refresh his memory. Then the examiner should look at them to see if they were sufficient to refresh the witness' memory.[7]

3.   Take the witness through each test: how it is performed, how long it takes, his expertise in performing the test, and how much of a sample he used.

4.   Get the witness to admit that other substances create the same result as does the substance he identified.

5.   Ask the witness if he ran a 'blank' test. This is a test to ascertain the purity of the chemicals used to administer the test and the correctness of the machines and instruments used for the analysis. It is performed by running the identical test on the identical machines without the unknown substance. The results should be zero or blank. Ask the

witness *when* he last performed a 'blank' test and when he last checked the machines. Imply that the laboratory instruments may not have been optimally efficient.

6. Ask the witness if he ran a 'control' test. This is a test using a known substance, e.g. heroin. Ask him about the source of the known substance. Ask him *when* the test was performed. If there is a time lag imply that there is no way of knowing whether the results of tests performed on the unknown substance really matched the paradigmatic standards set by the control tests on the known substance because of possibilities of changes in the instruments used.

7. Ask the witness to examine his laboratory data sheets and test charts. Ask him for 'blank' test results, 'control' test results, and the results of the tests on the unknown substance.

8. Ask him if the infrared analysis is the most specific means of drug identification. (It is.) Then ask him if he used it. (Few do.)

9. Ask the witness if the test is 'specific.' Since there are other substances which yield the same results, none are. Then force him to define the limits of the test. Then have him admit he did not perform other tests. Do not define their limitations.

10. Elicit from the expert an admission that the tests are empirically developed and that no one has any knowledge of *why* the substances react as they do. Force the witness to admit that the logic of the test is inferential, and that the specificity of the test is assumed on the basis of accumulated consistent data rather than on the basis of generally accepted scientific principles. Once all this is conceded, the tests can be shattered by showing other substances yielding results identical to the suspected narcotic.[8]

11. Find out if witness did formal tests to isolate out those substances which have *identical* test results as the suspected drug. Find out if witness did formal tests to isolate out those substances which interfere with the lucidity of tests that aim at identifying the sample as the suspected drug.

12. Get the witness to admit that even the pyramiding of many tests only results in a higher probability of an accurate analysis and not a certainty of accuracy.

IV. ATTACKING THE ANALYSIS OF HEROIN

A. Quantitative tests and eliminative tests

Ordinarily the chemist is only looking for *some* heroin in the unknown substance, and not for the exact amount. Since heroin, like most drugs, is not sold in pure form, but instead is 'cut' or diluted with some other substance such as milk, sugar or starch, the following series of questions may prove effective:

Q   Did you do a quantitative test on the substance? (Few do.)
Q   So you don't know what percentage, if any, of the substance is

heroin? (The exact percentage *can* be determined. It rarely exceeds 20%.)

Q   What other substances were present? (Chemists rarely test for other substances.)

Q   You do admit that some other substances give off the same reaction as heroin on the Marquis test? On the ultraviolet spectrography test?

Q   Did you test for quinine? For ephedrine? (Because sugar is the common 'cutting' agent and it has a sweet taste, and because heroin has a bitter taste, many drug sellers add quinine, which has a bitter taste, to the heroin to create an illusion of potency. Quinine can be tested for and it does interfere with the analytical tests for heroin. Ephedrine also may be added to the mix and it also interferes with the tests.)

B.   The Marquis test

The chemist is faced with an initial dilemma: he cannot identify the unknown drug until he extracts it and he may fail to extract it unless he knows its identity. As a result, he performs a general test to identify the substance. For heroin, this is the Marquis test.

The test consists of adding 40 drops of formaldehyde to 60 cc. of concentrated sulfuric acid and then applying this mixture to the unknown powder. A deep purple color indicates heroin.

The test is replete with limitations. First, it is a subjective test dependent on the description by the chemist. The chemist should be asked if he took a color photograph of the result. Few do. Second, many other drugs produce the same purple color. Ask the chemist about the reaction of the test to morphine, to codeine, to dionin, to caffeine, to cocaine, to strychnine. Third, the test is very sensitive. One microgram (i.e., one millionth of a gram) can yield the purple color. The chemist should be asked:

Q   You have done hundreds of these tests have you not?

Q   You use the same equipment do you not?

Q   Will you swear that one tiny trace of some other sample or some other substance was not present?

C.   Chromatographic separations

Once the chemist has made a preliminary hypothesis that the sample is heroin, he should attempt to separate that heroin from the remainder of the mixture. These tests are necessary to eliminate interfering substances such as quinine. These tests, however, are but a procedural stepping-stone to the more specific tests described in section D, below.

The most common test is paper chromatography. A strip of chromatographic paper is partially immersed in a suitable solvent. The unknown substance is then dissolved in the solvent and the various components of the unknown substance move to various heights on the paper. Presumably, different materials will move to different heights. The paper has a base line. The measuring standard from base line to substance is called the Rf value. The substances, mostly colorless, are detected by ultraviolet light at a wavelength of 254 millimicrons and are then sprayed with a chemical to help isolate them.

This test has limitations. The chemist should be closely examined as to whether or not he has conducted 'control' tests and 'blank' tests. The examiner should determine *when* these tests were performed; he should also obtain the charts of these tests. (See III [5] and [6].) The examiner should then find out what the Rf value of the unknown substance was. Heroin has a range of 56-64. Atropine (55-62), codeine (58-65), and strychnine (57-64) have similar ranges. Ask the examiner if they were separated. Ask the examiner if he did multiple separation tests.

The value of this test to the chemist is that it can isolate heroin from many interfering substances (i.e., quinine, Rf range of 76-83). Many hurried chemists fail to perform this test, which is the only reliable way to isolate heroin from quinine.

Thin-layer chromatography is another separation test. In principle, it is similar to paper chromatography, however substances such as silica gel, aluminium oxide, and cellulose power replace the paper. It is faster, but it can only be done with small quantities of a substance.

The gas chromatographic test is yet another separation test. The basic instrument consists of a tubular column containing a selected substance which is mixed with the unknown substance. The speed of flow is recorded on a graph and is dependent on the absorption of the components. Components are characterized by retention time. Quinine and codeine have the same retention time as heroin and thus have the same peaks.[9]

D.   Ultraviolet absorption spectrophotometry

The UV test is the most common test performed if heroin is suspected. A solution, either a basic water extract (water plus chloroform plus sodium or ammonium hydroxide) or an acidic water extract (water plus chloroform plus hydrochloric acid), is placed in a glass cell. A small amount of the unknown substance is placed in the solution, and the entire glass cell is placed into the ultravviolet spectrophotometer machine. A beam of light is passed through the cell, and a photocell, beneath the glass cell, measures the amount of light passing through the solution, recording these measurements on paper by way of a curve. Thus the wavelengths that are *not* absorbed are the wavelengths that are recorded. For heroin passing through the acidic water extract, the minimum peak is 250 and the maximum peak 275 microns: passing through the basic water extract the minimum peak is 278 and the maximum 295-300 microns.

The following cross-examination may be fruitful:

Q   What type of machine was used? Model? When manufactured? Aren't there newer machine? Aren't there more sensitive machines?

Q   When was the machine serviced before the test? How soon after?

Q   Ask to see the curve. If it is not full graph paper, with all the charts built in, ask him if there is more precise paper?

Then examine him on the need for control tests and blank tests.

Q   Did you run a control test?

Q   When?

Q   Where did you get the substance from?

Q   From whom was it received?

Q   Who has access to it?

Q   Did you adjust the machine so that the water extract had 0% transmittance?

Q   Did you run a 'blank' test?

Q   Quinine destroys the validity of the test, does it not? Isn't the curve for quinine too strong and too closely related to the heroin curve?[10]

Finally, remind the chemist (and the jury) that if the heroin is not isolated other substances may invalidate the test:

Q   Is it true that other materials have *different* wavelengths than heroin?

Q   It is possible for a material containing several materials with dissimilar wavelengths to collectively have the same peaks as heroin?

Q   It is true that other materials have the *same* wavelengths as heroin?

Q   It is true that some materials interfere with the analysis of heroin?

V.   ATTACKING THE ANALYSIS OF LSD

[Editors' note: this section has been cut, since it repeats strategies already discussed.]

VI. ATTACKING THE ANALYSIS OF MARIJUANA.[11]

A. Qualifications of expert

Because of some unique factors about marijuana, an additional series of questions may be effective in challenging the qualifications of the expert:

Q   The classification of vegetable substances is also a scientific discipline, is it not?
A   Yes.
Q   And the name of this scientific discipline is called plant taxonomy, is it not?
A   Yes.
Q   How many courses have you ever taken in plant taxonomy?
A   No one has ever taken courses in that except maybe the 20 guys who are plant taxonomists. So he's going to say 'none.'
Q   The 'L.' in *Cannabis sativa L.* stands for Linnaeus, doesn't it?
A   Yes, it does.
Q   Have you ever read Linnaeus' original taxonomy work?
A   No.
Q   Have you ever compared Linnaeus' taxonomy classification of cannabis with other taxonomy classifications? If he says 'yes,' ask: 'Name one.' He can't name one. If he says 'no,' you just shake your head.
Q   Oh, by the way, can you tell me when was Linneaus' work written? He doesn't know.
Q   Would 1694 refresh your recollection?
Q   This was the original classification of cannabis, wasn't it?
Q   Have you ever read J.B. Lamarck's work, published in 1783?
Q   Have you ever discussed this work with any learned person in the field of plant taxonomy?
Q   You are not now, nor do you pretend to be, a plant taxonomist?
Q   So you're really not qualified to give an opinion as to the classification of plants are you?

B. The tests

Some of the tests commonly used to detect marijuana are:
1. Examination under a microscope. 2. Duquenois or Reagent test. 3. Beam test. 4. Gas chromatography test. 5. Paper chromatography. 6. Gharmarcy. 7. Wasicky.

None of these tests are specific, i.e., all of them yield the same results when other substances are tested. Furthermore, there is no known reason why these tests produce the standard reactions. They are empirical support tests only. Thus, without theoretical support in scientific principle, one exception destroys the tests.

Usually a series of color and precipitate tests are run. Because the tests are inferential only (see section III [10]) the witness will

258

say all the tests were positive and therefore the substance is marijuana. The results are not certain.[12] The limitations of these tests can be adequately explored by proceeding in the manner described in section III above. Once the witness concedes that he did not run *all* the tests and that the ones he did run were inferential, ask him if he tested for the labiate culinary herbs: tea, sage, oregano, catnip, azmidor, rosemary, thyme. Any of the tests based on accumulated data are destroyed by the exceptions. The labiate culinary herbs have identical results on the color tests. The chromatography tests are discussed in section IV C.

The ultraviolet spectrographic tests are not specific enough because of the variability and low specificity of the absorption characteristics of cannabis. There is too broad a spectral shift—and, consequently, it is rarely used.

Cross-examination of the chemist is an art long ignored by practitioners. With a moderate effort and some basic knowledge it is a tool worth using. Its employment may lead to acquittals of otherwise hopeless cases.

NOTES AND REFERENCES

1. MGLA ch. III, §12.
2. MGLA ch. III, §13.
3. Commonwealth v. Harvard, 356 Mass. 452, 253 N.E.2d 346 (1969).
4. Ohio Revised Code §231 7.40.
5. State v. Tims, 9 Ohio State 2d 136, 224 N.E.2d 348 (1967).
6. The best sources on drug analysis are E.G.C. Clarke, *Isolation and Identification of Drugs* (London: Pharmaceutical Press, 1969); R.C. Sullivan, *Criminalistics: Theory and Practice* (Boston: Holbrook Press, 1971); Internal Revenue Service, *Methods of Analysis of Alkaloids, Opiates, Marijuana, Barbiturates and Miscellaneous Drugs* (Pub. No. 341, Rev. 2-0, Feb. 1960).
7. See Commonwealth v. Marsh, 354 Mass. 713, 242 N.E.2d 545 (1968).
8. N.D. Cheronis, *Tentative and Rigorous Proof in the Identification of Organic Compounds and Application of these Concepts to the Detection of the Active Principals of Marijuana*, 4 Microchemical Journal 558 (1960).
9. J. Grooms, *Quantitative Gas Chromatographic Determination of Heroin in Illicit Samples,* 51 Journal of the Association of Official Analytical Chemists 1010 (1968).
10. *Official Methods of Analysis of Association of Official Analytical Chemists 622* (11th ed.: Horwitz ed. Association of Official Analytical Chemists, 1970), has two tests for heroin, one if it is in presence of quinine and one if it is not mixed with quinine.
11. J. Shellow, *The Expert Witness in Narcotic Cases,* 6 Criminal Law Review 20 (1970) is a fine overview of the area.
12. L. Grlic, *Recent Advances in Chemical Research of Cannabis,* 16 United Nations Bulletin on Narcotics 34 (1964); L.T. Heaysman, E.A. Walker and D.T. Lewis, *The Application of Gas Chromatography to the Examination of the Constituents of Cannabis Sativa L.,* 92 Analyst 450 (1967).

14

Mary Douglas

Environments at risk

When the scientist has a very serious message to convey he faces a problem of disbelief. How to be credible? This perennial problem of religious creed is now a worry for ecology. Roughly the same conditions that affect belief in a denominational god affect belief in any particular environment. Therefore, in a series of lectures on ecology, it is right for the social anthropologist to address this particular question. We should be concerned to know how beliefs arise and how they gain support. Tribal views of the environment hold up a mirror to ourselves. Putting ourselves in line with tribal societies, we can try to imagine the figure we would cut in the eyes of an anthropologist from Mars. From our own point of view he would take an agnostic stand. But today, to do justice to this lecture's subject, we should ourselves attempt the difficult trick of letting go of what we know about our environment — not forgetting it, but treating it as so much science fiction. Like the alien anthropologist, let us suspend belief for a little while, so as to confront a fundamental question about credibility.

We are far from being the first civilization to realize that our environment is at risk. Most tribal environments are held to be in danger in much the same way as ours. The dangers are naturally not identical. Here and now we are concerned with overpopulation. Often they are worried by underpopulation. But we pin the responsibility in quite the same way as they do. Always and everywhere it is human folly, hate and greed which puts the human environment at risk. Unlike tribal society, we have the chance of self-awareness. Because we can set our own view in a general phenomenological perspective, just because we can compare our beliefs with theirs, we have an extra dimension of responsibility. Self-knowledge is a great burden. I shall be arguing that part of our current anxiety flows from loss of those very blinkering or filtering mechanisms which restrict perception of the sources of knowledge.

First, let us compare the ecology movement with others of

260

historical times. An example that springs to mind is the movement for the abolition of slavery of a century ago. The abolitionists succeeded in revolutionizing the image of man. In the same way, the ecology movement will succeed in changing the idea of nature. It will succeed in raising a tide of opinion that will put abuse of the environment under close surveillance. Strong sanctions against particular pollutions will come into force. It will succeed in these necessary changes for the same reason as the slavery abolition movement, partly because of its dedication and mostly because the time is ripe. In many countries in the nineteenth century slavery was becoming more costly than wage labour.[1] If this had not been the case, I doubt if that campaign would ever have got off the ground. Where locally it was not the case, all the arguments about brotherly love, Christianity and common humanity were of no avail. The Clapham Sect, I believe, abstained from sugar as a protest against plantation slavery. In the same spirit some of my friends have abstained from South African sherry. This is a less impressive sacrifice since there are other better sherries. But those of us who do not own a car will not end exhaust fume pollution any more than the Clapham Sect diminished by one whit the place of sugar in the native diet. The tide of opinion against slavery was not against industrial development. And the tide of opinion which will reduce the worst pollution effects will not stem industrialization. Here is the crunch of the environmental problem that leads it far beyond the nuisance of water and air pollution and increasing noise. The ecologists have had to raise their sights to the global level. Their gloomy forecasts for the imminent end of our planet put us, the laymen, in the role of the helpless hero of a thriller. Several nasty deaths are in store for us. Time will reveal whether the earth will be burnt up by the unbalancing of its radiation budget, or whether a film of dust will blank out the sun's rays, or whether it will explode in an atomic war. Overpopulation and overindustrialization are the twin causes. But herein lies the dilemma. The obviously overpopulated and starving masses are in the nonindustrial regions. Their hope of food lies in new technological developments. But these come from the already industrial countries. Must we stop the growth of science which may one day feed the existing hungry? How do we control population anyway? And which ones should we start with? On a giant hoarding over the Chicago Expressway is a notice which says: 'Think before you litter.' A rather coarse expression, I thought, when I first assumed it to be family planning advice. But if I understand Dr Paul Ehrlich aright the anti-litter campaign could do well to take on the double objective, especially in Chicago. The starving millions of Asia are not the ones who own two cars, whose factories discharge effluent into lakes, or whose aeroplanes give off loud bangs. Ehrlich says: 'The birth of every American child is fifty times more of a disaster for the world than the birth of each Indian child. If you take consumption of steel as a measure of overall consumption, you find that the birth of each American child is 300 times more of a disaster for the world than the birth of each Indonesian child.'[2]

At the top global level the scientists speak with different voices, and none has a clear solution. This is the level at which we are free to believe or disbelieve. The scientists would not wish to be treated as so many old sandwich-men bearing placards which say: 'The end is near.' Our disbelief is just as much a problem for them as our gullibility. Therefore, whether their message is true or false, we are forced to study the basis of plausibility.

Another movement of ideas which this current ecology question recalls is the growth of classical economics. A realization which transfixed thoughtful minds in the eighteenth century and onwards was that the market is a system with its own immutable laws. How can we appreciate the boldness of the illumination with which Ricardo discerned that system and its homeostatic tendencies? In our day and for this audience I can only hope to give an idea of the thrill of analyzing its complexity, and the power and even sheer beauty of the system as it revealed itself, by reminding you of the excitement engendered in linguistics by Chomsky's revelation of the structural properties of language. But that is a very pale analogy. Consider how few political decisions are affected by linguistics compared with the implications of economic science. For the sake of that system and its unalterable laws, many good men have had to harden their hearts to the plight of paupers and unemployed. They were deeply convinced that much greater misery would befall if the system was not allowed to work out its due processes. In the same way, the ecologists have perceived system. In fact, their whole science consists in assuming system, reckoning inputs and outputs and assessing the factors making for equilibrium. The pitch of excitement in the ecological movement rises when the analysis is lifted to the level of whole continents and even to the level of this planet as a whole. In exactly the same way as the old economists, the ecologists find themselves demanding a certain toll of human suffering in the name of the sytem which, if disturbed, will loose unimaginable misery on the human race. Sometimes there is a question of bringing water a thousand miles to irrigate a desert. The ecologists know how to do it. They can easily make a desert blossom and so bring food and life to starving people. They hesitate to answer for the consequences in the area from which the water has been diverted. Their professional conscience bids them consider the system as a whole. In the same spirit as Ricardo deploring the effects of the Poor Laws, ecologists find themselves unwillingly drawn into negative, even reactionary positions.

These digressions into economics and slavery suggest a way of restricting somewhat the problem of credibility which is altogether too wide. It will be rewarding to watch how belief is committed along the bias towards or away from restriction. For ultimately this is it — restriction or controlled expansion? Which of us tend to believe the experts who warn that our system of resources is limited? And which of us optimistically follow those who teach that they cannot possibly tell yet what resources may lie unknown beneath the soil or in the sea or even in the air? And the same question about our own bias may be raised for the bias of the experts themselves.

Phenomenology, as I understand it, is concerned with what it is we believe we know about reality and with how we come to believe it. An anthropologist's survey of tribal environments is different from the ecologist's survey. Ecology imports objective measures from a scientific standpoint and describes in those terms the effect of the system of cultivation on the soil and of the soil on the crop yield, etc. It is concerned with interacting systems of physical realities. The anthropologist, if he is not lucky enough to have access to an ecological survey in his research area, has to make a rough dab at this kind of assessment and then use it to check with the tribe's own view of their environment. In this sense an anthropologist's survey of tribal environments is an exercise in phenomenology. Each tribe is found to inhabit a universe of its own, with its own laws and its own distinctive set of dangers which can be triggered off by incautious humans. It is almost as if there were no limit on the amount of variation two tribes can incorporate into their views of the same environment. Some objective limits must apply. Nevertheless, if we were to rely entirely on tribal assessments, we would get wildly incongruous views of what physical possibilities and constraints are in force.

For example, I worked in the Congo on the left bank of the Kasai river, among the Lele. On the other bank of the same river lived the Bushong, where my friend Jan Vansina worked a little later on. Here were two tribes, next-door neighbours, who celebrated their cold and hot seasons at opposite points in the calendar. When I first arrived, green to Africa, the Belgians said how wise I had been to arrive in the cold season: a newcomer, they said, would find the hot rainy season unbearable. In fact it was not a good time to arrive, because all the Lele were working flat out to clear the forest and fire the dead wood, and then to plant maize in the ash. No one but the very aged and the sick had time to talk to me and teach me the language, until the rains arrived and ended their period of heavy work. When I knew the language better, I learnt of a total discrepancy between the European and native assessment of the weather. The Lele regarded the short dry season as unbearably hot. They had their sayings and rules about how to endure its heat. 'Never strike a woman in the dry season', for example, 'or she will crumple up and die, because of the heat.' They longed for the first rains as relief from the heat. On the other bank of the Kasai, the Bushong agreed with the Belgians that the dry season was pleasantly cool and they dreaded the onset of the first rains. Fortunately the Belgians had made excellent meteorological records, and I found that in terms of solar radiation, diurnal and nocturnal temperatures, cloud cover, etc, there was very little objective difference that could entitle one season to be called strictly hotter than the other.[3] What the Europeans objected to, apparently, was the humidity of the wet season and the absence of cloud which exposed them directly to the rays of the sun. What the Lele suffered from in the dry season was the increased radiant heat which resulted from the heavy screen of clouds. They recognized and hated the famous glass-house effect that we are told will result from an excessive carbon dioxide screen for ourselves. But above all, the Lele time-table required them to do all their agricultural work in one short, sharp burst, in the dry season. The Bushong, across the river, with a more complex agricultural system, worked away steadily the whole year round. They also distinguished wet and dry seasons, but they concurred with the Europeans about the relative coolness of the dry period. How did the Europeans arrive at their assessment, since objectively there was so little to choose? No doubt because the seasons were named and their attributes set at Leopoldville, the capital, where temperature readings showed a difference between the seasons that did not obtain in the interior. In this example, credibility derives from social usage. If the Lele could have changed their time-table, their perception of their climate could have been altered. But so would a great deal else in their life. They were relatively backward technologically, compared with the Bushong. A different time-table, spreading their work through the calendar and through the whole population, would have greatly bettered their exploitation of their environment. But for such a fundamental revolution they would have had to create a different society.

Time-tables are near the heart of our problem in the pheno-

menology of environments. Andrew Baring, in studying a Sudanese
people, tells one that their mythology is full of dynastic crises,
plots, unrest and revolution. Whenever discontent reaches boiling-
point in the myth cycle, a new king takes over the palace. The up-
start always shows his administrative flair by changing the times of
meals. Then the discontent simmers down and all is well until
something goes wrong again with the order of day in the royal
household. Then the stage is set for a new dynastic upheaval and a
new king to settle the time-table problem afresh. Largely, the
doom ecologists are trying to convince us about a kind of time-
bomb. Time is running out they say. Whenever I suggested to the
Lele small capital-intensive projects that would improve their
hunting or the comfort of their houses, they would answer 'No
time'. The allocation of time is a vital determinant of how a given
environment is managed. It is also true that the time perspective
held by an expert determines what answer he will give to a technical
problem. Therefore, we should start our discussion of how credibility is engendered by considering the time-dimension.

Among verbal weapons of control, time is one of the four final
arbiters. Time, money, God, and nature, usually in that order, are
universal trump cards plunked down to win argument. I have no
doubt that our earliest cave ancestress heard the same, when she
wanted a new skirt or breakfast in bed. First 'There wouldn't be
time', and then, 'We couldn't afford it anyway'. If she still seemed
to hanker: 'God doesn't like that sort of thing'. Finally, if she
were even prepared to snap her fingers at God, the ace card: 'It's
against nature, and what is more, your children will suffer'. It is a
strong hand when the same player, by holding all these cards, can
represent God and nature, as well as control the time-table and the
bank account. Then the time-scale, as presented by that player in
control, is entirely credible.

It is only just beginning to be appreciated how much the perception of time-scale is the result of bargaining about goals and
procedures. For a touching insight, read Julius A. Roth's little
book *Timetables, Structuring the Passage of Time in Hospital
Treatment and Other Careers*.[4] Here he describes the attempts of
long-term patients in a TB hospital to get some satisfactory response out of their doctors. Spontaneously and inevitably the
clash of interests expressed itself in a battle about the time-tabling
of diagnosis and treatment. For the patient, his whole life is held
in suspense, no plans can be made, no sense of progress enjoyed
until he knows when he can go home. His anxiety concentrates in
a passionate study of the time-scaling of the disease and its treatment. With no means of extracting from them any clue to his

central preoccupations, and no sanctions to employ to enforce their collaboration, the patient would try to impose on the medical staff a kind of natural time-table. 'I've been here six months now, doctor, by this time you ought to have decided whether I need surgery.' 'Massey got a pass after only three months here — why can't I?' 'Hayton got discharged three months after surgery — why should I have to wait four months?' The doctors' strategy is evasive. They consistently reject the very idea of a rigid time-table, and struggle to refute the culture of the ward where patients go on working out a clear set of phases by which they judge the competence of the doctors and the course of their illness. Sometimes if the doctor does not seem to accept these spontaneously emerging laws of disease and treatment, the patient discharges himself, feeling that the basis of mutual understanding has failed.

The Lele afford another example of how time-tabling is used as a weapon of control. They think that they can do something to bring on the rains. This technique of weather control was not a rain-dance, or a magic spell. It was something which any individual might do, with the effect of hastening the onset of the rainy season. The belief in its effectiveness became an instrument for mutual coercion. Laggard farmers would beg others to wait until they had had time to clear their fields and burn the wood. The punctual ones would warn them to hurry up, lest the rain be provoked by action in the north. It seemed to me, sceptical as I was of the value of their technique, exactly like the departmental head who creates artificial deadlines to hasten decisions or otherwise keep his staff on their toes. 'I can only vouch for the students' reaction if we get our policy settled by the next meeting.' Much later I learnt that the Lele techniques of seeding the clouds by the smoke of their burning forests might indeed be effective. Air pollution can increase condensation and precipitation of moisture. It seems that a district in Indiana, thirty miles downwind of South Chicago's smoky steelworks, tends to have 31 per cent more rain than other communities where the air is clearer.[5] The joke was on me that the Lele weather control turned out to be scientifically respectable. For the purpose of my present argument, its efficacy is irrelevant. All that matters is that time is a set of manipulable boundaries. Time is like all the other doom points in the universe. One and all are social weapons of control. Reference to their power sustains a view of the social system. Their influence is thoroughly conservative. For no one can wield the doom points credibly in an argument who is not backed by the majority view of how the society should be run. Credibility depends so much on the consensus of a moral community that it is hardly an exaggeration to say that a given community lays on for itself the sum of the physical conditions which it experiences. I give two well-known examples.

In many tribal societies it is widely agreed that wives should be faithful to their husbands. Women probably concur in that ideal, and they would perhaps like to add to it that husbands should be equally faithful to their wives. However, the latter view does not obtain whole-hearted male support. Therefore, since the men are the dominant sex, to sustain their view of sexual morality they need to find in nature a sanction which will enforce the chastity of wives without involving male infidelity. The solution has been to fasten on a natural danger to which only female physiology is exposed. Hence we find very commonly the idea that miscarriage is due to adultery. What a weapon that provides: the woman tempted to adultery knows that her unborn baby is at risk, and her own life too. Sometimes she is taught that the health of her older children will also suffer.[6] What a paraphernalia of confession and cleansing and compensation attends the guilty mother in her labour. What she has done is against nature and nature will retaliate.

For a different example — a warlike tribe of Plains Indians, the Cheyenne, believes that murder of a fellow tribesman is the ultimate wickedness. The tribe used to depend on wandering herds of bison for its food. The bison, it was thought, did not react to the murder of men of other tribes. The bison were not affected by ordinary homicide as such. But a fratricide emitted a putrid stink which frightened off the herd and so put the vital resources of the tribe at risk. This dangers from the environment justified special sanctions to outlaw the murderer.[7]

When homicide and adultery are seen to be triggering agents for danger points in a physical environment, the tribal view of nature begins to emerge as a coherent principle of social control. Red in tooth and claw, perhaps, but nature responds in a highly moral and avenging form of aggression. It is on the side of the constitution, motherhood, brotherly love, and it is against human wickedness.

When I first wrote *Purity and Danger* about this moral power in the tribal environment, I thought our own knowledge of the physical environment was different.[8] I now believe this to have been mistaken. If only because they disagree, we are free to select which of our scientists we will harken to, and our selection is subject to the same sociological analysis as that of any tribe.

We find that in tribal society certain classes of people are liable to be classed as polluters. These classes are not the same for all tribes. In some social structures the polluters will be one type, in others another. Imagine a tribal insurance company which set out to cover people against the risk of being accused of causing pollution. Their market researchers and surveyors should be able to work out where they should charge the highest premiums in any particular social system. In some societies the élite possessors of esoteric knowledge are certain to be charged with owning too much science and misapplying it to selfish ends. Sorcery charges against big operators in New Guinea or against cunning old polygamists among the Lele can be paralleled by our own charges against big business polluting lakes and rivers and poisoning the children's food and air for their commercial profit. The Lele shared with other Congolese tribes an acute anxiety about their population. They believed themselves to be dying out because of malicious attacks by sorcerers against the fertility of women and on the lives of babies. They continually said that their numbers had declined because of jealousy. 'Look around you,' a man said to me on this theme, 'do you see any people? Do you see children?' I looked about. A handful of children played at our feet and a few people sat around. 'Yes, ' I said, 'I see people and children.' 'Look again', he said, disgusted with my missing the cue. 'There are no people here, no children.' His question had been couched in the rhetorical style expecting the answer 'No'. It took me a long time to learn the tonal pattern which should have led our dialogue to run in this way.[9]

'Look around you! Do you see any people? Tell me — do you?' Emphatic answer: 'No!'

'Do you see any children? Tell me, do you?' Emphatic answer: 'No!' Then his answer would come: 'See how we have been finished off. No one is left now, we are destroyed utterly. It's jealousy that destroys us.' For him, the sorcerer's sinful lore was as destructive as for us the science that serves military and business interests with chemical weapons and pesticides.

In another type of society the probability of being accused of pollution will fall on paupers and second-class citizens of various types. Paupers I define as those who, by falling below a required level of achievement, are not able to enter into exchanges of gifts, services and hospitality. They find themselves not only excluded from the main responsibilities and pleasures of citizenship but a charge upon the community. They are a source of embarrassment to their more prosperous fellows and a living contradiction to any current theories about the equality of man. In tribal societies, wherever such a possibility exists, these unfortunates are likely to be credited with warped emotions. By and large they may be called witches, and risk being accused of causing the natural disasters which other people suffer. Somehow they must be eliminated, controlled and stopped from multiplying. If you want to intuit from our own culture how such accusations of witchcraft gain plausibility I recommend you to attend a conference of professional social workers. Over and over again you will hear the objects of the social workers' concern being described as 'non-coping'. The explanation given for their non-coping is widely attributed to their having 'inadequate personalities'. In something of this way, landless clients in the Mandari tribe are said by their patrons to have a hereditary witchcraft streak which makes them vicious and jealous.[10] The fact that they are emotionally warped justifies their fellow-men in withholding the liberties and protections of citizenship.

As for females, in all these societies and anywhere that male dominance is an important value, women are likely to be accused of causing dangerous pollution by their very presence when they invade the men's sphere. I have written enough about female pollution in *Purity and Danger*.

It should be clear now that credibility for any view of how the environment will react is secured by the moral commitment of a community to a particular set of institutions. Nothing will overthrow their beliefs if the institutions which the beliefs support command their loyalty. Nothing is easier than to change the beliefs (overnight!) if the institutions have lost support. If we could finish classifying the kinds of people and kinds of behaviour which pollute the tribal universes we would have performed an ecological exercise. For it would become clear that the view of the universe and a particular kind of society holding this view are closely interdependent. They are a single system. Neither can exist without the other. Tribal peoples who worship their dead ancestors often explicitly recognize that each ancestor only exists in so far as cult is paid to him. When the cult stops, the ancestor has no more credibility. He fades away, unable to intervene, either to punish angrily, or to reward kindly. We should entertain the same insight about any given environment we know. It exists as a structure of meaningful distinctions. In so far as it is only knowable by the powers attributed to it and the practical action taken in its regard, and in so far as its powers are evoked as techniques of mutual control, a known environment is as fragile as any ancestor. While a limited social reality and a local physical environment are meshed together in a single experience, there is perfect credibility for both. But if the society falls apart, and separate voices claim to know about different environmental constraints, then do credibility problems arise.

At this point I should correct some possible false impressions. I have tried to avoid examples which lend themselves to a skull-duggery, conspiracy theory of how the environment is constructed in people's minds. There is no possibility of one group conning another about what nature will stand for, and what is against nature. I certainly would not imply that fears of world overpopulation are spread by frustrated car-owners in the commuter belt who would like a clear run from Surrey to the City. I would not imply that residents on the south coast press for legislation to control population because they know there is no hope of legislating specifically to ban summer crowds from the seaside. But personal experience could make them lend a friendlier ear to those demographers who are most gloomy about population increase, and ignore the optimistic ones. It obviously suits Lele men and Cheyenne men to teach that the physiology of childbirth and the physiology of the wild bison respond to moral situations. But no one can impose a moral view of physical nature on another person who does not share the same moral assumptions. If Lele wives did not believe in married fidelity as a part of the social order they might be more sceptical about an idea which suits husbands extremely well. Because all Cheyenne endorse the idea that murder in the tribe is disastrous they can believe that bison are sensitive to its smell. The common commitment to a set of social meanings makes inferences about the response of nature plausible.

Another misunderstanding concerns the distinction between true and false ideas about the environment. I repeat the invitation to approach this subject in a spirit of science fiction. The scientists find out true, objective things about physical nature. The human society invests these findings with social meaning and constructs a systematic time-tabled view of the way that human behaviour and physical nature interact. But I fear that it is an illusion if scientists hope one day to set out a true, systematic, objective view of that interaction. And so it is also illusory to hope for a society whose fears of pollution rest entirely on the scientists' teaching and carry no load of social and moral persuasion. We cannot hope to develop an idea of our environment which has pollution ideas only in the scientists' sense, and none which, in that strict sense, are false. Pollution ideas, however they arise, are the necessary support for a social system. How else can people induce each other to cooperate and behave if they cannot threaten with time, money, God and nature? These moral imperatives arise from social intercourse. They draw on a view of the environment to support a social order. As normative principles they have an adaptive function. Each society adapts itself to its habitat by precisely these means. Telling each other that there is no time, that we can't afford it, that God wouldn't like it and that it is against nature and our children will suffer, these are the means by which we adapt our society to our environment, and it to ourselves. In the process our physical possibilities are limited and extended in this way and that, so that there is a real ecological interaction. The concepts of time, money, God and nature do for human society the adaptive work which is done non-verbally (but otherwise probably by very similar means) in animal society.

I have spoken of pollution ideas which are used by people as controls on themselves and on each other. In this light they seem to be weapons or instruments. However, there is another fundamental aspect of the subject. Pollution ideas draw their power from our own intellectual constitution. Impossible here to describe the learning process by which each individual works out a set of expectations, and derives rules which guide him in his behaviour. The beliefs that there are rules and that future experience is expected to conform to them underlie social intercourse. There is some fascinating experimental work on this aspect of behaviour by American sociologists. However faulty our probings and conclusions may be, we assume a rule-obeying, stable environment. We expect, as we learn a fairly satisfactory set of rules, that the *et cetera* principle can be left to look after the known. The *et cetera* principle is like the automatic pilot which, once the controls are fixed, will keep the plane on course. *Etcetera, etcetera.* Findings rules that work is satisfying as it allows us to suppose that more of the machine can be turned over to the automatic pilot to look after. Discovering a whole system that works is exciting because it suggests an even greater saving of *ad hoc* painful blundering. Hence the emotional response to discovering that market prices work as a system, or that language or mythology has a systematic element.

The deepest emotional investment of all is in the assumption that there is a rule-obeying universe, and that its rules are objective, independent of social validation. Hence the most odious pollutions are those which threaten to attack a system at its intellectual base. The system itself rests on a number of unchallengeable classifications. One of the well-known examples in anthropology is that of the Eskimo girl in Labrador who would persistently eat caribou meat after winter had begun.[11] A trivial breach of an abstinence rule, it seems to us. But by a unanimous verdict, she was banished in midwinter for committing what was judged to be a capital offence. These Eskimo have constructed a society whose fundamental category is the distinction of the two seasons. People born

271

in winter are distinguished from those born in summer. Each of the two seasons has a special kind of domestic arrangement, a special seasonal economy, a separate legal practice, almost a distinct religion. The regularity of Eskimo life depends upon the observances connected with each season. Winter hunting equipment is kept apart from summer equipment, summer tents are hidden in winter. No one should touch the skins of animals classed as summer game in winter. As Marcel Mauss put it, 'even in the stomachs of the faithful' salmon, a summer fish, should have no contact with the sea animals of the winter.[12] By disregarding these distinctive categories the girl was committing a wrong against the social system in its fundamental form. A lack of seriousness about the categories of thought was not the reason given for why she was condemned to die by freezing. She had to die because she had committed a dangerous pollution which set everyone's livelihood at risk. Contrast this sharp categorization and this cruel punishment for contempt with the Eskimo experience of the last millennium. Here is a civilization which has seen precisely what may well be in store for us, a slow but steady worsening of the environment. Are we going to react, as doom draws near, with rigid applications of the principles out of which our intellectual system has been spun? It is horribly likely that, along with the Eskimo, we will concentrate on eliminating and controlling the polluters. It might be a more worthwhile recourse to think afresh about our environment in a way which was not possible for the Eskimo level of scientific advance. Nor is it only scientific advance which lies in our grasp. We have the chance of understanding our own behaviour.

If the study of pollution ideas teaches us anything it is that, taken too much at face value, such fears tend to mask other wrongs and dangers. For example, take the population problem. A straight response to pollution fears suggests we should urgently try to control population. The question is treated in an unimaginative and mechanical way. It is made a matter of spreading information and making available contraceptive devices. At first, people's eagerness to use them is taken for granted, but soon the clinics report apathy, carelessness, lack of determination, lack of agreement between husbands and wives. Then the question moves from voluntary to automatic methods such as sterilization or compulsion by law. We are almost back to 'Think before you litter' and forcible control of animal populations. But if we were to learn from biologists, Wynne-Edwards' work on animal populations has many a moral for the demographer.[18] In some human societies social factors encourage the voluntary control of family size. Not all tribes tolerate unlimited expansion of their numbers. It cannot be ignored that the world demographic problem is an Us/Them situation. Even in Ricardo's day it was They, the labourers, whose improvident fecundity created the pressure on resources, while We, the rich, could be relied on to procreate more cautiously. And today it is We, the rich nations, who wag our fingers at Them, the poor ones, with their astronomical annual increase. Somewhere a social problem about the distribution of prestige and power underlies the stark demographic facts. Let us not miss the lessons of this confrontation with biology and anthropology.

In essence, pollution ideas are adaptive and protective. They protect a social system from unpalatable knowledge. They protect a system of ideas from challenge. The ideas rest on classification. Ultimately any forms of knowledge depend on principles of classification. But these principles arise out of social experience, sustain a given social pattern and themselves are sustained by it. If this guideline and base is grossly disturbed, knowledge itself is at risk.

In a sense the obvious risk to the environment is a distraction. The ecologists are indeed looking into an abyss. But on the other side another abyss yawns as frighteningly. This is the terror of intellectual chaos and blind panic. Pollution is the black side of Plato's good lie[14] on which society must rest: it is the other half of the necessary confidence trick. We should be able to see that we can never ask for a future society in which we can only believe in real, scientifically proved pollution dangers. We *must* talk threateningly about time, money, God and nature if we hope to get anything done. We must believe in the limitations and boundaries of nature which our community projects.

Here we return to the comparison with the classical economists and the slavery abolitionists. It would be good to know which of our experts is likely to take a restrictive view of the environment, certain that time, money, God and nature are against change, and which of them is likely to favour expansionary policies. It is easy to see why the laymen can't lift their noses above the immediate horizon. The layman tends to assimilate the total planetary environmental problems to his own immediate ones. His horizon is his back yard. For the scientists, as well as this same tendency, there is another source of bias. To understand a system — any system — is a joy in itself. The more that is known about it the more the specialist is aware of its intricacies, and the more wary he is of the complex disturbances which can result from ignorance. The specialist thus has an emotional investment in his own system. As Professor Kuhn has said, in his *The Structure of Scientific Revolutions,*[15] scientists rarely change their views, they merely retire or die away. If there are to be solutions to a grave problem, they will come from the fringes of the profession, from the amateur even, or from those areas of knowledge in which two or three specialisms meet. This is comforting. In the long run, if there is a long run, unless the man in the street specially wants to choose the pessimistic restrictionist view on any ecological problem, he can wait and see. The scientific establishment has its own structure of stability and challenge. Our responsibility as laymen and as social scientists is to probe deeper into the sources of our own bias. Suppose we are really set for the worst terrors that the ecologists can predict. How shall we comport ourselves?

Our worst problem is the lack of moral consensus which gives credibility to warnings of danger. This partly explains why we fail so often to give proper heed to the ecologists. At the same time, for lack of a discriminating principle, we easily become over-whelmed by our pollution fears. Community endows its environ-ment with credibility. Without community, unclassified rubbish mounts up, poisons fill the air and water, food is contaminated, eyesores block the skyline. Flooding in through all our senses, pollution destroys our well-being. Witches and devils ensnare us. Any tribal culture selects this and that danger to fear and sets up demarcation lines to control it. It allows people to live contentedly with a hundred other dangers which ought to terrify them out of their wits. The discriminating principles come from social struc-ture. An unstructured society leaves us prey to every dread. As all the veils are successively ripped away, there is no right or wrong. Relativism is the order of the day. I myself have tried to join the work of taking down some of the veils. We have adopted first this standpoint, then that, seen tribal society from within and from without, seen ourselves as the scientists see us or from the stance of the anthropologist from Mars. This is the invitation to full self-consciousness that is offered in our time. We must accept it. But we should do so knowing that the price is William Burroughs' *Naked Lunch*.[16] The day when everyone can see exactly what it is on the end of everyone's fork, on that day there is no pollution and no purity and nothing edible or inedible, credible or incredible, be-cause the classifications of social life are gone. There is no more meaning. Neither melancholic madness nor mystic ecstasy, the two modes in which boundaries are dispensed, can accept the other invitation of our time. The other task is to recognize each environ-ment as a mask and support for a certain kind of society. It is the value of this social form which demands our scrutiny just as clearly as the purity of milk and air and water.

REFERENCES

1. John Hicks, *A Theory of Economic History,* London, 1969, pp. 122—40.
2. Paul Ehrlich, *Listener,* 30 August 1970, 215.
3. Mary Douglas, 'The Lele, Resistance to Change' in *Markets in Africa,* ed. Bohannan and Dalton, Evanston, Ill. 1962.
4. Julius A. Roth, *Timetables, Structuring the Passage of Time in Hospital Treatment and other Careers,* New York, 1963.
5. Eric Aynsley, 'How air pollution alters weather', *New Scientist,* 9 October 1969, pp. 66—7.
6. Mary Douglas, *The Lele of the Kasai,* International African Institute, Oxford, 1963, p. 51.
7. E.A. Hoebel, *The Cheyénnes, Indians of the Great Plains,* New York, 1960, p.51.
8. Mary Douglas, *Purity and Danger, An Analysis of Concepts of Pollution and Taboo,* London, 1966.
9. Mary Douglas, 'Elicited responses in Lele language', *Kango-Over-zee, 1950,* xvi, 4, 224—7.
10. Jean Buxton, 'Mandari Witchcraft', in *Witchcraft and Sorcery in East Africa,* ed. John Middleton and Edward Winter, London, 1963.
11. E.A. Hoebel, *The Law of Primitive Man: A Study in Comparative Legal Dynamics,* Harvard, 1954.
12. Marcel Mauss and M.H. Beuchat, 'Essai sur les variations saisonnières des sociétés Eskimos', *L'Année Sociologique,* 9 (1904—5), 39—132.
13. V.C. Wynne-Edwards, *Animal Dispersion in Relation to Social Behaviour,* Edinburgh and New York, 1962.
14. Plato, *The Republic,* paras 376—92, Lindsay translation, Everyman.
15. T.S. Kuhn, *The Structure of Scientific Revolutions,* Chicago, 1962.
16. William Burroughs, *The Naked Lunch,* New York, 1962; London, 1965.

15

Dorothy Nelkin

Controversy as a
political challenge

Critics of science and technology perceive a vast distance between technology and human needs—indeed between the governing and the governed. They question the ability of representative institutions to serve their interests. They resent the concentration of authority over technology in private bureaucracies and public agencies responsible for technological change. And they challenge assumptions about the importance of technical competence as the basis for legitimate decision-making authority.

SOURCES OF OPPOSITION

Those who live in the vicinity of an airport, a nuclear waste disposal facility, or who work in a vinyl chloride plant, have practical reasons to protest. They are directly impacted by land appropriation, noise, immediate risks, local economic or social disruption, or by some encroachment on their individual rights. Others, including many environmentalists, creationists, and laetrile supporters, protest out of adherence to a 'cause.' And there are people not directly affected by a specific project or controversy who oppose science and technology because of global political concerns. Some nuclear critics, for example, see science as an instrument of military or economic domination and oppose technology for ideological reasons.

These conflicts over science and technology draw support from sharply contrasting social groups. Most active are middle-class, educated people with sufficient economic security and political skill to participate in decision-making. But conservative fundamentalists in California seeking to influence a science curriculum share concern about local control with citizens in a liberal university community seeking to influence the siting of a power plant.

Finally, an important source of criticism has been the scientific community itself. Many young scientists became politicized during the 1960s. At that time, they focussed on antiwar activities, on university politics, and on the issue of military research in universities. More recently, their attention has turned to the environment, energy, biomedical research, and harmful industrial practices. These scientists often initiate controversies by raising questions about potential risks in areas obscured from public knowledge.

THE POLITICAL ROLE OF EXPERTS[1]

Technical expertise is a crucial political resource in conflicts over science and technology. For access to knowledge and the resulting ability to question the data used to legitimize decisions is an essential basis of power and influence. The cases [in *Controversy*] suggest the important role of the activist-scientist in formulating, legitimizing, and supporting the diverse concerns involved in the controversies. Scientists were the first to warn the public about the possible risks of recombinant DNA research. They were the first to call the public's attention to the problems of developing nuclear power before solving the problem of radiation disposal techniques. As various controversies develop, scientists are called upon to buttress political positions with the authority of their expertise. The willingness of scientists to expose technical uncertainties and to lend their expertise to citizen groups constitutes a formidable political challenge.

The authority of scientific expertise rests on assumptions about scientific rationality. Interpretations and predictions made by scientists are judged to be rational because they are based on data gathered through rational procedures. Their interpretations therefore serve as a basis for planning and as a means for defending the legitimacy of policy decisions.

The cases demonstrate how protest groups also exploit technical expertise to challenge policy decisions. Power plant opponents have their own scientists who legitimize their concerns about thermal pollution. Environmentalists hire their own experts to question the technical feasibility of questionable projects. Laetrile supporters have their own medical professionals. Even fundamentalists seeking to have the Biblical account of creation taught in the public schools present themselves as scientists and claim 'creation theory' to be a scientific alternative to evolution theory.

Whatever political values motivate controversy, the debates usually focus on technical questions. The siting controversies develop out of concern with the quality of life in a community, but the debates revolve around technical questions—the physical requirements for the facility, the accuracy of the predictions establishing its need, or the precise extent of environmental risk. Concerns about the freedom to select a cancer therapy devolve into technical arguments about the efficacy of treatment. Moral opponents of fetal research engage in scientific debate about the precise point at which life begins.

This is tactically effective, for in all disputes broad areas of uncertainty are open to conflicting scientific interpretation. Decisions are often made in a context of limited knowledge about potential social or environmental impacts, and there is seldom conclusive evidence to reach definitive resolution. Thus power hinges on the ability to manipulate knowledge, to challenge the evidence presented to support particular policies, and technical expertise becomes a resource exploited by all parties to justify their political and economic views.[2] In the process, political values and scientific facts become difficult to distinguish.

The debates among scientists documented in these cases show how, in controversial situations, the value premises of the disputants color their findings. The boundaries of the problems to be studied, the alternatives weighed, and the issues regarded as appropriate—all tend to determine which data are selected as important, which facts emerge. The way project proponents or citizen groups use the work of 'their' experts reflects their judgments about priorities or about acceptable levels of risk. Whenever such judgments conflict, this is reflected in the selective use of technical knowledge. Expertise is reduced to one more weapon in a political arsenal.

When expertise becomes available to both sides of a controversy, it further polarizes conflict by calling attention to areas of technical ambiguity and to the limited ability to predict and control risks. The very existence of conflicting technical interpretations generates political activity. And the fact that experts disagree, more than the substance of their disputes, fires controversy. After hearing 120 scientists argue over nuclear safety, for example, the California State Legislature concluded that the issues were not, in the end, resolvable by expertise. 'The questions involved require value judgments and the voter is no less equipped to make such judgments than the most brilliant Nobel Laureate.'[3] Thus the role of expertise in these disputes leads directly to demands for a greater public role in technical decision-making.

THE PARTICIPATORY IMPULSE

Most of the controversies described in this book were provoked by specific decisions: to expand an airport; to ban a drug; to site a power plant. Those who propose these projects define the decision and the issues involved primarily as technical—subject to objective criteria based on energy forecasts, studies of environmental impact, accident statistics, or predictions of future needs. Opposition groups, on the other hand, perceive such decisions in a political light. They use experts of their own, but mainly for tactical purposes—to prove that technical data are at best uncertain and subject to different interpretations. They try to show that important questions involve political choices and that these can be obscured by technical criteria. In the end, they seek a role in making social choices.

The cases suggest that the nature of public opposition depends on a number of circumstances. If a project in question (for example, an airport) directly affects a neighborhood, local activists are relatively easy to organize on the basis of immediate interest. Many issues (such as recombinant DNA research or the automobile airbag) have no such natural constituency. Risks may be diffuse. Affected interests may be hard to define, or so dispersed as to be difficult to organize. Or the significant affected interests may be more concerned with employment than with their environment and therefore willing to accept certain risks.[4] In such cases, participation is limited and the controversies involve mainly scientists or professional activists. Even in these conflicts, however, political protest can be maintained only if the leadership can count on support among a wider group of people who, though generally inactive, will lend support at public hearings, demonstrations, and other key events.

What channels do these groups exploit as they try to influence technology policy? First, they seek to capture technical resources. But those who oppose a project must also organize their activities to develop maximum public support. Expanding the scope of conflict is necessary to push decisions commonly defined as technical out into the political arena. Thus dramatic and highly publicized media events often are important. Tactics in the controversies described in this book range from routine political actions like lobbying or intervention in public hearings to litigation, referenda, and political demonstrations. These channels of participation depend on the institutional framework or political system in which opposition takes shape (note the tactics of environmental politics in the Soviet Union). In the United States litigation has become a major means for citizens to challenge technology. The role of the courts has expanded through the extension of the legal doctrine of standing—private citizens without alleged personal economic grievances may bring suit as advocates of the public interest.[5] The courts have been used by citizens not only in environmental cases, but in challenging research practices as well, as we see in the litigation over fetal research.

The participatory impulse has also forced elected representatives seeking to maintain their popular support to consider technical issues that are normally beyond their political jurisdiction. For example, the Cambridge City Council and local government bodies in a number of other university communities claim the authority to judge the adequacy of safety regulations in biology laboratories. Local townships forbid the transport and disposal of radioactive materials within their jurisdiction. And referenda on specific technologies such as nuclear power are increasingly common.

Such events evidence a changing relationship between science and the public. But it should not be assumed that demands for participation imply antiscience attitudes. More often they suggest a search for a more appropriate articulation between science and those affected by it. In its report on recombinant DNA research, the Cambridge Review Board thoughtfully expressed a prevailing view.

> Decisions regarding the appropriate course between the risks and benefits of potentially dangerous scientific inquiry must not be adjudicated within the inner circles of the scientific establishment. . . . We wish to express our sincere belief that a predominantly lay citizens' group can face a technical, scientific matter of general and deep public concern, educate itself appropriately to the task, and reach a fair decision.[6]

This view sees science and technology policy as no different from other policy areas, subject to political evaluation that includes intense public debate. This, indeed, is the political challenge posed by the current state of controversy over science and technology, and it has profound implications for their resolution.

REFERENCES

1. For the changing role of experts see A. Teich (ed.), *Science and Public Affairs* (Cambridge, MA: MIT Press, 1974); J. Primack and F. Von Hippel, *Advice and Dissent: Scientists in the Political Arena* (New York: Basic Books, 1974).
2. G. Benveniste, *The Politics of Expertise* (Berkeley, CA: Glendessary Press, 1972).
3. Report on Hearings before the California State Committee on Resources, Land Use, and Energy, *New York Times,* June 2, 1976, 1: 1.

4.   See R. Neuhaus, *In Defense of People* (New York: Macmillan, 1971).

5.   For a review and bibliography on citizen litigation, see Joseph Dimento, 'Citizen Environmental Litigation and Administrative Process,' *Duke Law Journal* 22 (1977), pp 409–452.

6.   Cambridge City Council, Experimentation Review Board, 'Guidelines for the Use of Recombinant DNA Molecule Technology,' December 21, 1976.

16

Barry Casper and Paul Wellstone

Science court on trial
in Minnesota

The 'science court,' proposed in 1976 by a White House Task Force as an objective, apolitical, value-free forum in which the authority of scientific experts could appropriately contribute to the resolution of public policy disputes, recently received its first serious test in Minnesota in a controversy over a high-voltage transmission line. Governor Rudy Perpich put the full weight of his office behind a concerted effort to bring together in a science court setting the electric power cooperatives that are building the line and the militant protesting farmers whose land it crosses. The failure of that effort reveals the profound political naiveté in the original science court idea and the serious practical problems in instituting such a tribunal.

Those involved—the governor, the utilities, and the farmers—have viewed the proposed forum from quite different perspectives, guided by quite different considerations of self-interest. The net effect of their negotiations has been to reveal a crippling defect of voluntary participation in the procedures for initiating the forum and to surface very different conceptions of what kind of science court would be useful.

We shall focus particular attention on the viewpoint of one of the parties to these negotiations—the farmers, in part, because we have come to know them and sympathize with their position. More important, we believe the modified version of the science court the farmers have proposed is a significant advance over the original conception; they have recognized those aspects of the concept that might truly alter the politics of technology.

Physicist Arthur Kantrowitz headed the White House Task Force on Anticipated Advances in Science and Technology (*Science,* August 20, 1976, pp. 653—56), which first developed the science court concept. The task force proposal had three stages. First, the court would identify the significant questions of science and technology associated with a controversial public policy issue; it would leave political, ethical, and other questions for subsequent consideration in the decision-making process. Second, a panel of impartial scientist-judges would preside over an adversary proceeding in which scientific experts would testify and scientist-advocates would cross-examine them. Third, the judges would issue their decision on the scientific facts pertaining to the disputed technical questions.

When Jimmy Carter took office, the Kantrowitz task force was dissolved, and the science court was never tried.

THE CONTROVERSY IN MINNESOTA

In 1973, two electric utilities—Cooperative Power Association and United Power Association (the 'co-ops')—announced their intention to build a high-voltage (±400 kilovolts), direct-current transmission line across western Minnesota. The line was part of a billion-dollar project to stripmine coal in North Dakota, build a generating plant there, and then transmit electricity to a station on the outskirts of Minneapolis-St. Paul. The powerline with its gigantic 150-foot-high towers would cross 172 miles of farmland.

Many of the farmers are deeply upset; they do not want the powerline crossing their land. Most of them inherited their land from their parents and feel it is their responsibility to improve it and hand it down to their children. This powerline and the ones that are sure to follow (the Minnesota Supreme Court recently laid down a 'nonproliferation' principle, which will result in routing as many lines as possible through the same corridor) are a threat to everything they value.

The farmers have been particularly enraged by the imperious and impersonal way in which their land has been appropriated through the combined forces of the government and the utilities. Farmland became the choice route after the Minnesota Highway Department refused to allow the line to follow interstate highways (it would be 'unsightly'); the state Department of Natural Resources similarly refused to allow the line to cross wildlife areas. The utilities, exercising the right of eminent domain, chose paths diagonally across many farmers' fields, thereby interfering with center pivot irrigation and adversely affecting other farming practices. When requested to shift the route to skirt field boundaries, they refused, saying it would be too expensive.

The farmers are especially upset by the way the eminent domain rights were acquired. Due to unusual circumstances (a new
Minnesota law covering transmission-line siting was passed just as
this project was started), the utilities had the option of choosing
to obtain these rights from local zoning boards or from the
Minnesota Environmental Quality Board (MEQB). Assuming they
would receive routine approval, they chose the local option. However, when aroused farmers convinced the commissioners in Pope
County (home of the most militant farmers) not to grant eminent
domain rights, the utilities reversed themselves and asked the
MEQB to assume authority. The state supreme court later ruled
that this maneuver was legal, but the farmers feel the rules were
changed just when they had won.

The Energy Agency and the MEQB held public hearings in 1975,
in principle giving the farmers an opportunity to influence decisions
about the line. From the farmers' perspective, however, the hearings were a sham.

Consider, for example, the key question of need. The co-ops,
both members of the huge Mid-Continental Area Power Pool,
claimed that they could not meet expected electricity demand
without the proposed line. The farmers simply did not know
enough about this intricate subject to challenge the utilities'
figures at the hearings. Even the Energy Agency lacked the staff
and resources to make an independent assessment of need.
Looking back over this process, Justice Lawrence R. Yetka, in
concurring with the Minnesota Supreme Court decision allowing
the co-ops to move ahead on the project, nonetheless observed:

> . . . proceedings on the question of need for the facilities were con-
> ducted, but they were hastily organized and possible alternative
> sources of power do not appear to have been seriously considered.
> It is apparent from the transcripts of the public hearings held through-
> out the state that the finding of need was a foregone conclusion. . .
> (*Finance and Commerce*, Minneapolis, September 30, 1977, p. 18).

Thus, public hearings, ostensibly a forum for 'full public participation,' proved inadequate to the task of examining and assessing
the merits of the co-ops' statements of need. The mismatch between the resources available to the farmers and those available to
the co-ops was clearly the principal reason.

Another set of issues that has figured prominently in this dispute relates to alleged problems of health and safety. Among these
are corona effects, including production of ozone and nitrous
oxides in the air surrounding the conductors, and field effects, including shocks resulting from the normal electric field of the line
and from induced currents when transients (sudden voltage shifts)

occur. There is also concern about possible adverse biological effects resulting from long-term exposure to low-level electric and magnetic fields.

However, health and safety issues have not been the dominant reasons behind the farmers' opposition to the powerline. A series of interviews we have conducted over the past year with farmers involved in the protest makes this clear. Nevertheless, many of the farmers express a degree of concern about health and safety problems. And considerable attention has been paid to this issue in the protest, in part, because the institutions available to the protesters, such as environmental impact statements, have channelled them in this direction; and in part, because uncertain threats to health and safety make good organizing issues for a protest movement.

THE GOVERNOR AND THE SCIENCE COURT

Soon after Rudy Perpich became governor on December 29, 1976, he disappeared mysteriously from his office for two days. When he returned on January 11, he disclosed that he had driven out to western Minnesota to meet with the protesting farmers. Thus, from the beginning of his administration, Perpich committed himself publicly to resolving the issue.

At Perpich's urging, the co-ops reluctantly agreed in February 1977 to forego work on the line, pending attempts to mediate the dispute. Josh Stulberg, a professional mediator, was brought in from the American Arbitration Association. In March, Perpich first broached the idea of using a science court procedure and on July 7 he announced that the Ford Foundation had provided a start-up grant of $5,440. By this time officials of the National Science Foundation had indicated informally that, once preliminary agreement on a format had been reached, they would favourably consider a request for a larger grant to fund the science court.

Meanwhile, a suit against the co-ops by a number of protest groups had been moving through the courts, which had issued an injunction in March against any surveying or construction activities. On September 30, the Minnesota State Supreme Court ruled in favour of the co-ops and work resumed on the line, beginning in the far west at the North Dakota border. As survey and construction crews began working their way across the state, they encountered increasing resistance and threats of violence from protesters.

Stulberg attempted to mediate the conflict, but in meeting with the farmers and the co-ops, he encountered disagreements over several fundamental issues relating to the science court. The co-ops refused to declare a moratorium on construction while the science court met; the farmers insisted on a moratorium, arguing that it made no sense to build the line while holding a forum on the question of whether it *should* be built. The co-ops insisted on limiting the focus of the court to health and safety issues; the farmers wanted the focus widened to include the need for the line and the possibility of following alternate routes.

Finally Perpich intervened personally. He selected a group of the more moderate protest leaders and invited them to meet with him on November 15 at the Governor's Mansion in St. Paul, where he appealed to them to participate in a science court without insisting on a moratorium on construction. The group agreed, and newspapers and television across the state proclaimed that the science court was 'on.' Given the public relations climate created by the governor, it was assumed the co-ops would have to go along. Much to everyone's surprise and the governor's chagrin, the co-ops turned him down. They had won in the state agencies and in the courts; why should they participate in this *ad hoc* proceeding?

At this point the science court became a political football. On December 15, a general meeting of powerline opponents issued a statement demanding that Governor Perpich set up a science court by January 1, 1978. Four days later, Perpich held a private meeting with co-op officials and then announced that they had conditionally agreed to a science court restricted to health and safety issues, provided that there be no moratorium on construction and that 'no additional restrictions will be placed on the operation of the line should the Science Court fail to complete its deliberations.' Although construction would continue, Perpich urged that the protest stop.

At a large gathering the following day in the Lowry Town Hall in Pope County the farmers vehemently rejected the co-ops' conditions. This meeting signaled a significant step-up in the protest. Survey crews had reached Pope County. During the day carloads of militant farmers cruised the wintry country roads harassing survey and construction crews; at night survey stakes were removed and towers being built were taken down.

Local police were unable to cope with an insurrection of this magnitude, so the governor called in the state troopers. One hundred and seventy-five troopers, one-third of the state's highway patrol, moved into Pope County to confront the demonstrators and many farmers were arrested. Local sentiment strongly supported the farmers. The Pope County attorney resigned, refusing to prosecute those arrested, and no other lawyer in the county would take his place. Statewide public opinion polls favoured the farmers.

Under pressure to 'do something,' the governor again turned to the science court. His 'state of the state' message, delivered on January 24, 1978, devoted considerable attention to the powerline dispute, appealing to the farmers and co-ops to approve the science court. After his address he met with the farmers and again urged them to participate. The next move was up to them. Their response was a novel proposal—a modified science court quite different from what the governor and the co-ops had in mind.

The farmers' proposal to the governor was made on March 19, 1978. At a large meeting in the Lowry Town Hall the night before (which one of us [BMC] moderated), the farmers agreed to relax their demand for a moratorium on construction and to participate in a science court—provided certain conditions were met: the topics under consideration by the court would be broadened beyond health and safety to include need, alternate routes, eminent domain, and the impact of the line on an agricultural environment. The farmers would be guaranteed adequate funding to develop their case and to bring in their own experts. And the hearings would be directed to the public.

Given the expansion of the topics to include nontechnical issues, having scientific experts act as judges no longer made sense. Still, the proceedings needed some closure. The farmers noted that since this was really a political issue, an accountable public official should make the judgment; they demanded that the governor not delegate his responsibility to 'experts' and proposed instead that he himself be the judge. In the farmers' version of the science court, the governor's staff would monitor the adversary proceeding and then the governor would issue a public report stating and justifying his conclusions about the issues before the court.

When presented with this alternative proposal at the March 19 meeting with the farmers, Perpich expressed sympathy with including the question of need and with guaranteeing adequate funding for the farmers to make their case. He was taken aback by the idea that he should be the judge, but agreed to think about it. Two weeks later, however, after conferring with his aides and the co-ops, Governor Perpich issued a formal reply. He announced his refusal to broaden the scope of the science court beyond health and safety or to be its judge.

Replying to the governor a few days later the farmers repeated their demands that the science court include the broad range of issues of concern to them. There the matter stands. The utilities continue to build, and efforts are underway to make the powerline an election issue. The final outcome of the dispute is unclear, but the prospects for any kind of science court appear very dim.

UTILITY OF THE SCIENCE COURT

The contrast between Governor Perpich's, the co-ops' and the farmers' conceptions of the kind of science court that would be 'useful' shows clearly how inextricably such a forum is linked to politics.

While the health and safety question lent scientific overtones to this controversy, Governor Perpich was confronted basically with a political dispute. A politician's instinct in such circumstances is to hand over responsibility to someone else. And in that respect, the science court idea is a politician's dream—it focuses public attention on peripheral technical issues and delegates the decision to the 'experts.'

The co-ops initially decided that they had nothing to gain from the science court—they had already won in the state agencies and in the courts. Later they apparently saw the science court as a means for diverting the energies of the farmers away from opposition in the fields; with no moratorium on construction, the line could be completed while the science court deliberations dragged on.

The farmers saw the science court in quite a different light. If restricted to health and safety questions, it would have been a fitting climax to the chain of statutory and procedural restrictions that has inexorably channelled them away from their basic concern about the powerline—the sacrifice of their land without their consent for an allegedly greater social need whose validity they question—into a forum that would consider only quite peripheral technical issues.

The farmers' only hope of winning is to resist this channelling toward a narrow technical forum where experts decide and to keep the issues in the political arena. The critical element of the science court idea, from the point of view of the farmers, is that they would be provided with *sufficient funds* to develop and make their case effectively and an *adversary forum* to face the utilities on an equal footing before the public.

The impasse in Minnesota reveals two defects in the original science court proposal. First, there is an unrealistic expectation of voluntary participation by all parties on the same terms in an *ad hoc* proceeding. Each party to this dispute made essentially political calculations as to what format, what timing, and what issues would best serve its own interests. It is not surprising that they arrived at different conclusions. For the science court to work, it must be institutionalized—under terms chosen to promote public understanding of the full range of issues that bear on the dispute and not the perceived self-interests of the parties involved.

Second, the originators of the science court idea had an overly narrow conception of what is wrong with our present processes for dealing with technology disputes and this led them to an overly narrow conception of the kinds of new processes needed. In this dispute the farmers were channelled away from the issues they really cared most about. Furthermore, although there was nominally full participation of all parties in public hearings about the line, the co-ops had adequate resources to develop and present their case, while the farmers did not. As a consequence these hearings did not produce the kind of public discussion and debate of the powerline that would enable the people of Minnesota to reach informed independent judgments on the key issues that underlie this dispute.

What might make a difference is something along the lines the farmers have proposed: an adversary forum to deal with the full range of relevant issues, with adequate funding for all sides to present their cases effectively and cross examine each other's witnesses.

This kind of proceeding would represent a significant advance over the formats currently employed in agency and legislative hearings and the unsystematic and uneven information dissemination processes that currently accompany statewide referenda. Having public adversary processes with adequate funding for all participants could be a significant innovative step towards rectifying the built-in imbalances of political power that follow directly from the imbalances in economic resources in our society.

[For a more detailed discussion of this case, see Casper, B.M. and Wellstone, P.D., *Powerline: The First Battle of America's Energy War*. Amherst: University of Massachusetts Press, 1981.]

17

Donald Schon

The fear of innovation

Companies want new technology, new ideas. We all know this. Then why do they fight so hard to prevent anything new from ever happening?

The modern industrial corporation is required to undertake technological change—change that is destructive to the corporation's stable state. The corporation is ambivalent to innovation: on the one hand, it believes itself to be committed to it—it believes that technological innovation is essential to corporate growth—but on the other hand it fears innovation and it tries, in various ways, to prevent it.

Innovation, according to the 'rational view,' is similar to other major functions of an organization—such as sales, accounting, or production. Therefore, if we accept the rational view, we say that innovation is a *manageable process* in which risks are controlled by mechanisms of justification and review.

I want to show here that the rational view of innovation ignores or violates actual experience. In light of that experience, the notion of innovation as an orderly, goal-directed, risk-reducing process is a myth. I shall explore some of the ways in which it is a myth, then go on to show that when an organization lives by the myth it can discover that it has made innovation impossible.

Implicit in the rational view is the notion that skilled men can anticipate and control the risks of innovation. Phrases like 'the management of innovation' suggest that we can foresee and quantify the likely dangers and rewards of a technical project and weigh them against the likely risks and rewards of alternative efforts. By selecting only those projects whose benefits justify their anticipated costs, by playing risks off against one another—in short, by a process of justification, decision, and optimization—we can (it is assumed) keep the risk of innovation within bounds.

Risk has its place in a calculus of probabilities. It lends itself to quantitative expression—as when we say that the chances of finding a defective part in a batch are two out of 100. In the framework of benefit-cost analysis, the risk of an innovation is how much we stand to lose if we fail, multiplied by the probability of failure.

Uncertainty is quite another matter: A situation is uncertain when it requires action but resists analysis of risks. For example, a gambler takes a risk in an honest game of blackjack when, knowing the odds, he calls for another card. But the same gambler, unsure of the odds, or unsure of the honesty of the game, is in a situation of uncertainty.

How does this apply to technical innovation? Men involved in technical innovation in a corporation confront a situation in which the need for action is clear but the action itself is not. This situation is painful and full of anxiety for the individuals and, in a sense, for the corporation as a whole. To be in a corporation in a time of corporate uncertainty is to be engulfed in waves of anxiety. So long as this situation exists, the corporation cannot function effectively, because it is not designed for uncertainty—a situation in which there are no clear objectives to reach, no measures of accomplishment, and no proper concept of control. A corporation cannot operate in uncertainty, but it is beautifully equipped to handle risk. It is precisely an organization designed to uncover, analyze, evaluate, and operate on risks. Accordingly, *the innovative work of a corporation consists in converting uncertainty to risk.*

The work begins with more information than can be handled and operates on this information, at lower levels of the corporation, until clear alternatives of action, together with their probable benefits and risks, can be defined. At this point, management can play the investment game—the game of deciding where to put its bets. The game requires analysis of investment alternatives, estimating their markets, costs of technical feasibility, and investment decisions. The game is played with competitive corporations as opponents. The rewards and punishments of the game can be measured in dollars.

In the process of innovation, everything is done to permit decision on the basis of probable dollar costs and dollar benefits. In the process, the corporation converts the language of invention to the language of investment. Instead of talking about materials, properties, performances, experiences, experiments and phenomena, the corporation talks of costs, shares of market, investment, cash flow, and dollar return.

The conversion of uncertainty to risk takes varying forms, depending on the *kind* of uncertainty to be dealt with. Technical innovation involves many different kinds of uncertainty. Some of these spring directly from the non-rational character of the process of invention; others are only indirectly related to that process. I will be concerned here with the uncertainties springing from determination of technical feasibility, novelty, and market.

One focus of uncertainty is in the question: Can it be done? Is it technically feasible?

In the process of invention, the requirements to be met change continually in response to unexpected findings. At any time, these may turn a good risk into a poor one—or an indifferent idea into an idea of great promise. We can describe these changes in terms of curves of technical difficulty. For example, as an investigator works over a period of time, he may first encounter his most difficult problem and later on only minor ones whose solutions he knows he can attain if he works long enough. If we plot his progress over time against 'difficulty' or 'demands of the solution with respect to the state of the art', the curve may look like Figure 1.

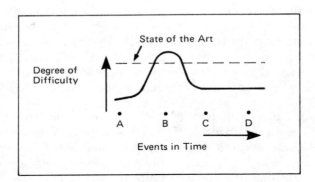

Figure 1

Here, the investigator must surpass the state of the art at B, whereas his subsequent tasks are well within it.

Or there may be a number of separate peaks. For example, in considering the development of new semiconductor materials, we would have to identify separate peaks for the development of high-purity materials and for the development of high-yield production techniques. The curve might look like Figure 2.

Such curves would not be a source of major uncertainty if it were possible always to identify ahead of time the problems and their degree of difficulty. Small improvements in the properties

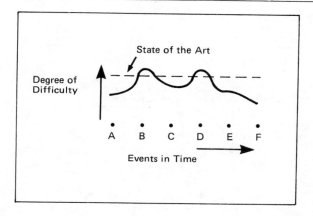

Figure 2

of materials—for example, raising by a few degrees the melting point of an alloy or increasing slightly the abrasion resistance of a plastic coating—may require no more than the continued application of known methods. On the other hand, achieving these improvements can turn out to be extremely difficult. A group working on the development of a new type of transformer, using piezoelectric principles, worked for weeks on what it thought was a materials problem, until it was able to make the calculations necessary to show that the performance requirements for which it was striving pressed to the limits both the energy capacity of the material and its ability to accept mechanical stress.

It is not always apparent, even to a skilled investigator, whether he is working on a minor problem of adjustment or a major problem of principle. He cannot place himself on the curve of technical difficulty. He may think he is here (Figure 3), when in fact he

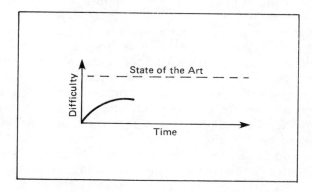

Figure 3

is here (Figure 4).

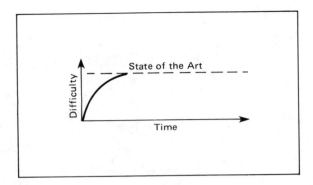

Figure 4

Technical feasibility, therefore, often resists the kind of definition required by the investment game. It may evade definition throughout the entire process of innovation.

Another focus of uncertainty concerns: Who has done it before? Who is doing it now? Although the patent literature offers a great deal of information about prior invention, a search may not uncover all relevant art. If a company undertakes a development, it can have no guarantee that its competitors are not doing similar things. Even if it knows that others are working in the same field, it cannot be sure how far they have gone. (The development of the automatic light meter for cameras is an example: The company that ultimately achieved commercial success had been spurred on by the belief that a competitor had the product, only to discover, later on, that its competitor did not have it after all.)

And questions of marketing: Who will buy it? How large will its market be? How much of a share will it get? How long will it last?

The marketing questions have become the subject matter of a new profession—a profession which aspires to be a science. Without attacking the question of whether or not a marketing science is now possible—whether there are inherent uncertainties about marketing which no amount of information (acquired ahead of time) will resolve—we can see that the process of answering marketing questions occurs over a period of time and that there are bound to be stretches of uncertainty which can be resolved only by the expenditure of time and money, if indeed they can be resolved at all. The uncertainties of marketing are particularly apparent in the case of consumer products. 'Scotch' tape, for example, was originally introduced as a way of mending books.

Before its introduction, its makers had no inkling of the huge market potential it offered.

Companies make marketing decisions under pressure of money and time, confronted with more information than they can handle. Although one may be able to predict, in principle, the effects of single-variable changes—prices, for example, or color—these variables never act alone. There are always many relevant variables, with multiple, interacting effects. There is no sure way of learning from past experience, for situations are never identical; differences that appear to be trivial may turn out to be critical.

Moreover, it costs money to find out. The introduction of a consumer product on a national scale is a major enterprise. A single regional market test, with appropriate preparation and follow-up, may cost hundreds of thousands of dollars—and yet the results may not permit formulation of the national picture. *In principle,* uncertainties may be resolvable, but the cost of resolution is high and the very process of resolving may cost more than can be justified.

To this we must add the fact that in the process of development there is interaction between need and technology: the market originally conceived for the product may be ruled out by technical limitations discovered in the process of invention. A different market may suggest itself—in this case, a more limited market. In short, the product for which a market must be anticipated is not an 'it' that remains constant throughout the process of development. Rather, the product changes through this process—and its possible market changes with it.

THE COST OF INNOVATION

Throughout the process of technical innovation, decisions about technical feasibility, novelty, and markets—among many other factors—must be made on insufficient evidence, by individuals who have more information than they can handle. The resolution of these uncertainties—the conversion of uncertainty to risk—takes time and money and requires justification in its own right. Its benefits must be balanced against its costs.

The process of innovation has a cost curve, as well as a curve of difficulty, which exhibits characteristic patterns. Let us consider the development of a new metal-to-metal adhesive. It begins in the laboratory when a chemist, working on another problem, notices an adhesive effect and reproduces it. Up to this point, the cost—in man-hours and materials—may be in the order of $5,000 to $10,000.

295

The chemist shows the effect to the research director, who finds it intriguing and authorizes a search of the report and patent literature, some further experimentation and a first examination of the markets. This may go on for a month or two and bring the total cost to $40,000.

At this point, the research director feels he has something worth presenting to management. Management likes the idea and launches a full-scale development project. The adhesive formulation has certain disadvantages, of course, and the development team makes efforts to improve it, as well as to explore variations of the chemistry involved, so as to 'cover all the ground.' A detailed analysis of the market—in the automotive, aircraft, and appliance industries—gets under way. It reveals the need for improved handling properties. At the same time, the company undertakes a thorough patent search which suggests that, although the formulation may be novel, the company must try certain other formulations as well, in order to cut off the possibility of later competition from an equally effective product. At the end of six months, the total cost of the effort has risen to $200,000.

Now the company files its patent applications and sets up a pilot production line. The line turns out to have a number of bugs in it; there are many more problems in producing large quantities of adhesive of reliable composition than in making small samples. Because of the scale of effort, modifications in the process become far more expensive than they had been before. Quality control becomes a problem. Shelf life has to be examined. Use tests, which had already been conducted on a small scale, are now undertaken on a large-scale, long-term basis. Meanwhile, word of the development has begun to spread outside the company and management decides to undertake a crash program in order to market the product before fall. A year and $600,000 have passed away.

By the time the first full production line has been set up, the marketing strategy has been settled, the sales force has been educated to the product, the use tests have been completed and evaluated, quality-control techniques have been established, two years have elapsed, and $1.2 million has been spent. Now the company can market the product—even though bugs in production, reliability, and quality still crop up. The product may or may not achieve the volume anticipated for it. The company may or may not make a profit.

The process described here takes no longer and is no more expensive than most product and process developments as they occur in medium-to-large firms. To be sure, there are instances of new products that have been invented, developed, and brought to the point of marketing for as little as $5,000. But at the other extreme are examples of products that took 10 years and $20 million. Regardless of scale, however, the shape of the development-cost curve remains relatively constant.

Often the first invention—the first demonstration of an effect— takes no more than a month or two and a few thousand dollars. As the company takes its first exploratory steps, the rise in the rate of expenditure is slow. As the company passes further checkpoints, the rate of expenditure increases. The total commitment grows exponentially. Each new major commitment requires complementary commitments. Only when the product or process reaches commercialization does the curve begin to level off (Figure 5).

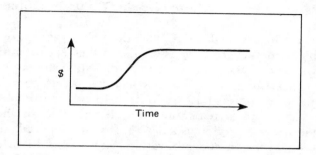

Figure 5

The development-cost curve takes on meaning when it is put in the context of the corporate investment game and the conversion of uncertainty to risk. Despite the careful efforts of many companies to establish checkpoints beyond which they will not go without adequate justification, they will find themselves having to make decisions on insufficient evidence in a general climate of uncertainty. As they begin to climb the slope of the curve, they rather quickly reach what appears to be a point of no return. At this point, the executive vice president can say to the president: 'We have put in so much. We may as well put in a little more and find out whether the investment is worthwhile.' To which the president can always reply: 'Don't throw away good money after bad.' But when this point is passed there comes another point: where the mistake—if the development is a mistake—is too big to admit. Large-scale developments, of the kind undertaken by supercorporations or the military, may proceed for months or years beyond the point where they should have stopped; they continue because of massive commitments to errors too frightening to reveal. They have their own momentum. In these cases, the personal commitment of the people involved in the development, the apparent logic of investment, and the fear of admitting failure, all combine to keep the project in motion until it falls of its own weight.

The problem of innovation within the corporation is, therefore, a problem of decision in the face of continuing uncertainty. A man must take leaps—not once, at the beginning of the process, but many times throughout the process—always in the face of uncertainty and on the basis of inadequate information. The need for such leaps of decision grows out of the uncertainty inherent in the process. A company cannot escape it by careful planning or by gathering exhaustive data. The uncertainties *resist* resolution—and the process of attempting to resolve them is itself a form of commitment.

Then why the rational view? Why the belief that innovation is manageable? One key to the answer lies in the very nonrationality and uncertainty of the process. A man may state the rational view when he is describing a process in which he has no part, or when he is trying to tell others how to do it, or to exhort others to do it, or, again, when he is reassuring himself about it. In short, the rational view may function as a device. It is an idealized, after-the-fact view of invention and innovation—as we would like them to be, so that they can be controlled, managed, justified. It is a view designed to calm fears, gain support, or give an illusion of wisdom.

The attraction of the rational view is easy to understand. Uncertainty is frightening. If the development of new products and processes is unavoidable—as growing numbers of corporations believe—it is cold comfort that most new products fail, that invention is full of unanticipated events, that innovation is inherently uncertain and subject to a treacherous cost curve. It is far more cheering, even apparently necessary, to believe that invention and innovation are rational, deliberate processes in which success is assured by intelligent effort. As a result, uncertainty becomes taboo, unmentionable—especially in the context of important corporate decisions. It is necessary to have a *clear, rational view* of where we have been and where we are going. It is necessary to believe that the future is essentially predictable and controllable, if only we gather the right facts and draw the right inferences from them. We suppress the surprising, uncertain, fuzzy, treacherous aspects of invention and innovation in the interest of this therapeutic view of them as clear, rational, and orderly. Armed with this myth, managers make decisions and mobilize resources. They and their subordinates must then live out the actual uncertainty of this process.

But there is another answer to the question: Why the rational view? This answer focuses on its partial truth and on its utility.

The rational view of invention and innovation is more nearly correct for more nearly marginal inventions. The less significant the invention, the more the process tends to be orderly and predictable. The more radical the invention, the less rational and predictable.

I represent the universe of new products and processes with a target-like diagram (Figure 6).

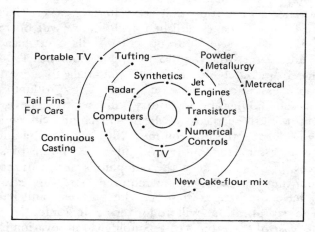

Figure 6

In the outer rings are those inventions whose acceptance requires least change. Indeed, these are hardly inventions at all. They require little or no change in the technology. They do not significantly advance the state of the art. They require no modification of scientific theory. Production equipment can be made to handle them with only minor change. Little or no rethinking of the corporate organization or approach is required.

As we move toward the center of the diagram, we encounter inventions that carry with them *major* changes in technology: transistors, synthetic fibers, etc. Their introduction goes hand in hand with change in scientific theory. They command new concepts in marketing and new marketing organization, as well as radically different production equipment and major new investment. They force their corporate organizations to undergo major change.

Let us take up the question of degree of novelty—how does this affect the organization within which innovation occurs? We can state a general rule here: The more peripheral the development, the less change required for acceptance, the more the development will tend to conform to the rational view. But the question of degree of novelty is more complex than this simple rule would suggest. For example, a technically insignificant invention—such as one that provides individually packaged food products—may require major changes in marketing. On the other hand a major technological invention, as was true of nylon, may require surprisingly little change in textile machinery. There is an unusual optical effect: what looks like trivial change from afar may close-up be monumental.

For example, consider the replacement of a natural material by a synthetic—a disruption that may threaten several levels of a company at once. The established technology will be rendered obsolete. Men who have built up craft skills over decades will suddenly find them irrelevant to the new problems. Production will change from batch processing of materials to a continuous chemical process. New means of controlling product quality, production scheduling, and inventory will have to be devised. The far greater productive capacity of the new machines will require new marketing concepts: the old volumes will no longer be adequate. It may be necessary to double or triple the volume of sales, at lower prices, in order to keep the company's net profit constant. New accounting methods will be required. The very strategy of sales will have to change. In short, the business—and hence the nature of the corporation—will no longer be the same.

Not the least of the effects of technological innovation is on top management itself. For top management, the cumulative effects of all these changes may be overwhelming. If the president came up through the business and draws his confidence from his intimate knowledge of the details of the present operation, technological innovation threatens to throw him onto completely unfamiliar ground. He understood the old business, but he does not understand the new one. How can he manage if he does not understand the business he is in? Where is he to draw the resources of experience he needs in order to trust his own judgements? Is he to become completely dependent on the proponents of the new product or process? He is faced with a crisis of both self-confidence and trust in others.

In brief, technological innovation disrupts the stable state of corporate society. On every level, it affronts the society's vigorous and continuing efforts to stay as it is. To this onslaught, the corporate society responds in a variety of ways. Some of the more evident are these:

- □ *It rejects the effort at innovation. It puts down the idea, fires or demotes the men associated with it, and makes its further discussion taboo. Many corporate pasts are littered with discarded efforts of this sort—begun, then stopped, then relegated to corporate limbo.*
- □ *It allows the effort to continue, but isolates it from the rest of the corporation. In this way, research projects—or whole research departments—may function in a vacuum, cut off from the contacts with the corporation, contacts that are essential to their ever coming to reality.*
- □ *It contains the threat, allowing it to proceed, but on a level so much reduced that the innovation is always far short of critical mass.*
- □ *It seeks to convert the threat to an activity acceptable within the corporate society. Efforts at radical innovation become product improvements or service to production or sales, which can be carried on without disrupting effect.*

These are straightforward negative responses. But the corporation cannot respond to innvoation in an exclusively negative way. It cannot because it believes itself to be committed to technological innovation as essential to corporate growth. Innovation is something the corporate society must both espouse and resist.

- □ *It may compartmentalize innovation, permitting it to occur in one part of its business while preventing it from occurring in others.*
- □ *It may oscillate between support and resistance, confusing corporate members by an on-again, off-again approach to change.*
- □ *It may resist innovation while not appearing to do so, while in fact proclaiming the official doctrine of innovation. For example, it may encourage the development of new product ideas only to find consistently that none of them meets the stringent criteria laid down in advance (just as a mother may want her son to get married but never approves of the girls he brings home); or it may foster a research effort, only to reject or ignore its results.*

These strategies do not usually represent a conscious pose or a management deception. They are responses of a corporate society to two requirements which are equally legitimate—even necessary—but unfortunately in conflict.

The society of a corporation attempts to maintain a stable state. This effort is not inertia, but conservative dynamism. The various forms of corporate resistance to change which reflect themselves as obstacles to technological innovation are processes of *conservation*—processes which are essential to the survival of any social or biological organism. Active conservatism is the natural state of organisms. It is nonsense to say that companies should throw off this old-fashioned habit. But it is the crisis of the modern industrial corporation that it is required to undertake technological change—change that is destructive of its stable state—in order to survive. This is the paradox which accounts for its ambivalence to innovation, and yet forces it to adopt new forms and styles of change.

18

Brendan Gillespie, Dave Eva and Ron Johnston

Carcinogenic risk assessment in the USA and UK: The case of Aldrin/Dieldrin

The nature and scale of the hazards associated with new tech-nologies have required governments to make difficult decisions about the form of control they should exercise over new products and processes. Although governments have received advice from a large and growing body of scientists, this has not always simplified the decision-making process nor rendered it more rational and ob-jective. In this paper, we try to explain why two governments reached contradictory conclusions to the same problem on the basis of the same scientific evidence. More specifically, we will try to explain why two chemical pesticides, Aldrin and Dieldrin (A/D), were judged to be carcinogenic in the US but not in Britain when the same data were available to public authorities in both coun-tries. The issue here is not one of differing assessments of risk or of balancing risks and benefits. Rather, what requires explanation is the different regulatory statuses assigned to the same products by the US and British governments.

The observation that scientific experts disagree and oppose each others' views is hardly new—witness the appearance of a sizeable literature on the subject.[1] However, much of the literature incor-porates weaknesses which inhibit our understanding of the role played by science in risk determination and, more generally, in controversies among experts in matters of public policy. Firstly, much of it focuses on one country, usually the US, whilst assum-ing that the roles of scientific advisers, and the problems which their use entails for governments, are everywhere the same. The comparative approach adopted in this paper should offset this, as well as a second, deficiency:[2] namely, the roles of science and scientists are generally analyzed in terms of universalistic categories which obscure the effects of different political and cultural settings on the type of science providing the basis for policy decisions. One frequently used variant holds that science provides the facts, and their evaluation from divergent value and ideological perspectives results in contrary interpretations. Conflicts among scientists are then explained in terms of the political views, bias or irrationality of one or more of the disputants, and little attention is paid to the content of the conflicting positions.

This approach has merit to the extent that it accepts that scientists become integrally involved in political conflicts. Nevertheless, two weaknesses severely limit its explanatory power, and call into question the fact-value distinction on which it is based. Firstly, it does not provide an accurate description of actual controversies. Nowotny, for instance, has argued that the traditional division of labour between science and society is, in practice, breaking down in complex technical spheres of decision-making like risk assessment.[3] Secondly, the essentially positivist view of science, on which the distinction is based, is overly restrictive in limiting explanations of scientific controversies to factors 'extrinsic' to science: values, bias, and the like. When science is examined as a form of organized, intellectual production, a much more complex relationship between scientific concepts, theories and methodologies, on the one hand, and ideological and value commitments, on the other, emerges, which also allows explanations of controversies in terms of factors 'intrinsic' to scientific development.[4]

Through a cross-national analysis of the contradictory A/D decision, we hope to set the pertinent features of the science-public policy relationship into sharper focus than before. We begin our analysis by looking at the differing interpretation of the evidence about A/D provided for the British and US authorities. Significant differences are found to exist in the criteria used to infer carcinogenicity, in the type of scientists providing the authoritative interpretations, and in the choices with which the respective public authorities were faced. This examination of the provision of scientific advice is complemented by an analysis of the factors affecting its reception. We identify those institutional, legal, cultural and economic features of the decision-making contexts which made them more or less receptive to the advice they were proffered.

BRITISH AND US DECISIONS[5]

Aldrin and Dieldrin are two organochlorine pesticides that were widely used in agriculture in the 1960s. Workers applying A/D and consumers eating residues in and on treated crops were exposed to

304

short- and long-term toxic hazards. (Aldrin rapidly degrades to Dieldrin in plants and animals, so when we discuss the hazards of A/D we are really talking about the hazards associated with Dieldrin.)

The carcinogenic risk of A/D had been reviewed in a variety of national and international settings since the mid 1950s.[6] While several expressed uncertainty and requested more evidence, none of the reviews had been prepared to declare A/D carcinogenic. None, that is, until September 1974, when the US Environmental Protection Agency (EPA) found that A/D posed an unacceptable carcinogenic hazard.[7] Not surprisingly, the decision was viewed as controversial in the US, and was criticized in other countries.

EPA had initiated administrative hearings in August 1973 to determine whether A/D's registrations should be *cancelled*—that is, effectively banned.[8] These hearings had been in progress for a year when EPA held more hearings to determine whether A/D's registrations should be *suspended*—that is, their sale temporarily banned. Congress had provided EPA with these statutory alternatives to ensure that the agency could eliminate 'unreasonable' risks that might arise in the course of lengthy cancellation hearings. The two decisions are administratively distinct, and a decision to suspend does not automatically prejudice the cancellation decision.

It was during the suspension hearing that the question of A/D's carcinogenicity came to the fore and, in fact, dominated the proceedings. An environmental group, the Environmental Defense Fund (EDF), had petitioned EPA for A/D's cancellation and suspension. Their case was supported, and effectively prosecuted, by EPA's Office of General Counsel (OGC). Shell, A/D's manufacturer, opposed these groups and was supported by the Department of Agriculture (USDA). The case was argued before Administrative Law Judge Perlman who was paid by EPA but independent of the agency for career advancement and tenure.[9] He recommended that A/D be suspended because of the carcinogenic risk the chemicals posed, and this was endorsed by EPA's Administrator, Russell Train. When the decision was upheld in the Appeals Court, Shell withdrew from the cancellation hearings.[10]

The following month, in October 1974, a group of British experts reviewed EPA's decision. They concluded 'that there was no reason to recommend any change in the UK action on Aldrin and Dieldrin as decided at the time of the 1969 reviews of organochlorine pesticides'.[11] No new evidence had been produced in the interim and the British experts had frequently discussed the evidence on A/D with Shell toxicologists over the years.[12]

The British experts were members of a Panel of the Advisory

Committee on Pesticides and Other Toxic Chemicals. The Committee is an expert body that advises UK government departments on the hazards associated with pesticides. The departments then negotiate what action to take with manufacturers of the product in question. The negotiations are conducted within the framework of a voluntary (that is, non-statutory) agreement between government and industry — the Pesticides Safety Precautions Scheme (PSPS).[13]

It is quite clear, then, that the experts and decision-makers in PSPS and EPA reviewed the same experimental evidence of A/D's possible carcinogenicity and came to contradictory conclusions. Furthermore, although each group of experts and decision-makers was aware of A/D's status in the other decision-making forum, this did not change their conclusion. We therefore have a genuine paradox to explain.

LIMITATIONS OF TOXICOLOGY

The first possible explanation of the different decisions is the uncertainty in determining carcinogenic risk. Toxicology, the field most directly concerned with the harmful effects of chemicals, has developed a fragmentary and incomplete understanding of the cancer-causing process.[14] It is conceivable, then, that a number of competing and equally plausible interpretations of the same evidence could co-exist. The toxicologists' uncertain knowledge of cancer might make it difficult to choose between rival interpretations, or to completely rule one out.

Although this argument could be applied to all chemicals tested for carcinogenicity, it seems particularly relevant in this case. Firstly, other chemicals have been shown to be more definitely carcinogenic than A/D.[15] Secondly, even those scientists who thought A/D to be carcinogenic regarded them as considerably less potent than other known carcinogens.[16] Thirdly, Shell had spent $10 million producing evidence of the hazards associated with A/D, so there was a much better data base for these than for most other chemicals.[17]

Briefly, the evidence was of three types:

1. *Epidemiology* — that is, correlations between incidence of human cancer and exposure to particular substances. Although ingestion of A/D-contaminated foodstuffs had resulted in British and US consumers storing Dieldrin in their body fat, there were no control populations which could be used to evaluate the effects of these exposures. The best available epidemiological data derived

from the medical records of workers employed in the production of A/D.[18] About 1,000 workers had been occupationally exposed, but only 69 of this pre-selected group had been exposed for 10 years or more. Since the latency period for cancer can be 20, 30 or even 40 years, it was difficult to know what significance to attribute to the two cases of cancer that had occurred in this group by 1974. Finally, Shell argued that indirect epidemiological evidence was available. Their scientists believed that humans metabolized A/D analogously to the drug phenobarbitone. They argued, therefore, that epidemiological data on phenobarbitone were relevant to the determination of A/D's carcinogenic hazard.

2. *Animal models.* These were experiments in which specific doses of A/D were administered to populations of laboratory animals, and the incidence of cancer tumours in exposed populations were compared with those in controls. The significance of results from these experiments for humans is highly controversial. By 1969, experiments with A/D in mice, rats, dogs and monkeys had been performed.

3. *Biochemical tests.* In recent years, a variety of so-called 'quick tests' have been developed to predict carcinogenic hazards more cheaply and quickly than animal tests.[19] Shell scientists developed several *in vitro* tests which they used on A/D. Essentially, they tested for certain biochemical properties of chemicals that Shell claimed were correlated with carcinogenic activity.

Clearly, there are general problems in determining carcinogenicity, and these were exacerbated by the particular biological properties of A/D. The uncertainty gave rise to the possibility of different interpretations and hence different decisions. Even so, uncertainty cannot explain why the British and US decisions went one way rather than the other; that is, it cannot explain why A/D were judged to be carcinogenic in the US but not in Britain, and vice versa. Since the uncertainty of the evidence was common to both EPA and PSPS, it cannot be a sufficient explanation of their difference.

EVALUATION OF THE EVIDENCE

The different interpretations of the A/D evidence could have arisen in at least two ways, depending on the degree of consensus that existed over the standards for inferring carcinogenic risk. If there were consensus over the standards, the explanation of the differing interpretations would be different than if there were competing standards of carcinogenicity. Thus, the way in which the two decisions were reached is important to our overall argument; clarification of just how the decisions differed must logically precede any account of why they differed.

We will outline the arguments that Shell and EPA presented to Judge Perlman in the first part of this section, and the evaluation of the A/D data by British experts in the second.

(A) US Evaluation

At the suspension hearing, the Shell lawyers argued that a scientific approach requires *all* the evidence to be examined and that the results should be reproducible.[20] In a similar vein, the Director of Shell's Tunstall Laboratories argued that carcinogenicity could be inferred only when five criteria were met:

1. The exposed animals experience a higher incidence of tumors.
2. Tumors develop in more than one species.
3. The development of these tumors can be proven to be compound-related.
4. The animals have proven to be an adequate model for extrapolating to man.
5. Human data are available proving at least one incidence of cancer.[21]

Shell conceded the first criterion and accepted that five experiments—Davis and Fitzhugh (1962), Davis (unpublished), McDonald et al. (unpublished), Walker et al. (1973) and Thorpe and Walker (1973)—demonstrated that Dieldrin had increased the incidence of liver tumours in mice.[22] Shell then argued that this finding could not be extrapolated to humans because A/D did not meet the other four criteria.

Evidence was produced by several Shell witnesses that the mouse liver was not a valid predictor of human cancer (Criterion 4). They argued that Dieldrin had not induced tumours at sites other than the liver, and that the induction of hepatomas (liver tumours) in mice was highly dependent on a variety of genetic and environmental factors. More critically, Shell witnesses argued that the response of the mouse liver was unique and quite different from other species. Thus, they argued that evidence from rat, monkey and dog studies failed to indicate a carcinogenic response (Criterion 2). Next, they advanced a five-stage process whereby liver tumours were formed, and argued that whilst all five stages occurred in mice, they did not occur in humans.

This latter argument was quite subtle, and drew on evidence from a variety of sources. The indirect epidemiological evidence, the results from the *in vitro* biochemical experiments and the medical histories of occupationally-exposed workers, were all marshalled to support an argument for a mechanism which, Shell argued, showed how A/D did *not* induce human cancer (Criterion 3). Finally, Shell argued that the direct epidemiological evidence indicated that A/D did not meet their fifth criterion: the demonstration of one A/D-induced human cancer.

The EPA case against A/D was quite different in form. Their lawyers argued that a number of principles had been well established in the scientific community for assessing carcinogenic hazards, and that on this basis A/D must be considered carcinogens.[23] Nine cancer principles were presented in the A/D case. The most relevant ones were as follows:

1. A carcinogen is any agent which increases tumor induction in man or animals.
2. Well-established criteria exist for distinguishing between benign and malignant tumors; however, even the induction of benign tumors is sufficient to characterize a chemical as a carcinogen. . .
7. The concept of a 'threshold' exposure level for a carcinogenic agent has no practical significant because there is no valid method for establishing such a level.
8. A carcinogenic agent may be identified through analysis of tumor induction results with laboratory animals exposed to the agent, or on a *post hoc* basis by properly conducted epidemiological studies.
9. Any substance which produces tumors in animals must be considered a carcinogenic hazard to man if the results were achieved according to the established parameters of a valid carcinogenesis test.[24]

We may demonstrate that EPA was using different standards than Shell by considering the agency's response to Shell's five criteria. EPA's witnesses had re-examined the pathological and statistical analysis of the animal data, and it was this which helped elicit Shell's concession of their first criterion. With this established, EPA strongly objected to Shell's 'two species' criterion (Criterion 2). They argued that the negative evidence from the rat, dog, and monkey studies was the result of poorly designed experiments. Further, the uncertainties surrounding the use of animal models and the dictates of 'prudent policy,' meant that, for them, positive evidence should supercede negative results.[25] In this case, however, evidence on the mouse derived from five different strains, and was therefore sufficient, in their opinion, to indict A/D as carcinogens.

Although EPA could hardly disagree with Shell that the induction of tumours in test animals be 'compound-related' (Criterion 3), there was scope for disagreement on the meaning of this term. EPA witnesses insisted that they were not required to produce causal mechanisms, as Shell had demanded. Indeed, Epstein argued that such a requirement 'would define away the entire field of chemical carcinogenesis.'[26] EPA's ninth cancer principle declared that a positive result from a 'valid carcinogenesis test' was sufficient to consider that chemical carcinogenic. Since A/D fulfilled this condition, EPA concluded that they were carcinogens.

There were also objections to the mechanism Shell had presented. Farber argued that

> it is evident that many chemicals require metabolic conversion to active derivatives before they can initiate the development of cancer. However, the specifics of the metabolic processes which result in cancer in various test animals are not clear, to say nothing of the metabolic processes in man. No one as yet can draw any valid correlation between a particular pattern of metabolism and the induction of cancer in any species, and any judgments concerning carcinogenicity or lack thereof based on metabolic patterns have no scientific basis at this time.[27]

Similarly, the EPA witnesses rejected the analogy on which Shell's indirect epidemiological evidence was based and the validity of the biochemical tests as indicators of carcinogenicity.

If EPA were to demonstrate A/D's carcinogenicity, the agency had to accept Shell's fourth criterion, and justify the mouse as a valid indicator of human cancer. Shell's attack on the mouse had important regulatory implications.

> Health agencies in many countries recommend mice for routine testing of pesticides, food additives and drugs. If the argument that 'mice are of no value for carcinogenicity testing' were acceptable, the existing regulatory system would have been pitched into chaos.[28]

By the same token, the regulatory status of other pesticides would have been threatened if A/D were suspended. Aware of these implications, the EPA witnesses strongly objected to doubts about the reliability of the mouse liver, and argued that well-designed experiments controlled any genetic or environmental susceptibilities. They pointed out that the mouse was a standard test animal in national and international regulatory systems, and that it was used in Shell's own studies — Walker et al. (1973) and Thorpe and Walker (1973). Finally, evidence was produced suggesting that chemicals 'adequately' tested in mice had been shown to be carcinogenic when 'adequately' tested in other species.

EPA also objected to Shell's fifth criterion, requiring the demonstration of at least one A/D-induced human cancer. They argued that since animal tests were sufficient to predict carcinogenic risk, it was ethically unjustifiable to wait for the demonstration of human harm. Moreover, since the direct epidemiological data for A/D failed to meet methodological standards, it was not 'prudent' to reach a negative conclusion, which might subsequently be reversed by the accumulation of more evidence.

On all of the critical points, EPA convinced both Perlman and Train that theirs was the superior argument. Perlman had few doubts about his decision and justified his findings that A/D were unacceptable carcinogenic hazards as follows:

> We believe that this conclusion represents established traditional and 'conventional wisdom'. The Shell Chemical Company has strenuously and with sophistication attempted to demonstrate that 'this truth' does not apply to Aldrin and Dieldrin for the reasons we have detailed above. We do not believe that traditional wisdom or science has been overcome thereby. Shell's presentation with respect to the shortcomings of the mouse as an appropriate test animal and its lack of significance for man is based, in part, on matters far from established in the scientific community, speculation and surmise. In reality, our knowledge with respect to cancer is very limited. Many, many years would be required to pursue the theories, hypotheses and correlations advanced by witnesses for Shell without any confidence that they could be proven.[29]

(B) British Evaluation

For reasons that will become clearer below, there is no public record of the deliberations preceding the British decision. Nevertheless, we may infer the standards employed by British advisers from several sources, and we will do so in chronological order. The first source of evidence derives from several Editorials in the principal British medical journals: *The Lancet* and the *British Medical Journal*.[30] It seems unlikely that they would have been written without at least consulting the PSPS advisers, and their 'anti-EPA tone' would tend to confirm this. The Editorials were dubious about a lay person/judge's ability to evaluate such complex issues and of the validity of using 'cancer principles' to make the decision. In dealing with the evidence, they showed a greater willingness to accept the available epidemiological evidence than the EPA witnesses. Moreover, they shared Shell's scepticism that human carcinogenic risk could be inferred from mouse data.

These themes are echoed in the views of several British committees that met in the 1960s. In 1969, PSPS's Advisory Committee reviewed the evidence and placed great stress on the negative epidemiological evidence derived from observations on Shell's workers.[31] In contrast, the Committee considered that evidence from animal experiments could only provide 'presumptive evidence.' They reported that:

> In consultation with a number of cancer experts and pathologists we were

311

unable to obtain common agreement as to whether these lesions represented malignant tumours which would have indicated that Dieldrin had had an undoubted carcinogenic effect on these mice. Certain additional experiments were then suggested which might help yield results which would help to decide whether these liver lesions were capable of autonomous growth and had the accepted characteristics of malignant tumours.[32]

In 1967, a different Committee reviewed the evidence, but merely requested more data to clarify the issue.[33] PSPS's Advisory Committee had also reviewed the available evidence in 1964, and concluded in much the same way as the 1969 Committee.[34] They could obtain no consensus on whether the tumours were malignant or benign, and they hinted that their induction might be unique to the mouse.

The only other source that illuminates the British position is a report issued by the Panel on Carcinogenic Hazards in 1960.[35] The report outlines several principles to guide the interpretation of animal tests, and is significant because it is still recommended in the PSPS agreement.[36]

The Panel defined a carcinogen as a substance that increased the incidence of *malignant* tumours in animals or humans. The Panel was also prepared to differentiate carcinogens from co-carcinogens and initiators that were not carcinogenic in themselves, but which enhanced or augmented the effect of a carcinogen.[37] On the assumption that a carcinogenic response was specific to some test animals, the Panel recommended that two species be tested. The report does not make clear whether positive findings are required in both species, but it does suggest that: 'Negative results will reinforce confidence in the safety of the material; the significance of positive findings needs careful consideration in each case.'[38]

Putting these pieces of evidence together, we may conclude that British cancer experts were reluctant to label Dieldrin carcinogenic because of:

1. the negative epidemiological evidence available;
2. the lack of consensus as to whether Dieldrin had induced *malignant* tumours;
3. concern about the mouse liver as a valid indicator of carcinogenicity; and
4. the failure of Dieldrin to induce tumours in species other than the mouse.

Our review of the interpretation of the A/D evidence indicates that EPA and Shell witnesses employed different standards for the determination of carcinogenic risk. Similarly, EPA and PSPS experts required the evidence to satisfy quite different conditions

312

before labelling A/D 'carcinogenic', and this was the most immediate reason for the different British and US decisions. Moreover, insofar as we can compare the British evaluation with that of Shell, there are some striking similarities. Indeed, the only noticeable difference is that the British authorities required evidence of malignant tumours in test animals, whereas Shell's argument made no distinction between malignant and benign tumours. It seems reasonable to take as a working hypothesis, then, that the British authorities used the same, or very similar, standards as Shell to evaluate the A/D evidence.

ADVISERS AND THEIR ADVICE

Since different standards were applied to the same evidence, we must turn our attention to the scientists who provided EPA and PSPS with the authoritative interpretation of the evidence, and the nature of their advice. How can we account for the existence of such disparate standards, and what features of those standards, and their adherents, made them more or less compelling to decision-makers in PSPS and EPA?

In an earlier paper, Johnston and Robbins developed a model relating the type of occupational control exercised over scientists to the type of knowledge they produce, the specialties they work within, and the relations between those specialties.[39] These authors had previously analyzed one environmental controversy, and found that the disputants not only evaluated the same data differently, and derived contrary policy implications from the same, or similar, evidence, but that also, motivated by different social and scientific commitments, they were predisposed to produce different 'facts'. We may consider whether similar forces are operating in this case, and if they had any bearing on the British and US decisions, by examining the work of J. M. Barnes and Umberto Saffiotti, the principal advisers (respectively) to PSPS and the EPA lawyers. Both scientists occupied important organizational positions — Barnes as head of the British Medical Research Council's Toxicology Unit, and Saffiotti as head of the National Cancer Institute (NCI)'s Chemical Carcinogenesis Program. Their organizational roles meant not only that they could determine the type of work supported by these key organizations, but also that they could provide authoritative advice to those seeking it. Thus Barnes was a member of every British Committee that has reviewed the toxic hazards of pesticides since the 1950s.[40] Similarly, Saffiotti had served on many US governmental committees, and was centrally involved in the development of EPA's cancer principles. Perlman described him as

> . . . a world renowned expert whose initial testimony was cleared and approved by [the NCI] and whose demeanour and knowledge during his several days of cross-examination especially impressed us.[41]

Whereas the British Toxicology Unit has maintained close links with the chemical industry,[42] the NCI has been the leading agency in the US crusade against cancer.[44] Not surprisingly, then, the social commitments expressed by these institutional leaders in their approach to carcinogenic hazards are quite different:

> *Saffiotti:* That we take a position of caution and prudence in the matter of exposing the entire population to the potential hazards of chemical carcinogens is dictated by the tragic knowledge that — with the present trends in cancer mortality in the United States — out of 200 million Americans now living, 50 million will develop cancer and 34 million will die of it. Yet most cancers appear to be caused by environmental factors and therefore could be preventable.[44]

> *Barnes:* The safety of man from hazards presented by pesticide residues will not necessarily be increased by carying 'wolf' on every conceivable occasion that some direct or indirect carcinogenic activity can be detected in a substance filling a valuable role as a pesticide. Without pesticides many people would die for other reasons long before they reached the age at which they might develop cancer. Cancer was widespread long before modern pesticides were synthesized. If chemical carcinogens are responsible for any significant fraction of human cancer of unknown origin it is probable that such carcinogens will be of natural origin. With aflatoxin, cyasin, and the pyrroloziding alkaloids before us as examples of carcinogens found widespread in nature, it would be unwise if not irrational, to try to create undue alarm about carcinogenic hazards from pesticides that display no carcinogenic activity even faintly comparable with that of the compounds listed above.[45]

These quite different assessments of the significance of industrial chemical carcinogenic hazards support, and are supported by, competing scientific accounts of the carcinogenic process:

> *Saffiotti:* . . . extremely small amounts of chemical energy are required for the critical chemical interaction, which can trigger off a permanent change in the regulation of cell growth which in turn, as cells replicate, will become manifested as cancer. Such a trigger effect of carcinogens is basically and completely different from most other toxic effects of chemicals, which require the continued presence of the agent to produce the effect, or a large number of target cells to be affected, to reveal a physiologic or pharmacologic effect, as is the case with hormones or vitamins. This is a fundamental point — well borne out by all modern molecular biology — which the 'traditional' toxicologist, used to the study of 'compound-dependent effects' rather than 'trigger effects,' is

at times found to misunderstand. Because of this trigger effect, carcinogenesis is not like most other forms of toxicity and is not explained by the same generalizations.[46]

Barnes: There is already enough evidence to support the belief that for many carcinogens the latent period before the cancer develops bears some inverse relation to the size of the dose. If there are data on the response to high doses it may not be difficult to calculate that at some lower dose the latent period of response will exceed the life-span of the host. Thus, if the dose of a carcinogen is low enough, the response in an exposed person will be manifest only in the hereafter. What is needed to examine this belief is not more molecular biology but more studies on the response of the whole animal. . . . A critical look at the limited amount of information that exists on the whole-animal response to carcinogens should help to disperse the prevailing gloom that there is neither a practical safe dose of a chemical carcinogen nor a reasonable experimental basis for attempting to derive one.[47]

Thus we can see how commitment to a 'trigger' mechanism of carcinogenesis, whereby a single molecule can initiate a cancerous response, legitimates the 'prudent' policy advocated by Saffiotti — whilst Barnes' commitment to the traditional toxicological mechanism, which requires prolonged contact between cells and chemicals, supports a more 'permissive' policy. This analysis suggests that a more detailed examination of the social and scientific commitments of the PSPS toxicologists and the EPA witnesses would reveal similar, systematic differences on issues such as the pathological classification of tumours, the evaluation of epidemiological evidence and the significance of mouse data.

It is beyond the scope of this paper to document these claims further, but we think the evidence presented is sufficient to conclude that the advice and the advisers that EPA and PSPS accepted were very different in the two cases. Moreover, the social and scientific commitments embedded in the advice were consistent with the two decisions: the more agriculturally-oriented advice with the British decisions, and the more health-protective advice with the US decision. But having clarified those features of the advice that seem most relevant to the decisions, we must examine those features of the two decision-making contexts that made them more or less receptive to the advice they received. This would not be necessary if the British and US decision-makers had received only one set of arguments. However, the decision-makers in EPA had to choose between two rival arguments presented in a court-like setting, whilst British decision-makers asked a committee of experts to review their previous decision in the light of the opposed US decision. We must therefore determine what led the British and US decision-makers to favour differing interpretations of the same evidence.

Before doing so, we should mention one other way in which the two types of advice differed, and which may have affected their reception. As we indicated above, the type of occupational control exercised over scientists has implications not just for the content but also for the form of the knowledge produced. On Johnston and Robbins' model, the collegiate form of occupational control operating in basic research tends to result in universalistic and theoretically-oriented forms of knowledge that quickly diffuse into other scientific fields. In contrast, a greater degree of patron control, as occurs in more mission-oriented research establishments, is related to particularistic, atheoretical concepts and techniques, which are only locally intelligible. Whereas molecular biology, the approach underlying Saffiotti's 'trigger mechanism', has become the conceptual foundation for a significant part of contemporary biological sciences,[48] 'traditional toxicology', the approach underlying Barnes' 'continuous contact mechanism', is hardly represented in university scientific research. This evidence, therefore, tends to support the prediction of Johnston and Robbins' model, and suggests that there were differences not only in the social values of EPA advisers and the content of their advice, but also in the character of the science forming the basis of their advice.

DECISION-MAKING INSTITUTIONS

In general, the location of regulatory responsibility in different types of agencies will have direct implications for the evaluation of the benefits of pesticides and the hazards associated with their use. The considerable discretion inevitably vested in government agencies allows the bureaucracy to define its mandate in close relation to its own interests and goals. We need to evaluate, therefore, how the institutional missions of the British and US governmental agencies with primary responsibility for the regulation of pesticides influenced the reception of the rival interpretations of the A/D evidence.

In Britain, several government departments are party to the PSPS Agreement, but there is little doubt that the Ministry of Agriculture, Fisheries and Food (MAFF) plays the leading role. The Minister of Agriculture, Fisheries and Food answers Parliamentary questions about pesticides, and MAFF provides most of PSPS's supporting staff. Eighty per cent of the British usage of pesticides is agricultural, so MAFF is the agency with the most direct interest in pesticides, and particularly in their contribution to agricultural production. Indeed, there is good evidence that this interest tends to dominate MAFF's regulation of pesticides.[49]

It was just such a conflict of interest that led President Nixon to transfer responsibility for the regulation of pesticides from USDA to EPA in 1970.[50] Critics of USDA had argued that pesticides would be more effectively regulated by an agency with sole responsibility for environmental protection, and with no role in the promotion of agricultural production. When the agency was established, many of the activists who had supported the move were attracted by EPA's mandate, and found work in quite senior positions. The vigorous manner in which EPA has subsequently endeavoured to protect the US environment has led critics to accuse the agency of 'capture' by environmentalists.[51]

There are good grounds for believing, then, that MAFF's and EPA's institutional missions have led these agencies to regard pesticides in quite different ways, with MAFF emphasizing their contribution to agricultural efficiency and EPA predominantly concerned with the hazards that their use entails. Moreover, we can see how these missions would have made them receptive to different types of scientific advice: EPA to the more health-protective type and MAFF to the more agriculturally-oriented advice. Still, the organization of the two decision-making processes suggests that any connections of this sort are less direct than, at first, they would appear. Both processes are formally designed to attenuate the political pressure that government agencies can apply to the providers of technical advice: in the US, by insulated, quasi-legal proceedings; and, in Britain, by the operation of an expert-based committee. What role, if any, can we therefore attribute to EPA's and MAFF's institutional responsibilities in explaining the British and US decisions?

In the US, the lawyer who led the OGC team has opined '. . .that there simply would not have been an Aldrin/Dieldrin case to review. . .if the matter had been left up to the scientists designated within EPA's Office of Pesticides.'[52] When the OGC lawyers were preparing their brief for the A/D suspension hearing they naturally consulted with EPA's principal source of expertise: the Office of Pesticide Programs (OPP). This Office had been formed from the scientific divisions of EPA's regulatory predecessors — USDA, the Food and Drug Administration (FDA) and the Department of the Interior.[53] Until 1970, these scientists had effectively controlled the regulation of pesticides in the US. Now Karch has argued that, despite being employed by EPA,

the views of many OPP scientists (on cancer testing) were generally along the lines of those in industry and USDA, who felt, for example that evidence of advanced stages of malignancy were necessary to characterize tumors as carcinogenic. Also many scientists in OPP had extensive toxicological experience in industry, USDA or FDA with substances that induce reversible effects below a certain 'threshold' of action. Thus they believed 'no-effect' levels should be set for carcinogens as they are for substances posing acute and certain other chronic hazards. Furthermore, many in OPP believed rodents (mice especially) were inappropriate animal models for cancer testing.[54]

Not surprisingly, the OPP scientists did not believe that the available data would support a case alleging the carcinogenicity of A/D. The OGC lawyers were not satisfied by the response from OPP and a bitter conflict developed between the two divisions within EPA. The intensity of the disagreement can be gauged by the fact that Kent Davis, who had transferred to EPA from FDA, and who had co-authored two of the A/D mouse studies, offered to support Shell's defence of A/D.[55]

The OGC lawyers, however, were not compelled to rely exclusively on EPA's in-house expertise. They were aware of the controversial nature of the toxicological field, and the diversity of institutions engaged in research. They therefore sought, and found, scientists who were prepared to oppose the OPP/Shell interpretation of the A/D evidence. In fact, they did not have to look very hard. Several of the scientists who subsequently appeared as EPA witnesses, such as Saffiotti and Epstein, had been highly critical of the way in which toxicologists in FDA and OPP had discharged their regulatory duties.[56] The suspension hearing therefore provided an excellent opportunity for them to carry on this attack and establish the superiority of their arguments within an important regulatory forum.

What the transfer of regulatory responsibility achieved, then, was to shift the critical regulatory locus away from the scientists in OPP to the lawyers in OGC. This, in turn, allowed the introduction of a new group of scientific advisers into the decision-making process, whose advice was more in keeping with the interpretation which the OGC lawyers had given EPA's mandate. The lawyers could not be certain that their advisers' evaluation of the evidence would be accepted by Perlman and Train, but, by introducing the rival interpretation, they were attacking the hegemony that OPP scientists had enjoyed until that time, and providing a possible replacement.

Turning to the British case, we need to clarify how the decision-making process operates before we can evaluate how MAFF's institutional goal influenced the reception of the rival interpretations of the A/D evidence. By 1974, the PSPS Agreement had been in operation for nearly 20 years, and, in this time, a considerable amount of decision-making authority had been delegated to the Advisory Committee. Although no industrial representatives sit on the Committee or its Sub-Committee, a company may send a representative to attend any meeting at which its products are being discussed. Considerable efforts are made to achieve a consensus between government advisers and industrial scientists. The success of this process in generating consensus has meant that government departments have effectively been able to delegate their authority, and rely on PSPS advisers to negotiate a satisfactory settlement.

In this way, PSPS advisers have developed a long tradition of working closely with pesticide manufacturers and advising a government department with a predominant interest in the promotion of agricultural production. Neither of the groups with a direct interest in the regulation of potentially carcinogenic substances — workers and consumers — are represented in PSPS's decision-making process. It is hardly surprising, then, that after working in a regulatory system like PSPS, advisers like Barnes develop a very positive evaluation of pesticides. But we have also seen that this commitment informed the scientific principles used to determine carcinogenic risk. We are now in a position to understand why PSPS advisers were so receptive to Shell's argument and, as we suggested earlier, shared the same or similar standards for inferring carcinogenicity. The tradition of close cooperation between PSPS advisers and industrial toxicologists within a context that placed a high value on agricultural production and the achievement of consensus meant that PSPS advisers had been actively involved in the development of Shell's case and in the standards they applied to their evidence.[57] In so far as the PSPS advisers had become the *de facto* decision-makers, they were highly receptive to an argument which they had helped develop, and which was congruent with their social and scientific commitments.

The toxicologists in OPP had developed a similar regulator-regulatee relationship before 1970, and they employed standards for determining carcinogenic risk that were much closer to those of Shell's and the PSPS advisers than they were to those of EPA's witnesses.[58] Indeed, it seems likely that the activities of multinational companies like Shell, and the need for international regulation of foodstuffs contaminated with pesticides, provided the opportunity for these scientists to develop a good measure of consensus among themselves on the scientific basis for the regulation of potentially carcinogenic pesticides. Moreover, since little toxicological work was performed independently of government and industry, there were few scientists to challenge the approach they developed, especially outside the US. When seen in this light, EPA's finding that A/D were carcinogenic emerges as a radical departure from the 'regulatory orthodoxy' that had prevailed in the US until that time, and that probably persists in most other national regulatory systems.[59]

The closed nature of the PSPS decision-making process is also relevant to the reception of the rival interpretations of the A/D evidence. Since there was virtually no way for 'outsiders' to contribute to the British decision, there was no way that EPA's witnesses' case could be presented systematically. But even if there were, it is not clear whether any British scientific equivalents to Saffiotti and his colleagues existed, or whether they would contribute to the decision-making process if they did. So there was neither the demand, nor the supply, nor the opportunity to play a role in the British decision-making process for scientists advocating an interpretation of the A/D evidence like EPA's witnesses. The result was that the PSPS advisers had little difficulty in dismissing the EPA decision and the rival approach on which it was based.

To sum up, the British institutions with responsibility for regulating pesticides — MAFF and PSPS — were motivated by a commitment to agricultural production that resulted in a highly favourable reception of Shell's argument. For the US decision, we concluded that EPA's institutional mission of environmental protection could be afforded only an indirect role. Through the agency of the OGC lawyers, a new group of scientists was introduced into the decision-making process who challenged Shell's and OPP's interpretation of the evidence. Once there, EPA's advisers and Shell's witnesses had to convince two laypeople that theirs was the better argument. Perlman's and Train's evaluations of the arguments were, in turn, constrained by the statute governing the regulation of pesticides. We need to determine, therefore, how far this statutory framework affected the reception of the rival arguments in EPA, and whether the absence of a similar statutory authority in Britain helps to account for the different British and US decisions.

REGULATORY STANDARDS

In general, systems of pesticide regulation employ two types of standard that influence the determination of carcinogenic risk — substantive standards and standards of proof. Taken together, these standards specify what evidence is required for the public authorities to classify a pesticide as a carcinogenic hazard. It is important, threfore, to determine how the standards used in EPA and PSPS affected the reception of the rival interpretations of the evidence on A/D.

The enactment of the Federal Environmental Pesticide Control Act (FEPCA) by Congress in 1972 established a strict, new standard whereby a pesticide could only be marketed as long as there were 'no unreasonable adverse effects on the environment.'[60] This standard was defined in FEPCA to mean 'any unreasonable risk to man or the environment, taking into account the economic, social and environmental costs and benefits of the use of any pesticide. '[61] Particular types of risk, such as carcinogenic ones, were not explicitly mentioned in FEPCA, but they were, nevertheless, covered by this standard. The Act also specified that, whenever required, manufacturers had to demonstrate that the use of their product conformed to this standard — in other words, at all times, the burden of proof of safety remained with the manufacturer.

Even though the PSPS Agreement is non-statutory, it does declare that '. . .if a chemical is known or shown to be a carcinogen it will not be permitted to occur as a residue in food.'[62] Presumably, this standard applies both before and after products are marketed, though manufacturers will not necessarily have to provide the evidence in the latter case.[63] The burden of proof, therefore, is more ambiguous in the British system of regulation than in the American.

Insofar as these standards do not define what is meant by 'carcinogen', 'unreasonable risk', 'burden of proof', and so on, they are too broad to account for the differing US and British decisions on A/D. To evaluate the significance of these standards, we need to see how they were interpreted in the two decision-making contexts and whether they were influenced in this by their statutory or voluntary form.

(A) US Standards

The EPA suspension decision apparently did not turn on the distribution of the burden of proof. Perlman stated in his opinion that

> . . .the respondent [i.e. EPA], who has the burden of going forward to present an affirmative case for suspension, but not the ultimate burden of persuasion as to safety, has in fact satisfied the burden of proof, which is not his, that the chemicals in question pose a high risk of causing cancer in man.[64]

More substantively, Perlman's and Train's interpretation of the

FEPCA standard did not require EPA to unequivocally demonstrate A/D's carcinogenicity: '. . .suspension is to be based upon potential or likely injury and need not be based upon demonstrable injury or certainty of future public harm.'[65] They were supported in this by similar interpretations of other environmental statutes. A number of precedents had been established in the US courts so that if a hazard was thought to be sufficiently serious, the decision-maker was required to make a policy determination of the likely effects of the hazard-producing activity, rather than waiting for definite evidence of harm to accumulate before deciding whether the risk was acceptable or not.[66] For serious hazards, then, rather less evidence would be required before treating an activity as if it were hazardous, than would be the case if the strictest standards of scientific causality were to be met. Both Perlman and Train believed that cancer fell into this category of 'serious hazard' and that it should be dealt with accordingly.

The strong Congressional aversion to carcinogens in the US food supply, expressed in the Delaney Clause, supported Perlman's and Train's view of the seriousness of carcinogenic hazards, and provided additional justification for a cautious approach to decision-making on pesticides. The clause was an amendment to the Federal Food, Drug and Cosmetic Act and stated that

> . . . no [food] additive shall be deemed safe if it is found to induce cancer when ingested by man or animal, or if it is found, after tests which are appropriate for the evaluation of the safety of food additives, to induce cancer in man or animals.[67]

Train was aware of the fuzzy distinction between 'intentional' food additives and 'unintentional' contaminants and argued that

> . . . since the Delaney Amendment does prohibit the setting of safe levels/tolerances of carcinogenic food additives, and since Aldrin-Dieldrin is present as a residue in processed foods, the Administrator has a particular burden to explain a basis for a decision permitting continued use of a chemical known to be carcinogenic in laboratory animals.[68]

It is likely, then, that Perlman's and Train's interpretations of the FEPCA standard would have made them more receptive to EPA's argument than Shell's. EPA's use of cancer principles as guides to decision-making was tailored to their needs. The principles assumed the necessity of making policy judgments when determining carcinogenic risk, and expressed a preference for minimizing carcinogenic risk. In contrast, Shell's argument seemed to reflect an insufficient appreciation of the novel regulatory situation in which they found themselves. Their stress on causality and the demonstration of harm was precisely the approach that had been rejected in recent court decisions.

Nevertheless, Perlman did not justify his decision in these terms. As we saw earlier, he thought his decision represented the conventional wisdom of the scientific community. But if this were so, why didn't experts in other countries, and in the US before 1974, adopt the same view? Even if one accepts that these other decision-making forums were dominated by pro-pesticide interests, this still leaves the problem of explaining why Perlman and Train preferred EPA's interpretation of the evidence. Should we accept Perlman's argument that the greater scientific merit of EPA's case was the decisive factor? Or should we afford the legislative tradition, within which Perlman evaluated the rival interpretations, the greater role, despite Perlman's indications to the contrary?

Unfortunately, it is not so easy to disentangle these factors. Nevertheless, we think that the legal framework undoubtedly helped structure the way in which Perlman evaluated the evidence. Perlman's and Train's acceptance of EPA's cancer principles, and the policy judgments associated with them, can only be understood in terms of their interpretations of their statutory responsibilities. Moreover, FEPCA determined that two laypeople, rather than scientists, were in the key decision-making roles.[69] So, unlike the PSPS advisers, EPA's decision-makers were not committed to a particular way of assessing carcinogenic risk, and they were unaware of the 'regulatory orthodoxy' that had prevailed until that time. At the same time, we think it is possible to identify some reasons, independent of immediate legislative concerns, that may have influenced Perlman's decision.

Firstly, EPA's witnesses were generally highly-qualified scientists, working in prestigious institutions rather than in private industry or government regulatory agencies. The claim that they represented 'the most advanced research findings and policy of both national and international cancer experts and agencies',[70] appeared, therefore, to have some substance. In contrast, Shell's witnesses had the disadvantage of representing a narrower, sectional interest. Furthermore, Shell's attempt to bolster the universality of their argument by relying on witnesses associated with international agencies backfired somewhat when Perlman discovered that the views of the Joint FAO/WHO Committee on Pesticide Residues were largely determined by one of Shell's witnesses, Francis Roe.[71] As the report was being used to support a controversial argument — Shell's 'two species' criterion — this discovery was quite damaging to a case purporting to represent the 'state-of-the-art' views of the scientific community.

Secondly, the evidence that EPA's witnesses presented within the framework of the cancer principles was generally coherent and consistent. The only disagreement was on an issue that, as it happened, was not critical to the outcome of the case.[72] Shell, too, had disagreements among their witnesses and their position shifted during the course of the proceedings. More significantly, their demand for evidence of causality contrasted unfavourably with their use of poorly corroborated biochemical evidence, and emphasized an apparently narrow dimension to their argument. The quality of Shell's work was also brought into question by EPA's reinterpretation of existing data in a way which suggested that Shell had systematically underestimated A/D's carcinogenic risk. Furthermore, the scientists who developed the methodology that Shell employed in their two mouse studies testified that Shell's analysis was 'almost guaranteed to give non-significance for even the strongest carcinogens.'[73]

Finally, the form of the two arguments may have influenced Perlman's and Train's judgment as to which approach would provide the better scientific basis for future regulation. We saw in an earlier section how the scientific approach of the EPA witnesses was generally more theoretical and universal in character than Shell's. In the context of the suspension hearing, this may have influenced Perlman's assessment of the relative scientific merit of EPA's and Shell's arguments.

We may conclude, then, that both the legal framework within which the A/D evidence was considered, and various features of the arguments themselves, led Perlman and Train to prefer EPA's interpretation of the A/D evidence to Shell's. Considerably more research would be needed to clarify which of these factors was the more important. Nevertheless, we would emphasize the fundamental importance of the legal framework in this connection. Whereas it is not clear that other laypeople, or uncommitted scientists, would be persuaded by EPA's case alone, it seems more likely that they would reach the same conclusion as Perlman and Train when they evaluated the evidence within the framework of FEPCA, the Delaney Clause, and US environmental law. Thus, we would argue that the motivation for the US decision on A/D is to be found in the American legal system rather than the 'conventional wisdom of the scientific community.'

(B) British Standards

The interpretation that PSPS advisers gave the PSPS cancer standard was subject to none of the same constraints as the interpretation of the FEPCA standard. To the extent that there was no statutory framework governing the interpretation of the cancer standard, the institutional constraints we considered in the last section were correspondingly more important in determining the outcome of the A/D review.

The PSPS Agreement did not express a strong commitment to the prevention of cancer, and if Barnes' views are typical, British decision-makers did not think that banning pesticides would do much to prevent cancer anyway. Similarly, no new tradition had emerged in British legal, or political, or scientific thinking to guide the specification of serious but uncertain risks. As a result, PSPS and other British regulatory systems have expected the traditional requirements of scientific causality to be satisfied before labelling a chemical carcinogenic, just as OPP scientists had done when USDA exercised regulatory responsibility for pesticides. This approach was clearly more congenial to the PSPS scientific advisers and their industrial counterparts. Not only was it more in keeping with their professional training than any other, but also the way in which the balance of doubt was distributed (in favour of pesticides) was more consistent with the ethos of PSPS.

Thus, the very 'weaknesses' of Shell's argument in the suspension hearing became their 'strengths' in PSPS. For example, the demands for causal mechanisms and demonstration of harm were all quite reasonable, and compelling, in the PSPS context. Similarly, PSPS advisers had been closely involved in FAO and WHO activities, and the Shell witness who had largely determined the views of the report that Perlman rejected had served on a number of British advisory committees. It is not surprising, therefore, that British medical opinion was irritated by Perlman's treatment of Roe's evidence, nor that the EPA decision was regarded as 'unscientific' and the unfortunate result of leaving such decisions to laypeople.

Finally, if we speculate on the decision that would have been reached if PSPS were statutorily based, there is no guarantee that the British decision would be reversed. On the contrary, it is quite likely that A/D would not be considered carcinogenic as long as the following conditions pertain: the assessment is based on the traditional notion of scientific causality; the burden of proof of safety is not unequivocally with Shell; the PSPS decision-making organization (with its strong commitment to agricultural production) remains intact; and there is an absence of scientists and citizens advancing more health-protective arguments. In Britain, as in the US, the critical point is not the existence of a statute, but the type of framework, legal or otherwise, that guides the determination and assessment of uncertain risks.

CULTURAL AND ECONOMIC FACTORS

In complex policy issues such as this, where the outcome is determined by the interaction of several causal factors, the roles of cultural and economic factors are diffuse and difficult to evaluate. Their influence is generally mediated through specific institutions and practices, but is, nonetheless, real. For our purposes, we need consider only four aspects of the British and US cultural and economic contexts: environmental movements and their definition of the pesticide issue; economic well-being; availability and type of expertise; and styles of government — particularly their use of law and science in environmental decision-making.

There is little doubt that the environment was a much more significant issue, politically, in the US than in Britain.[74] For whatever reasons, the environment became one of the major foci for the unrest that characterized American society in the late 1960s. The movement that mobilized around environmental issues became sufficiently strong to elicit a major response from US political and legal institutions. The creation of EPA, the enactment of FEPCA, and the reformulation of the principles on which technological risks were managed — all important factors in the US A/D decision — must be seen within the context of this movement.

In Britain, political conflicts did not develop around environmental issues in the same way, nor with the same intensity. The British environmental movement — if one may call it that — did not secure the same support, nor develop the same strength, as its US counterpart. Thus, in contrast to the US, pesticides were not a political issue in Britain after 1969.[75] There was no group lobbying for the transfer of primary regulatory responsibility from MAFF to a less compromised agency, for the enactment of more effective statutory controls nor for the removal of A/D from the British market. Finally, there was no constituency that could provide senior administrators for a new regulatory agency, nor monitor their activities, as groups like EDF did for EPA.

It is also interesting to contrast the types of concern expressed about pesticides in the two countries. In Britain, this concern was largely confined to wildlife conservation, an issue that traditionally has mobilized strong support. In the US, public health issues, and particularly cancer hazards, have also concerned critics of pesticides.[76] Indeed, environmentalists have generally capitalized on the widespread fear of cancer among the US population, and have tended to redefine environmental protection as health protection, and health protection as prevention of cancer. It was no accident, therefore, that the OGC lawyers based their suspension case on the

326

carcinogenic hazard that A/D posed, rather than some other 'unreasonable adverse effect on the environment'.

Fear of cancer has not carried the same political weight in Britain as in the US, especially in policy circles. British decision-makers and their advisers have not been convinced that cancer rates would be much reduced by the costly regulation of technologies like pesticides. It is not surprising, then, that there has been virtually no expression of public concern about A/D in Britain either before, during, or after the suspension hearing.[77]

One reason why the US political system has been more responsive to environmental demands than the British is the relative strengths of their national economies. The US has a much larger and wealthier economy than Britain's, and this has allowed it to absorb the costs of environmental and health protection more readily. Technologies have been regulated on the assumption that American industry is sufficiently robust to operate within these new constraints. In contrast, Britain's weaker economic position has legitimated a continued stress on the promotion of economic development through technological change. British decision-makers have been very reluctant to encumber industry with unnecessary regulation, and this interest has continued to inform the British management of technological risks.

Taken together, the different national resources and evaluations of dread diseases such as cancer have affected the availability, and type, of expertise for the assessment of carcinogenic risk.[78] In both countries, toxicology has been, historically, most closely associated with the private sector and the government agencies responsible for the regulation of toxic hazards. This has been reinforced by the failure of toxicology, as a scientific field, to develop a unique identity within national university systems. Still, the greater resources and commitment to health protection in the US has resulted in the development of a larger, more diverse biomedical research system, in which more basic research institutions, independent of regulator-regulatee interests, operate. These, in turn, have proved to be appropriate locations for the development of alternative approaches to the assessment of carcinogenic risk — an important pre-condition for the US A/D decision. This contrasts with the British biomedical research system, which is smaller, more closely tied to regulatory work, and apparently lacking the diversity of approaches to the assessment of carcinogenic risk that exists in the US.

In this connection, the different scientific traditions that the cultural and economic contexts of the two countries supported may also have contributed to the different decisions. McGinty, for instance, has argued that the British emphasis on epidemiology owes much to the influential tradition associated with Sir Richard Doll.[79] Doll and his co-workers had carefully demonstrated how cigarette smoking was the main cause of increasing rates of lung cancer in Western countries. This experience, and the authority it conferred on British epidemiology, may, therefore, have structured the way in which British scientific advisers regarded 'environmental' cancer hazards and the best means for their detection. This contrasts with the US situation, in which the NCI has tended to emphasize animal rather than human studies because of their relative costs.[80] Considerably more research would be needed to evaluate the contribution of these differing traditions to the British and US A/D decisions. Nevertheless, they are consistent with the respective decisions and would at least have supported their rationales.

Turning, finally, to the styles of government, we can see how the British and US systems for regulating pesticides are guided by the norms that generally inform the execution of public policy in these two countries. In Britain, consensus is generally achieved by restricting both access to, and information of, the decision-making process, whilst allowing maximum flexibility for negotiation among the most directly involved parties. This contrasts with the American pluralist tradition in which great importance is attached to the clash of conflicting ideas and the evolution of policy through adversarial processes.

The different approaches are clearly illustrated by contrasting the British and US attitudes to the legal process. In Britain, the law is often thought to breed inflexibility and rigidity, and to induce confrontation, polarization and irrationality — all of which hinder the achievement of consensus. In policy fields like the environment, therefore, a voluntary approach is preferred, and the law is used only as a last resort. Within the American pluralist tradition, however, the courts are the archetypal mechanism for conflict resolution. The participation, openness, and confrontation that they allow have served to assimilate disaffected groups back into the system. Thus, in contrast to Britain, the courts have been a critical locus for the development of US environmental policy and the management of technological risks.

The British and US styles of government have also structured the roles that science and scientists played in such decisions as that on A/D. While toxicology and its practitioners have generally been 'politicized' by their involvement in government decision-making[81] (for instance, by governments intervening in what was traditionally the autonomous sphere of the scientist to prescribe 'good laboratory practice,' and endorsing particular experimental designs and interpretive principles to be used in risk assessment), the respective styles of government have politicized them in different ways. In the US, toxicologists have been enlisted by conflicting social groups, especially industrialists and environmentalists, to support their arguments, and, as a result, have been drawn into adversarial political processes. Indeed, the political and legal forums in which US toxicologists confront each other have become important new battlegrounds for competing groups of scientists trying to establish the superiority of their approach. Pfizer has described this changed environment in which the toxicologist works as

> . . . at once both frustrating and stimulating; frustrating because he doesn't have the answers to many of the perplexing questions being asked; frustrating to some toxicologists because their time-honoured methods for making judgments about safety evaluations are being challenged; stimulating because suddenly there are a multitude of people interested in his professional activities. . . He can expect his data to be scrutinized by non-scientists, to be interpreted in the newspapers and in legal hearings, to face requirements for exactness and statistical validity with increasing rigour. The life of the toxicologist will never be the same.[82]

British toxicologists have not escaped politicization by their contact with government, but it has assumed a less obvious form. Rather than publicly confronting each other, toxicologists have been enlisted by the British government to generate a consensus and legitimate political decisions. In contrast to the conflicts among experts that characterize many American decisions in this field, British decisions emerge from a closed decision-making process with the apparently uncontroversial and authoritative support of science. Whereas US decision-making institutions depend upon, and, to some extent, generate conflicts among experts, British institutions tend to rely upon singular sources of expertise. In the A/D case, the result was the availability of an adversarial forum in the US, and scientists eager to enter it, whereas in Britain there was neither the forum nor, apparently, the scientists.

CONCLUSION

This case study should, at the very least, rid readers of the notion that the relationship of scientific knowledge to public policy is straightforward. In order to provide a reasonable account of why two countries derived different conclusions from the same scientific evidence, we had to invoke a set of interacting, and mutually supporting factors for each decision: the uncertainty inherent in the relevant scientific field; the application of different scientific standards, motivated by different scientific and social commitments; the bureaucratic politics of the agencies with responsibility for regulating pesticides; the way in which standards are defined in particular systems of regulation; and, finally, a series of contextual factors. Whilst many interesting questions were raised by the analysis, we will conclude this paper by reconsidering, in the light of this case study, some features of the science-public policy relationship outlined in the introduction.

Firstly, although this study has highlighted some important differences between the British and US approaches to the regulation of potentially carcinogenic pesticides, it is significant that both decisions were ultimately justified by recourse to science — in the British case by the traditional notion of scientific causality, and in the US case with reference to the consensus of the scientific community. Both decisions were, thus, presented as in some sense 'springing directly from the facts'. This testifies to the considerable cultural authority of science in these countries, particularly as a means of solving policy problems.

Our analysis of the A/D case, however, calls into question any exclusively scientific justification for such decisions. Thus, the British demand for evidence of causality, and the hostility towards the EPA decision expressed in British medical journals, both tend to conceal that the decision to wait for definite evidence of harm to accumulate is just as much an ethical and political choice as the decision to treat risk determination as a policy issue. Whether or not one agrees with the ethical and political commitments informing the US decision, the acknowledgement of this dimension is surely preferable to presenting decisions as the result of a methodological imperative.

Despite this insight, Perlman's treatment of the A/D decision as a policy issue, and his justification of it as the conventional wisdom of the scientific community, was rather like having one's cake and eating it. It may well be, as we have tried to show, that Perlman found 'good reasons' for preferring EPA's argument to that of Shell. But it is not clear whether they would have been compelling without the supporting *statutory* framework. Moreover, it is difficult to see in what sense the views of EPA's witnesses were more representative of the scientific community. The notion of 'community' seems inappropriate for a collection of scientists working largely in industrial and governmental laboratories, and, if anything, Shell's argument was more widely accepted in these circles than was EPA's. In cases such as this, then, science can only inform but not determine policy decisions. The attempt to ground policy decisions exclusively in science, for whatever motive, is inappropriate and ill-conceived.

Secondly, we found that the fact-value distinction was of little use beyond a trivial level for analyzing either the decisions or the roles of the scientific advisers. The apparently factual determinations of A/D's regulatory status involved considerably more than 'the facts', as did the advice on which they were based. Positivist-inspired views of science, which incorporate the fact-value distinction in their analysis, are thus inherently limited by their failure to examine the context of scientific advice and the way in which it can transmit social commitments.

Similarly, this case study supports the view that the traditional division of labour between science and public policy, or at least the way it is usually perceived, is breaking down in complex areas of decision-making like risk assessment. Scientists do not operate in an exclusively factual arena, and decision-makers play a more active role than is usually realized in determining what is to count as a fact. If both parties are to operate effectively, and in good faith, they will have to face up to this, and adjust their roles accordingly. We would argue that the minimum requirements for this are for scientists to accept responsibility for the commitments they inevitably make when devising approaches for the study of poorly understood phenomena. Moreover, they should be prepared to elaborate and clarify the implications of those commitments, especially in relation to the ethical and political considerations that arise in risk assessment. Lay participants in risk decision-making, in their turn, will need to develop a more subtle understanding of both the strengths and limitations of the contributions that science can make to the clarification of these difficult issues. Unless they do, it seems likely that the social control of technological change will continue to be impaired by the manipulation of outmoded ideologies of science by competing or dominant social groups.

NOTES AND REFERENCES

1.   S. Hadden, 'Technical Advice in Policy Making', in J. Haberer (ed.), *Science and Technology Policy* (Lexington, Mass.: Lexington Books, 1977), 81–98, and J. Milch, 'Technical Advice and the Democratic Process', paper presented to AAAS meeting, Washington, DC, 12-17 February 1978, provide good reviews of this literature.

2.   The need for comparative studies in the social studies of science field has been most recently argued by R. Brickman and A. Rip, 'Science Policy Advisory Councils in

France, The Netherlands and the United States: A Comparative Study', *Social Studies of Science,* Vol. 9 (1979), 167-98.

3. H. Nowotny, 'Scientific Purity and Nuclear Danger: The Case of Risk Assessment', in E. Mendelsohn et al. (eds), *The Social Production of Scientific Knowledge* (Dordrecht/Boston, Mass.: D. Reidel Publishing Co., 1977), 243-64.

4. R. Johnston and D. Robbins, 'The Development of Specialties in Industrialised Science', *Sociological Review,* Vol. 25 (1977), 87—108; and D. Robbins and R. Johnston, 'The Role of Cognitive and Occupational Differentiation in Scientific Controversies', *Social Studies of Science,* Vol. 6 (1976), 349-68.

5. The initial analysis of these decisions appeared in B. Gillespie, *British Control of Pesticide Technology* (unpublished PhD thesis, University of Manchester, 1977), Chapter 8.

6. M. Sloan, 'A US Historical Account — Aldrin/Dieldrin Registration, Uses and Sales', evidence presented at EPA's cancellation hearing (no date).

7. L. McCray, 'Mouse Livers, Cutworms and Public Policy: EPA Decision Making for the Pesticides Aldrin and Dieldrin', in R. Burt et al. for the Committee on Environmental Decision Making, *Decision Making in the Environmental Protection Agency,* Vol. IIa (Washington, DC; NAS, 1977), 58-118, has provided a good discussion of this decision.

8. See P. Spector, 'Regulation of Pesticides by the Environmental Protection Agency', *Ecology Law Quarterly,* Vol. 5 (1976), 233-63, for a relevant discussion of the statute governing the US regulation of pesticides.

9. McCray, op. cit. note 7, 65.

10. Environmental Defense Fund, Inc. v. Environmental Protection Agency, 210 F2d (1975), 1292.

11. Personal communication (BG), MAFF, 17 June 1976.

12. 'Insecticides and Cancer' (Editorial), *British Medical Journal* (25 January 1975), 170.

13. For a more detailed discussion of the British regulation of pesticides, see Gillespie, op cit. note 5, and B. Gillespie, 'British "Safety Policy" and Pesticides', in R. Johnston and P. Gummett (eds), *Directing Technology* (London: Croom Helm, 1979), 202-24.

14. T. Maugh, 'Chemical Carcinogens: I. The Scientific Basis for Regulation,' *Science,* Vol. 201 (29 September 1978), 1200-05, and 'II. How Dangerous are Low Doses?', *Science,* Vol. 202 (6 October 1978), 37-41; and D. Eva, *Toxicology and Society* (unpublished MSc thesis, University of Manchester, 1975).

15. Maugh, ibid., I, 1200, lists 26 'accredited chemical carcinogens'.

16. Shell Chemical Company, 'Brief. Aldrin/Dieldrin Consolidated Suspension Hearing' (Washington, DC, 13 September 1974), III.

17. Shell Chemical Company, 'Objections and Supplemental Brief to the Administrator on Suspension' (Washington, DC, 24 September 1974), I-9.

18. See K. Jager, *Aldrin, Dieldrin, Endrin and Telodrin* (London/Amsterdam/New York: Elsevier, 1970).

19. Maugh, op. cit. note 14, I.

20. Shell Chemical Company, op. cit. note 16, III.

21. 'Consolidated Aldrin. Dieldrin Hearing', *Federal Register,* Vol. 39, No. 102 (18 October 1974), 37269. We shall refer to this below as 'EPA hearing'.

22. K. Davis and O. Fitzhugh, 'Tumorigenic Potential of Aldrin and Dieldrin for Mice', *Toxicology and Applied Pharmacology,* Vol. 4 (1962), 187-89; K. Davis, 'Pathology Report on Mice for Aldrin, Dieldrin, Heptachlor Epoxide for Two Years', internal FDA memo (1965); W. McDonald et al., 'The Tumorigenicity of Dieldrin in the Swiss Webster Mouse', unpublished (1972); A. Walker et al., 'The Toxicology of Dieldrin (HEOD) I: Long Term Oral Toxicity Studies in Mice', *Food and Cosmetic Toxicology,* Vol. II (1973), 415-32; E. Thorpe and A. Walker, 'The Toxicology of Dieldrin (HEOD)

II: Comparative Long Term Oral Toxicology Studies in Mice with Dieldrin, DDT, Phenobarbitone, β-BHC and γ-BHC', *Food and Cosmetic Toxicology,* Vol. II (1973), 433-42. These and other experiments involving A/D were reviewed in S. Epstein, 'The Carcinogenicity of Dieldrin. Part 1', *Science of the Total Environment,* Vol. 4 (1975), 1-52.

23.   Office of General Counsel, Environmental Protection Agency, 'Respondent's Brief. Proposed Findings and Conclusions on Suspension' (Washington, DC, 16 September 1974).

24.   N. Karch, 'Explicit Criteria and Principles for Identifying Carcinogens: A Focus of Controversy at the Environmental Protection Agency', in Burt et al., op. cit. note 7, 134.

25.   'Prudent policy indicates that every possible measure should be taken to eliminate human exposure to chemical compounds as soon as their carcinogenic nature is identified': Office of General Counsel, op. cit. note 23, 41.

26.   Epstein, op. cit. note 22, 207.

27.   EPA hearing, 37258-59.

28.   I. Nisbet, 'Measuring Cancer Hazards', *Technology Review,* Vol. 78 (1975), 8-9.

29.   EPA hearing, 37259.

30.   'Insecticides and Cancer', *British Medical Journal* (25 January 1975), 170; 'Does This Chemical Cause Cancer in Man?', *The Lancet* (14 September 1974), 629-30; and 'Seventeen Principles About Cancer, Or Something', *The Lancet* (13 March 1976), 571-73.

31.   *Further Review of Certain Persistent Organochlorine Pesticides Used in Great Britain.* Report by the Advisory Committee on Pesticides and Other Toxic Chemicals (London: HMSO, 1969), para 79-83. This was the evidence that Shell conceded was methodologically flawed at the suspension hearing.

32.   Ibid., para 81. There is no evidence that these experiments were ever performed, which is perhaps not surprising in view of Shell's research expenditure on A/D.

33.   *Aldrin and Dieldrin Residues in Food.* Report by the Food Additives and Contaminants Committee (London: HMSO, 1967).

34.   *Review of the Persistent Organochlorine Pesticides.* Report by the Advisory Committee on Poisonous Substances Used in Agriculture and Food Storage (London: HMSO, 1964), para 77.

35.   Panel on Carcinogenic Hazards, 'Carcinogenic Risks in Food Additives and Pesticides', *Monthly Bulletin of the Ministry of Health,* Vol. 19 (1960), 108-12.

36.   Ministry of Agriculture, Fisheries and Food, *Pesticides Safety Precautions Scheme Agreed Between Government Departments and Industry* (London: MAFF, Pesticides Branch, revised March 1971), para 1.4. We shall refer to this as 'PSPS Agreement'.

37.   In accepting the arguments of EPA witnesses on this point, Perlman ruled: 'Whether the agent actually is a *sine qua non* of the observed response or merely enhances a virus or some other factor found in the host animal is irrelevant unless and until we know that similar factors are not also found in man': EPA hearing, 37257.

38.   Panel on Carcinogenic Hazards, op. cit. note 35, Appendix, para 7.

39.   Johnston and Robbins (1977), op. cit. note 4.

40.   Gillespie, op. cit. note 5, 283-84.

41.   EPA hearing, 37255.

42.   Since its foundation in 1947, the starting point for all its research projects had been 'substances of potential interest to industry and agriculture about whose toxicity little was known': *MRC Annual Report,* April 1968-March 1969, HC 299, 82.

43.   R. Rettig, *Cancer Crusade: The Story of the National Cancer Act of 1971* (Princeton, NJ: Princeton University Press, 1977).

44.   U. Saffiotti, 'Comments on the Scientific Basis for the "Delaney Clause",' *Preventive Medicine,* Vol. 2 (1973), 128-29.
45.   J. Barnes, 'Carcinogenic Hazards from Pesticide Residues', *Residue Reviews,* Vol. 13 (1966), 79.
46.   Saffiotti, op. cit. note 44, 127.
47.   J. Barnes, 'Assessing Hazards from Prolonged and Repeated Exposure to Low Doses of Toxic Substances', *British Medical Bulletin,* Vol. 31 (1975), 198-99.
48.   Although the NCI is a government laboratory, its employees tend to be well-integrated into the open, peer review processes characteristic of collegiate forms of occupation control. On molecular biology, R. Young, in 'Evolutionary Biology and Ideology: Then and Now', *Science Studies,* Vol. 1 (1971), 179, has written:

> . . . the fundamental paradigm of explanation — the goal of all science — has been to reduce or explain all phenomena in physico-chemical terms. The history of science is routinely described as a progressive approximation to this goal. This is the metaphysical and methodological explanation for the fact that molecular biology is the queen of the biological sciences and the basis on which other biological (including human) sciences seek, ultimately, to rest their arguments.

49.   This argument is documented in Gillespie, op. cit. note 13.
50.   J. Blodgett, 'Pesticides: An Evolving Technology', in S. Epstein and R. Grundy (eds), *The Legislation of Product Safety,* 2 Vols. (Cambridge, Mass.: MIT Press, 1974), Vol. 2, 197-287.
51.   P. Weaver, 'Regulation, Social Policy and Class Conflict', *Public Interest,* Vol. 50 (Winter 1978), 45-63.
52.   J. Kolojeski, 'Federal Administrative Trial of a Carcinogen: the EPA Aldrin/Dieldrin Pesticide Case', in H. Hiatt et al., *Origins of Human Cancer,* Book C (Cold Spring Harbor, Mass.: The Laboratory, 1977), 1719.
53.   See Spector, op. cit. note 8, for further details of this regulatory reorganization.
54.   Karch, op. cit. note 24, 161.
55.   McCray, op. cit. note 7, 104.
56.   The chairperson of the Mrak Commission on Pesticides and Their Relationship to Environmental Health (Washington, DC: US GPO, 1969) reported 'Pathologists of the (National) Cancer Institute made it known that they did not think much of the pathologists representing the Food and Drug Administration'. See E. Mrak, 'Some Experiences Relating to Food Safety', in F. Coulston and F. Korte (eds), *Environmental Quality and Safety: Global Aspects of Chemistry, Toxicology as Applied to the Environment,* Vol. 15 (Stuttgart: George Thieme, 1976).
57.   Shell's '. . .research workers discussed the significance of their findings on many occasions with the scientists advising the [PSPS] committee': *BMJ* Editorial, op. cit. note 12.
58.   Political scientists are familiar with the idea of regulators being 'captured' by their regulatees: see, for example, Weaver, op. cit. note 51. We would argue that this notion can be extended to encompass the scientific approaches that are developed for regulatory purposes.
59.   This argument differs from Karch, op. cit. note 24, who accepts that EPA's witnesses' position was the better established in the scientific community. Still, we would argue that this underestimates the novelty of the EPA position, and overestimates the 'unscientificity' of their opponents.
60.   PL 92-516, Section 3. See also Spector, op. cit. note 8.
61.   Ibid., Section 2 (bb).
62.   PSPS Agreement; Appendix F, para 1.1
63.   Personal communication (BG), MAFF, 28 April 1977.

64. EPA hearing, 37259.

65. Ibid., 37265.

66. For a more detailed discussion of this difference, see M. Gelpe and A. Tarlock, 'The Uses of Scientific Information in Environmental Decision Making', *Southern California Law Review*, Vol. 28 (1974), 371-427.

67. 21 USC Section 348 (c) (3) (A) (1970).

68. EPA hearing, 37267.

69. We are using the term 'layperson' in the sense of non-scientist. Of course, Perlman's training as a lawyer would also have predisposed him to approach the evidence from a particular perspective which, in this context, cannot be regarded as neutral.

70. Office of General Counsel, op. cit. note 23, 29-31.

71. EPA hearing, 37256.

72. Some of EPA's witnesses disagreed over whether the rat experiments indicated a carcinogenic hazard associated with A/D. This was also the only matter that Perlman and Train disagreed on. But it was not important because they both accepted that the evidence from the mouse was sufficient for their purposes.

73. EPA hearing, 37268, fn. 55.

74. C. Enloe, *The Politics of Pollution in Comparative Perspective* (New York: David McKay Company, 1975).

75. Gillespie, op. cit. note 5, Chapters 6 and 7.

76. See, for instance, R. Carson, *Silent Spring* (Boston, Mass.: Houghton Mifflin, 1962), Chapter 14.

77. With the exception of Gillespie et al., 'A Tale of Two Pesticides', *New Scientist* (9 February 1978), 350-52.

78. S. Strickland, *Politics, Science and Dread Diseases* (Cambridge, Mass.: Harvard University Press, 1972), and J. Sadler, *Elites in Science: the Case of Cancer Research* (unpublished MSc thesis, University of Manchester, 1976).

79. L. McGinty, 'Controlling Cancer in the Workplace', *New Scientist* (22/29 December 1977), 758, and 'Human Guinea Pigs?', *New Scientist* (10 August 1978), 386.

80. Kolojeski, op. cit. note 52, 1722.

81. S. Blume, *Toward a Political Sociology of Science* (New York: Academic Press, 1974), 212-14.

82. E. Pfizer, 'Toxicology', in L. Cralley (ed.), *Industrial Environmental Health* (New York: Academic Press, 1972), 71-72.

Bibliographical notes

All items are referred to in these Notes by the numbers assigned to them in the Bibliography.

General

Items 73 and 329 are works of reference: both cover wide ranges of relevant literature in history, sociology, philosophy and political science, and both contain some excellent review essays and useful bibliographies. Item 245 is the best survey of current sociology of science: for further insights into East European work, see also 139 and 203; 227 is also relevant. For a good review in French, see 195; in German, 332. Item 317 is an excellent survey of recent historical research, the scope of which allows us to omit several important references from this list; see also the 'Essay on Sources' in 167.

Teaching Readers in the sociology of science are rare: 6 and 161 are dated, though still useful; 105 contains several important papers; but 8 is now out of print. Item 241 is the nearest to a 'set textbook' for the area: but see also 373. Several recent collections of research papers are available: 13, 26, 54, 104, 109, 128, 172, 196, 356, 362 and 396, and the annual *Sociology of the Sciences Yearbooks* (see 171, 177, 223 and 224), all contain important material. Items 18, 31, 63 and 345 are useful Readers in the sociology and politics of technology. Other notable collections, in their respective areas, are 115 and 265.

Supplementary and background texts include: in sociology of science and of scientific knowledge, 9, 10, 12, 23, 79, 176, 214, 226, 288, 322, 335 and 383; in political sociology of science, 25, 75, 114 and 295; in the social history of science and technology, 16, 42, 66, 167, 234, 269 and 283; and in contemporary problems of technology, 53, 83, 220, 365 and 367. At a more elementary level, 46, 152, 236, 249, 275, 384 and 400 are all useful.

337

Relevant material is published in many 'mainstream' sociological, historical, philosophical and political science journals. Of the more specialized journals, *Social Studies of Science* (see 54, 325–328), *Social Science Information* and *Minerva* are the most productive, with *Science and Public Policy* and *Research Policy* (and, in the history of science, *Isis, Studies in History and Philosophy of Science* and *History of Science*) worth special attention. *Scientometrics* covers quantitative methods. *Isis* and *Technology and Culture* contain comprehensive book reviews; *Isis,* and *Science, Technology and Human Values,* and also the Science, Technology and Society Association (STSA) *Newsletter,* offer running bibliographies. For a range of useful items, especially on contemporary problems, *Science, Nature, New Scientist, Scientific American, Technology Review* and the *Bulletin of the Atomic Scientists* all repay study.

There is a plethora of American works on the inter-relationships between science and politics, and their effects both on scientific and on political life. From a wide range, note especially 20, 27, 114, 122, 123, 167, 183, 277, 285 and 344. For a British perspective, see 120 and 366; for samples of the experiences of 'insider' scientists in the political process, see 74, 157, 168, 381 and 389; and for samples which reflect, more generally, the experience of engagement in scientific research, see 36, 79, 159, 187, 341 and 357. Aspects of a Marxist perspective can be found in 66, 75, 132, 206, 269, 296 and 297, and in the periodicals *Radical Science Journal* (London) and *Science and Nature* (New York).

Part One

For samples of work carried out within the broad terms of the Mertonian tradition, see 51, 61, 62, 102, 103, 131, 230, 231, 247, 248, 370, 387 and 388. Other studies of the development of specialties include 79, 190 and 196; for summaries, see 49, 79 (Chapter 10) and 244; for recent, related work in 'applied' contexts, see 156 and 336. Additional references on scientific élites and their political roles include 186, 388, and the 'science and politics' literature summarized above.

Item 331 reviews the copious literature (mainly of the 1970s) which criticizes Merton's formulation of the norms of science and their operation: note also 84, 137, 230 and 243. For a Mertonian perspective on the control of deviance in science, see 387; for interesting case studies, see 108, 378 and 392. Summaries and criticism of studies of communication in science are contained in 78, 101 and 221; see also 115. On the nature and effects of competition in science, see 79 (esp. Chapters 7 and 9), 102, 110, 126 and 338. Surveys and criticism of citation, co-citation and other quantitative methods can be found in 78, 82, 115, 127, 197 and 326. Item 92 adds to the discussion on mobility.

For a review of 'scientific internationalism', see 329; discussion of the problem raised by relations with German scientists following World War I is in 94 and 309; 265 touches on scientific internationalism since 1945, in the context of the Pugwash movement.

For surveys of psychological research on scientists, see 212, 329 and 394. For discussion of the role and status of women in science, see 138, 144, 165 and 210. For the 'incorporation' of science, see 297.

Part Two

Items 281 and 282 contain the classic statements of Polanyi's notion of 'tacit knowledge': the concept is developed and applied in 288 and 383; of particular interest is Ravetz's elaboration of the importance of 'craft skills' in science.

Relationships between the sociology of science and the sociology of knowledge, explored earlier by Merton (see 226), have recently been revived: see 24, 143 and 333.

Recent (mainly European) work, much influenced by ethnomethodology, has produced empirical data on the methods by which scientific knowledge is socially constructed and similarity relations established in scientific sub-cultures: for exemplifications, see 54, 169, 187, 224, 287, 390 and 398. In this connection, 38 offers an interesting additional case study of a scientific controversy. Criticism and discussion of instrumentalist 'interests explanations' in the sociology of science can be found in *Social Studies of Science*, Vol. 11, Nos. 3 and 4 (August and November 1981).

Part Three

One byproduct of the Mertonian analysis of the norms of science was the claim that academic norms were carried by scientists into applied contexts, where they 'clashed' with management values. For statements of this view, see 58 and 174; for criticisms of these claims, see 7, 84 and 137; for summaries, see 240 and 331.

For discussions of the importance of new instrumentation in the development of science, see 79 (esp. Chapter 4) and 342. As to the process of technical innovation in general, 153, 185 and 294 offer a rich collection of case studies; excellent discussions can be found in 35 and 298. Item 60 is a classic case study of the relationship between technical and social change, and 234 a stimulating historical commentary on that relationship; see also 367. Studies of the historical relationship between science, technology, industrialization and economic growth generate controversy: see 184, 217, 284, 290 and 301 for aspects of this debate.

Part Four

For reviews of historical material relating to this section, see 316 and 317. For additional criticisms of 'finalization', see 154 and 343. For further case studies of the relationship between goal-orientation and the development of scientific theory, see 324 and 392.

Social control can be mediated through school curricula. Basil Bernstein offers an important sociological analysis of curricula: for sample statements, see 382; for a gloss, see 76. Items 192, 318 and 319 offer historical analyses of the development of science curricula: for contemporary studies of science education which raise sociological issues, see 165 and 385.

As the 'science and politics' literature (see above) demonstrates, political control of science is a pervasive phenomenon. Items 95 and 350 contain specific historical analyses of the mechanism of control in Germany: 16 sets this in a wider context. The cases of Lysenko and Oppenheimer exemplify aspects of political control, and have been extensively documented. For Lysenko, see 158, 222 and 296; 204, 328 and 393 discuss Soviet science in general. Literature on Oppenheimer is surveyed in 112 and 167: see also 183. The growth of political awareness among physicists, precipitated by involvement in the Manhattan Project (see 167), is part of a wider trend: again, the 'science and politics' literature refers; for early British experience, see 209 and 359; and several case studies annotated below are relevant (e.g. r-DNA; sociobiology; Jensen and Race/IQ). For a range of literature on scientists' response to war and military research, see 112, 167, 209, 235, 252, 265, 277, 289, 363, 366, 381 and 389.

Historically, science has often been used as a form of cultural expression by those whose interests it can serve. Items 3, 162, 313 and 346 document this phenomenon in Britain: see also 316. Changing relationships between scientific and religious thought are an aspect of this: 47 and 52 contain excellent source material; see also references in 317.

Two other topics can only be touched on here. For some discussion of the 'anti-science' literature (Ellul, Marcuse, Roszak, etc.) and associated movements, see 274, 329, 365 and 367. And for analysis of 'Appropriate Technology' (just one aspect of the much wider problem of 'Rich North/Poor South' relationships), see 66, 202 and 368.

Part Five

For discussions of the 'participatory impulse', and of the problems raised by attempts at 'democratic involvement' in technical decisions, see 17, 43, 44, 211, 256, 262, 263, 267, 323 and 402. For the growth of 'Public Interest Science', see 353. For further discussion of the Science Court proposal, see 5, 18, 220 and 260.

'Scientism' is introduced and discussed in 41, 87 and 307. The historical background to the rise of 'technocracy' is outlined in 4 and 121; the latter discusses Taylorism, as does 213. Item 253 is, in effect, a case study of a scientist approach to a social problem. Item 349 is a seminal attack on scientist pretensions. For discussions of the status and social function of 'systems theory', see 28, 50, 141 and 200. For exposition and discussion of Technology Assessment (TA), see 18, 31, 45, 63, 142, 155, 199, 345 and 374. Item 88 is a cautionary tale of the application of TA to environmental decisionmaking.

Aspects of the growth of the environmental movement are discussed in 88, 132, 272, 280, 297, 302 and 323; 251 is an excellent source for the historical background. For the Race/IQ debate, 21 and 151 (esp. the Preface: 1–67) are basic sources; 133 and 160 offer analysis, as do items in 13, 96, 235 and 297. For the sociobiology debate, 228 is a comprehensive bibliography; 1 offers both analysis and references; and 150 is in a feminist perspective. For the ABM debate, see 71, 129, 205, 285 and 344. For the r-DNA debate, see 118, 155, 194, 235, 260, 261, 312 and 320; for a very interesting account of this debate in Russia, see 204.

The basic references in the debate over risk accounting and assessment are contained in 19: see, in particular, the popular accounts by Inhaber and Rothschild, and Lowrance's book. For samples of recent discussions of risk, see 57, 113, 117, 140, 147, 164, 273 and 311. Many of these specifically discuss the risks of nuclear energy, and so merge with the nuclear power debate in general. Again, 19 contains many basic references in this debate (note especially the important articles by Gillette, fn. 125); see also 33, 59, 65, 140, 155, 163, 164, 223, 254, 261, 262, 264, 273, 285, 292, 323, 340 and 404. Disputes over the effects of low-level radiation figure prominently: see 250, 271 and 377; for the relation between these disputes and the test ban debates, see 67, 173, 183 and 265. The problems of effective disposal of radioactive waste also raise political issues: see 19, 111, 149, 163, 261, 300 and 386. For discussion of the UK Windscale Inquiry, see 19, 276 and 380. For the Three Mile Island (TMI) event, see 93, 148, 166, 215, 220 and 303.

Several items in the Bibliography are useful compilations of case studies of controversies and decisions involving technical uncertainties and/or public involvement: 33, 85, 155, 164, 235, 261, 285, 320, 323 and 397. Omitting the topics already mentioned above, the coverage is as follows: 33 offers notes and references on DDT, Concorde, thalidomide, water pollution by nutrients, road and rail transport policy, the Aswan High Dam, and car production technology; 85 is a standard reference on carcinogens — asbestos, vinyl chloride, benzene, tobacco, saccharin, pesticides, Aldrin/Dieldrin, etc.; 155 discusses environmental impacts and pesticides; 164 includes the ozone layer dispute, mercury pollution, chemicals in the work environment, and pollution of drinking water (by iron ore and asbestos); 235 touches on anti-vivisection, and environmental disputes; 261 discusses airport extensions, diethylstilbestrol (DES) in food, vinyl chloride, the automobile airbag, laetrile, smallpox immunization, fetal research, and creationism; 285 includes the supersonic transport (SST), 2,4,5-T, cyclamates, persistent pesticides, and chemical and biological warfare; 320 discusses The Pill, carcinogenicity (the Delaney amendment), DES, asbestos, and laetrile; 323 discusses asbestos; and 397 covers no less than 45 cases. In addition, there are two collections of case studies of 'marginal' sciences: 218 discusses acupuncture, continental drift, and parapsychology; 356 covers acupuncture, sea serpent sightings, UFOs, parasychology, astrology, mesmerism, creationism, and spiritualism. For polywater claims, see 406.

Supplementary references on some of these topics are as follows: DDT and pesticides, 72, 351; Concorde/SST, 198, 299; thalidomide, 339; vinyl chloride, 69; tobacco/smoking, 100, 183; the ozone layer, 70, 214, 337; chemicals at work, 36, 69, 330; anti-vivisection, 98; laetrile, 278, 403; creationism, 233, 258; 2, 4, 5-T and Seveso, 361; The Pill, 352; carcinogenicity, 268; acupuncture, 64, 405; continental drift, 216, 391, 395, 399, 400; UFOs, 360. In addition, material on the following topics and disputes are listed as follows: birth control, 68; fluoridation (the case against), 354;

low-level lead, 53, 293; medicine and public health, 145, 253, 291; microwave radiation standards, 37; oil and gas reserves, 364; PBB contamination, 48, 81; Santa Barbara oil spill, 232, 334; swine flu, 266; weather modification, 310; 'whistle blowing' by scientists, 19 (fn. 125), 80. Items 232 and 303 discuss the role of the mass media.

Bibliography

This Bibliography is not intended to be a comprehensive compilation of all relevant writings. The items listed below contain or cite a very wide range of important work. They have been selected primarily for their potential usefulness as 'ways in' to major topic areas, and as source material for case studies.

Items marked with an asterisk are cited in the Editors' introductory essays.

Each individual Reading contains its own references. Work cited in the Readings is not necessarily included in this final Bibliography.

1. Albury, W.R. (1980). 'Politics and Rhetoric in the Sociobiology Debate', *Social Studies of Science*, **10** : 519–36.
*2. Alfvèn, Hannes (1972). 'Energy and Environment', *Bulletin of the Atomic Scientists*, **28**, 5 (May) : 5–7.
3. Allen, David Elliston (1976). *The Naturalist in Britain : A Social History*. London : Allen Lane.
4. Armytage, W.H.G. (1965). *The Rise of the Technocrats : A Social History*. London & Boston, Mass.: Routledge and Kegan Paul.
5. Banks, R.S. (1980). 'The Science Court Proposal in Retrospect : A Literature Review and Case Study', *CRC Critical Reviews in Environmental Control*, **10**, 2 : 95–131.
6. Barber, Bernard and Hirsch, Walter (eds) (1962). *The Sociology of Science*. New York : The Free Press; London : Collier-Macmillan.
7. Barnes, S.B. (1971). ' "Making Out" In Industrial Research', *Science Studies*, **1** : 157–75.
8. Barnes, Barry (ed.) (1972). *Sociology of Science : Selected Readings*. Harmondsworth, Middx.: Penguin Books. Contains item 86, and extracts from items 84 & 125.
*9. Barnes, Barry (1974). *Scientific Knowledge and Sociological Theory*. London & Boston, Mass.: Routledge and Kegan Paul.
*10. Barnes, Barry (1977). *Interests and the Growth of Knowledge*. London & Boston, Mass.: Routledge and Kegan Paul.
*11. Barnes, Barry (1981). 'On the Conventional Character of Knowledge and Cognition', *Philosophy of the Social Sciences*, **11** : 303–33. German translation in Stehr & Meja (1981) : 163–90.
*12. Barnes, Barry (1982). *T.S. Kuhn and Social Science*. London : Macmillan.
*13. Barnes, Barry and Shapin, Steven (eds) (1979). *Natural Order : Historical Studies of Scientific Culture*. Beverly Hills, Calif. & London : SAGE Publications. Contains items 208 & 315.
*14. Ben-David, Joseph (1960). 'Roles and Innovations in Medicine', *American Journal of Sociology*, **65** : 557–68.

*15. Ben-David, Joseph and Collins, Randall (1966). 'Social Factors in the Origins of a New Science : The Case of Psychology', *American Sociological Review*, **31** : 451–65.

16. Ben-David, Joseph (1971). *The Scientist's Role in Society : A Comparative Study*. Englewood Cliffs, NJ : Prentice-Hall.

*17. Benveniste, Guy (1972). *The Politics of Expertise*. Berkeley, Calif.: Glendessary Press; London : Croom Helm.

18. Bereano, Philip L. (ed.) (1976). *Technology as a Social and Political Phenomenon*. New York & London : John Wiley & Sons. Contains items 43, 206 & 353.

19. Bickerstaffe, Julia and Pearce, David (1980). 'Can There Be a Consensus on Nuclear Power?', *Social Studies of Science*, **10** : 309–44.

20. Blissett, Marlan (1972). *Politics in Science*. Boston, Mass.: Little, Brown.

21. Block, Ned and Dworkin, Gerald (eds) (1976). *The IQ Controversy : Critical Readings*. New York : Random House/Pantheon Books; London : Quartet Books (1977).

*22. Bloor, David (1973). 'Wittgenstein and Mannheim on the Sociology of Mathematics', *Studies in History and Philosophy of Science*, **4** : 173–91.

*23. Bloor, David (1976). *Knowledge and Social Imagery*. London & Boston, Mass.: Routledge and Kegan Paul. Chapter 1 reprinted *in* Gebhardt (1981).

24. Bloor, David (1982). 'Durkheim and Mauss Revisited : Classification and the Sociology of Knowledge', *Studies in History and Philosophy of Science*, **13** : in press. German translation *in* Stehr & Meja (1981) : 20–51.

*25. Blume, Stuart S. (1974). *Toward a Political Sociology of Science*. New York : The Free Press; London : Collier-Macmillan.

26. Blume, Stuart (ed.) (1977). *Perspectives in the Sociology of Science*. New York & London : John Wiley & Sons.

27. Boffey, Philip M. (1975). *The Brain Bank of America : An Inquiry into the Politics of Science*. New York : McGraw-Hill.

28. Boguslaw, Robert (1965). *The New Utopians : A Study of System Design and Social Change*. Englewood Cliffs, NJ : Prentice-Hall. Extracts from Chapter 8 *in* Cross et al. (1974) : 297–301.

*29. Böhme, Gernot, van den Daele, Wolfgang and Krohn, Wolfgang (1976). 'Finalization in Science', *Social Science Information*, **15** : 307–30. Reprinted *in* Gebhardt (1981).

*30. Bourdieu, Pierre (1975). 'The Specificity of the Scientific Field and the Social Conditions of the Progress of Reason', *Social Science Information*, **14** : 19–47. Reprinted *in* Gebhardt (1981).

31. Boyle, Godfrey, Elliott, David and Roy, Robin (eds) (1977). *The Politics of Technology*. London : Longman, in association with The Open University Press. Contains item 255, and extracts of items 83 & 322.

*32. Brannigan, Augustine (1979). 'The Reification of Mendel', *Social Studies of Science*, **9** : 423–54.

33. Braun, Ernest, Collingridge, David and Hinton, Kate (1979). *Assessment of Technological Decisions – Case Studies*. London & Boston, Mass.: Butterworths/ SISCON.

*34. Braun, Ernest and MacDonald, Stuart (1978). *Revolution in Miniature : The History and Impact of Semiconductor Electronics*. Cambridge : Cambridge University Press.

35. Bright, James R. (1964). *Research, Development, and Technological Innovation : An Introduction*. Homewood, Ill.: Richard D. Irwin.

36. Brodeur, Paul (1971). 'A Reporter at Large : The Enigmatic Enzyme', *The New Yorker* (16 January 1971): 42–74.

37. Brodeur, Paul (1977). *The Zapping of America : Microwaves, Their Deadly Risk, and the Cover-Up.* New York : W.W. Norton.

38. Burchfield, Joe D. (1975). *Lord Kelvin and the Age of the Earth.* London : Macmillan; New York : Science History Publications.

*39. Burns, Tom (1969). 'Models, Images and Myths', *in* William H. Gruber and Donald G. Marquis (eds), *Factors in the Transfer of Technology* : 11–23. Cambridge, Mass. & London : The MIT Pess.

*40. Callaghan, James (1978). Speech to the Parliamentary and Scientific Committee Annual Lunch, Savoy Hotel, London, 22 February, unpublished mimeo.

*41. Cameron, Iain and Edge, David (1979). *Scientific Images and their Social Uses : An Introduction to the Concept of Scientism.* London & Boston, Mass.: Butterworths/SISCON.

42. Cardwell, Donald S.L. (1957). *The Organisation of Science in England.* London : Heinemann, Revised edition, 1972. Chapters 1 & 8 reprinted *in* Kaplan (1965) : 30–38, 86–105.

43. Carroll, James (1971). 'Participatory Technology', *Science,* **171** (19 February) : 647–53. Reprinted *in* Bereano (1976) : 492–506; and *in* Teich (1977) : 336–54.

44. Casper, Barry M. (1976). 'Technology Policy and Democracy', *Science,* **194** (1 October) : 29–35.

45. Casper, Barry M. (1978). 'The Rhetoric and Reality of Congressional Technology Assessment', *Bulletin of the Atomic Scientists,* **34**, 2 (February) : 20–31.

46. Chalmers, Alan F. (1976). *What is this Thing called Science? An Assessment of the Nature and Status of Science and Its Methods.* Atlantic Highlands, NJ : Humanities Press; Milton Keynes, Bucks. : The Open University Press (1978).

47. Chant, Colin and Fauvel, John (eds) (1980). *Darwin to Einstein : Historical Studies on Science and Belief.* Harlow, Essex : Longman Group, for The Open University; New York : Longman.

48. Chen, Edwin (1979). *PBB : An American Tragedy.* Englewood Cliffs, NJ : Prentice-Hall.

49. Chubin, Daryl E. (1976). 'The Conceptualization of Scientific Specialties', *The Sociological Quarterly,* **17** : 448–76.

50. Churchman, C. West (1979). *The Systems Approach and Its Enemies.* New York : Basic Books.

*51. Cole, Jonathan R. and Cole, Stephen (1973). *Social Stratification in Science.* Chicago & London : The University of Chicago Press.

52. Coley, Noel G. and Hall, Vance M.D. (eds) (1980). *Darwin to Einstein : Primary Sources on Science & Belief.* Harlow, Essex : Longman Group, for The Open University; New York : Longman.

*53. Collingridge, David (1980). *The Social Control of Technology.* London : Frances Pinter.

*54. Collins, H.M. (ed.) (1981a). 'Knowledge and Controversy : Studies of Modern Natural Science', *Social Studies of Science,* **11** (Special Issue) : 3–158. Contains item 348.

*55. Collins, H.M. (1981b). 'The Place of the "Core-Set" in Modern Science : Social Contingency with Methodological Propriety in Science', *History of Science,* **19** : 6–19.

*56. Collins, H.M. and Pinch, T.J. (1979). 'The Construction of the Paranormal : Nothing Unscientific is Happening', *in* Wallis (1979) : 237–70.

57. Conrad, J. (ed.) (1980). *Society, Technology and Risk Assessment.* New York &

London : Academic Press. Contains item 379.

58. Cotgrove, Stephen and Box, Steven (1979). *Science, Industry and Society*. London : George Allen and Unwin.

59. Cottrell, Sir Alan (1981). *How Safe is Nuclear Energy?* London : Heinemann.

60. Cottrell, W.F. (1951). 'Death by Dieselization : A Case Study in the Reaction to Technological Change', *American Sociological Review*, **16** : 358–65.

*61. Crane, Diana (1969). 'Social Structure in a Group of Scientists : A Test of the "Invisible College" Hypothesis', *American Sociological Review*, **34** : 335–52. Reprinted *in* Griffith (1980) : 10–27.

*62. Crane, Diana (1972). *Invisible Colleges : Diffusion of Knowledge in Scientific Communities*. Chicago & London : The University of Chicago Press.

63. Cross, Nigel, Elliott, David and Roy, Robin (eds) (1974). *Man-Made Futures : Readings in Society, Technology and Design*. London : Hutchinson, with The Open University Press. Contains extract from item 28.

64. Davis, Devra Lee (1975). 'The History and Sociology of the Scientific Study of Acupuncture', *American Journal of Chinese Medicine*, **3** : 5–26.

65. Del Sesto, Steven L. (1979). *Science, Politics and Controversy : Civilian Nuclear Power in the United States, 1946–1974*. Boulder, Col.: Westview Press.

*66. Dickson, David (1974). *Alternative Technology and the Politics of Technical Change*. London : Collins/Fontana Books.

67. Divine, Robert A. (1978). *Blowing on the Wind : The Nuclear Test Ban Debate 1954–1960*. New York : Oxford University Press.

68. Djerassi, Carl (1979). *The Politics of Contraception*. New York & London : W.W. Norton.

69. Doniger, David D. (1978). *The Law and Policy of Toxic Substances Control : A Case Study of Vinyl Chloride*. Baltimore, Md. & London : The Johns Hopkins University Press, for Resources for the Future.

70. Dotto, Lydia and Schiff, Harold (1978). *The Ozone War*. Garden City, NY : Doubleday.

71. Doty, Paul (1976). 'Science Advising and the ABM Debate', *in* Frankel (1976) : 185–203.

72. Dunlap, Thomas R. (1981). *DDT : Scientists, Citizens and Public Policy*. Princeton, NJ : Princeton University Press.

73. Durbin, Paul T. (ed.) (1980). *A Guide to the Culture of Science, Technology and Medicine*. New York : The Free Press; London : Collier-Macmillan.

74. Dyson, Freeman (1979). *Disturbing the Universe : A Life in Science*. New York : Harper and Row.

75. Easlea, Brian (1973). *Liberation and the Aims of Science : An Essay on Obstacles to the Building of a Beautiful World*. London : Chatto and Windus, for Sussex University Press; Toronto : Clarke, Irwin.

76. Edge, David (1975). 'On the Purity of Science', *in* W.R. Niblett (ed.), *The Sciences, the Humanities, and the Technological Threat* : 42–64. London : University of London Press.

*77. Edge, David (1977). 'The Sociology of Innovation in Modern Astronomy', *Quarterly Journal of the Royal Astronomical Society*, **18** : 326–39.

78. Edge, David (1979). 'Quantitative Measures of Communication in Science : A Critical Review', *History of Science*, **17** : 102–34.

*79. Edge, David O. and Mulkay, Michael J. (1976). *Astronomy Transformed : The Emergence of Radio Astronomy in Britain*. New York & London : John Wiley & Sons. German translation of Chapter 10 (abridged), entitled 'Fallstudien zu wissenschaftlichen Spezialgebieten', *in* Stehr & König (1975): 197–229.

80. Edsall, John T. (1981). 'Two Aspects of Scientific Responsibility', *Science*,

212 (3 April) : 11–14.

81. Eggington, Joyce (1980). *Bitter Harvest*. London : Secker and Warburg.

82. Elkana, Yehuda, Lederberg, Joshua, Merton, Robert, Thackeray, Arnold and Zuckerman, Harriet (eds) (1978). *Toward a Metric of Science : The Advent of Science Indicators*. New York & London : John Wiley & Sons.

83. Elliott, David and Elliott, Ruth (1976). *The Control of Technology*. London & Winchester : Wykeham Publications. Excerpt from Chapter 5 reprinted *in* Boyle et al. (1977) : 17–24.

84. Ellis, N.D. (1969). 'The Occupation of Science', *Technology and Society*, Bath University Press, **5** : 33–41. Extract reprinted *in* Barnes (1972) : 188–205.

85. Epstein, Samuel S. (1979). *The Politics of Cancer*. Garden City, NY : Anchor Press/Doubleday. Revised and expanded edition.

*86. Ezrahi, Yaron (1971). 'The Political Resources of American Science', *Science Studies*, **1** : 117–33. Reprinted *in* Barnes (1972) : 211–30.

87. Ezrahi, Yaron (1974). 'The Authority of Science in Politics', *in* A. Thackray and E. Mendelsohn (eds), *Science and Values* : 215–51. New York : Humanities Press.

88. Fairfax, Sally K. (1978). 'A Disaster in the Environmental Movement', *Science*, **199** (17 February) : 743–48. Also Letters, *ibid.*, **202** (8 December) : 1034–41; *ibid.*, **203** (2 February (1979) : 400.

*89. Ferguson, Eugene S. (1977). 'The Mind's Eye : Non-verbal Thought in Technology', *Science*, **197** (26 August) : 827–35.

*90. Feyerabend, Paul (1975). *Against Method : Outline of an Anarchistic Theory of Knowledge*. Atlantic Highlands, NJ : Humanities Press; London : New Left Books.

*91. Fleck, Ludwick (1935). *Entstehung und Entwicklung einer Wissenschaftlichen Tatsache*. English translation published as *Genesis and Development of a Scientific Fact* (1979). Chicago & London : The University of Chicago Press. New German edition (1980), with an introduction by Lother Schäfer and Thomas Schnelle. Frankfurt am Main : Suhrkamp.

92. Fleming, D. (1968). 'Emigré Physicists and the Biological Revolution', *Perspectives in American History*, **2** : 152–89.

93. Ford, Daniel (1981). 'A Reporter at Large : Three Mile Island – Part I : Class Nine Accident', *The New Yorker* (6 April) : 49–120; 'Part II : The Paper Trail', *ibid.* (13 April) : 46–109.

94. Forman, Paul (1973). 'Scientific Internationalism and the Weimar Physicists : The Ideology and Its Manipulation in Germany after World War I', *Isis*, **64** : 151–80.

95. Forman, Paul (1974). 'The Financial Support and Political Alignment of Physicists in Weimar Germany', *Minerva*, **12** : 39–66.

96. Frankel, Charles (ed.) (1976). *Controversies and Decisions : The Social Sciences and Public Policy*. New York : Russell Sage Foundation. Contains item 71.

*97. Freeman, Christopher (1974). *The Economics of Industrial Innovation*. Harmondsworth, Middx.: Penguin Books.

98. French, Richard D. (1975). *Antivivisection and Medical Science in Victorian Society*. Princeton, NJ & London : Princeton University Press.

*99. Friedkin, Noah (1978). 'University Social Structure and Social Networks among Scientists', *American Journal of Sociology*, **83** : 1444–65.

100. Friedman, Kenneth Michael (1975). *Public Policy and the Smoking-Health Controversy*. Lexington, Mass. & London : D.C. Heath.

101. Garvey, William D. (1979). *Communication : The Essence of Science*. Elmsford, NY & Oxford : Pergamon Press.

102.   Gaston, Jerry (1973). *Originality and Competition in Science : A Study of the British High Energy Physics Community.* Chicago & London : The University of Chicago Press.

103.   Gaston, Jerry (1978a). *The Reward System in British and American Science.* New York & Chichester, Sussex : John Wiley & Sons.

104.   Gaston, Jerry (ed.) (1978b). *The Sociology of Science.* San Francisco, Calif. & London : Jossey-Bass. Contains item 331.

105.   Gebhardt, Eike (ed.) (1981). *Science as Behavior : Sociological Approaches to the Sciences.* New York : Continuum. Contains items 29, 30 & 238, and extracts of 23 & 383.

*106.   Geison, Gerald L. (1981). 'Scientific Change, Emerging Specialties, and Research Schools', *History of Science,* **19** : 20–38.

*107.   Gibbons, Michael and Johnston, Ron (1974). 'The Role of Science in Technological Innovation', *Research Policy,* **3** : 220–42.

108.   Gibbons, Michael and King, Philip (1972). 'The Development of Ovonic Switches – A Case Study of a Scientific Controversy', *Science Studies,* **2** : 295–309.

109.   Gieryn, Thomas (ed.) (1980). *Science and Social Structure : A Festschrift for Robert K. Merton.* Transactions of the New York Academy of Sciences, Series II, **39**. Contains item 243.

110.   Gilbert, G. Nigel (1978). 'Competition, Differentiation and Careers in Science', *Social Science Information,* **16** : 103–23.

111.   Goodin, Robert E. (1978). 'Uncertainty as an Excuse for Cheating our Children : The Case of Nuclear Wastes', *Policy Sciences,* **10** : 25–43.

112.   Gowing, Margaret and Arnold, Lorna (1979). *The Atomic Bomb.* London & Boston, Mass.: Butterworths/SISCON.

113.   Green, C.H. and Brown, R.A. (1980. 'The Acceptability of Risk', *Science and Public Policy,* **7** : 307–18.

*114.   Greenberg, Daniel S. (1967). *The Politics of Pure Science : An Inquiry into the Relationship between Science and Government in the United States.* New York : New American Library. Revised paperback reprint as a Plume Book (1971). Published in Britain as *The Politics of American Science.* Harmondsworth, Middx.: Penguin Books (1969).

115.   Griffith, Belver C. (ed.) (1980). *Key Papers in Information Science.* White Plains, NY : Knowledge Industry Publ., for the American Society for Information Science. Contains items 61 & 116.

*116.   Griffith, Belver and Mullins, Nicholas C. (1972). 'Coherent Social Groups in Scientific Change', *Science,* **177** (15 September) : 959–64. Reprinted *in* Griffith (1980) : 52–57.

117.   Griffiths, Richard F. (ed.) (1981). *Dealing with Risk : The Planning, Management and Acceptability of Technological Risk.* Manchester : Manchester University Press.

118.   Grobstein, Clifford (1979). *A Double Image of the Double Helix : The Recombinant-DNA Debate.* San Francisco, Calif.: W.H. Freeman.

*119.   Groeneveld, Lyle, Koller, Norman and Mullins, Nicholas C. (1975). 'The Advisers of the United States National Science Foundation', *Social Studies of Science,* **5** : 343–54.

120.   Gummett, Philip (1980). *Scientists in Whitehall.* Manchester : Manchester University Press.

121.   Haber, Samuel (1964). *Efficiency and Uplift : Scientific Management in the Progressive Era.* Chicago & London : The University of Chicago Press.

122.   Haberer, Joseph (1969). *Politics and the Community of Science.* New York &

London : Van Nostrand Reinhold.

123. Haberer, Joseph (1972). 'Politicalization in Science', *Science*, **178** (17 November) : 713–24.

*124. Habermas, Jürgen (1971). 'The Scientization of Politics and Public Opinion', *in Toward a Rational Society* : 62–80. Translated by Jeremy J. Shapiro. London : Heinemann.

*125. Hagstrom, Warren O. (1965). *The Scientific Community*. New York & London : Basic Books. Reprinted as an Arcturus Paperback (1975). Carbondale, Ill.: Southern Illinois University Press; London & Amsterdam : Feffer & Simons. Extracts reprinted *in* Barnes (1972) : 102–25.

126. Hagstrom, Warren (1974). 'Competition in Science', *American Sociological Review*, **39** : 1–18.

127. Hahn, Roger (1980). *A Bibliography of Quantitative Studies on Science and its History*. Berkeley, Calif.: The University of California. *Berkeley Papers in History of Science* No. 3.

128. Halmos, Paul (ed.) (1972). *The Sociology of Science*. Keele, Staffs.: *Sociological Review Monograph* No. 18.

129. Halperin, Morton H. (1972). 'The Decision to Deploy the ABM : Bureaucratic and Domestic Politics in the Johnson Administration', *World Politics*, **25** (Autumn) : 62–95.

130. Hardin, Garrett (1969). *Science and Controversy : Population, a Case Study*. San Francisco, Calif.: W.H. Freeman. A commentary on readings in G. Hardin (ed.), *Population, Evolution and Birth Control : A Collage of Controversial Ideas*. San Francisco, Calif.: W.H. Freeman.

131. Hargens, Lowell L., Mullins, Nicholas C. and Hecht, Pamela K. (1980). 'Research Areas and Stratification Processes in Science', *Social Studies of Science*, **10** : 55–74.

132. Harvey, David (1974). 'Population, Resources, and the Ideology of Science', *Economic Geography*, **50** : 256–77.

133. Harwood, Jonathan (1976, 1977). 'The Race-Intelligence Controversy : A Sociological Approach. I : Professional Factors', *Social Studies of Science*, **6** : 369–94; 'II : "External" Factors', *ibid.*, **7** : 1–30.

*134. Hesse, Mary (1970). 'Is There an Independent Observation Language?', *in* R.G. Colodny (ed.), *The Nature and Function of Scientific Theories* : 35–77. Pittsburgh, Pa.: University of Pittsburgh Press. Reprinted *in* Hesse (1974) : 9–44.

*135. Hesse, Mary (1974). *The Structure of Scientific Inference*. London : Macmillan. Contains item 134.

*136. Hesse, Mary (1980). *Revolutions and Reconstructions in the Philosophy of Science*. Brighton, Sussex : The Harvester Press.

137. Hill, Stephen (1974). 'Questioning the Influence of a "Social System of Science" : A Study of Australian Scientists', *Science Studies*, **4** : 135–63.

138. Hinton, Kate (1976). *Women and Science*. SISCON Pamphlet. Manchester : Department of Liberal Studies in Science, Manchester University.

139. Hoffmann, Erik P. (1979). 'Contemporary Soviet Theories of Scientific, Technological and Social Change', *Social Studies of Science*, **9** : 101–13.

140. Hohenemser, Christoph, Kasperson, Roger and Kates, Robert (1977). 'The Distrust of Nuclear Power', *Science*, **196** (1 April) : 25–34.

141. Hoos, Ida R. (1972). *Systems Analysis in Public Policy : A Critique*. Berkeley, Calif. & London : University of California Press.

142. Hoos, Ida R. (1979). 'Societal Aspects of Technology Assessment', *Technological Forecasting and Social Change*, **13** : 191–202.

143. Horton, Robin and Finnegan, Ruth (eds) (1973). *Modes of Thought : Essays on Thinking in Western and Non-Western Societies*. London : Faber and Faber.

144. Høyrup, Else (1978). *Women and Mathematics, Science and Engineering : A Partially Annotated Bibliography with Emphasis on Mathematics and with References on Related Topics*. Roskilde, Denmark : Roskilde University Library.

145. Hughes, Richard and Brewin, Robert (1979). *The Tranquilizing of America : Pill Popping and the American Way of Life*. New York & London : Harcourt Brace Jovanovich.

*146. Hughes, Thomas Parke (1976). 'The Science-Technology Interaction : The Case of High-Voltage Power Transmission Systems', *Technology and Culture*, 17 : 646–62.

147. *IAEA Bulletin* (1980). 'The Respective Risks of Different Energy Sources' (7 Symposium Papers), *International Atomic Energy Agency Bulletin*, 22, 5/6 (October) : 35–128.

148. *IEEE Spectrum* (1979). Special Issue on 'Three Mile Island and the Future of Nuclear Power', *IEEE Spectrum*, 16, 11 (November).

149. Jakimo, Alan and Bupp, Irvin C. (1978). 'Nuclear Waste Disposal : Not in My Backyard', *Technology Review*, 80, 5 (March/April) : 64–72.

150. Janson-Smith, Deirdre (1980). 'Sociobiology : So What?', *in* Brighton Women and Science Group, *Alice Through the Microscope : The Power of Science over Women's Lives*, Chapter 3 : 62-86. London : Virago.

151. Jensen, Arthur R. (1972). *Genetics and Education*. London : Methuen.

152. Jevons, F.R. (1973). *Science Observed : Science as a Social and Intellectual Activity*. London : George Allen and Unwin.

153. Jewkes, John, Sawers, David and Stillerman, Richard (1969). *The Sources of Invention*. London : Macmillan. Second edition.

154. Johnston, Ron (1976). 'Finalization : A New Start for Science Policy?', *Social Science Information*, 15 : 331–36.

155. Johnston, Ron and Gummett, Philip (eds) (1979). *Directing Technology : Policies for Promotion and Control*. London : Croom Helm.

*156. Johnston, Ron and Robbins, Dave (1977). 'The Development of Specialties in Industrialised Science', *Sociological Review*, 25 : 87–108.

157. Jones, Reginald V. (1978). *Most Secret War*. London : Hamish Hamilton. Published in the USA as *The Wizard War : British Scientific Intelligence, 1939–1945*. New York : Coward, McCann and Geoghegan.

158. Joravsky, David (1970). *The Lysenko Affair*. Cambridge, Mass.: Harvard University Press.

159. Judson, Horace Freeland (1979). *The Eighth Day of Creation : Makers of the Revolution in Biology*. New York : Simon and Schuster.

160. Kamin, Leon J. (1974). *The Science and Politics of I.Q.* Potomac, Md.: Lawrence Erlbaum Associates, and John Wiley & Sons.

161. Kaplan, Norman (ed.) (1965). *Science and Society*. Chicago : Rand McNally. Includes extracts from item 42.

162. Kargon, Robert H. (1977). *Science in Victorian Manchester : Enterprise and Expertise*. Manchester : Manchester University Press.

163. Kasperson, Roger E., Berk, Gerald, Pijawka, David, Sharaf, Alan B. and Wood, James (1980). 'Public Opposition to Nuclear Energy : Retrospect and Prospect', *Science, Technology and Human Values*, No. 31 (Spring) : 11–23.

164. Kates, Robert W. (1978). *Risk Assessment of Environmental Hazard*. Chichester, Sussex & New York : John Wiley & Sons.

165. Kelly, Alison (1979). *Girls and Science. Monographs of the International Association for the Evaluation of Educational Achievement* No. 9. Stockholm : Almqvist &

Wiksell International.

166. Kemeny, John G. (1980). 'Saving American Democracy : The Lessons of Three Mile Island', *Technology Review*, **83**, 7 (June/July) : 64–75.

167. Kevles, Daniel J. (1978). *The Physicists : The History of a Scientific Community in Modern America*. New York : Alfred A. Knopf.

168. Kistiakowsky, George B. (1976). *A Scientist at the White House : The Private Diary of President Eisenhower's Special Assistant for Science and Technology*. Cambridge, Mass. & London: Harvard University Press.

169. Knorr-Cetina, Karin D. (1981). *The Manufacture of Knowledge: An Essay on the Constructivist and Contextual Nature of Science*. Oxford & New York : Pergamon Press.

*170. Knorr-Cetina, Karin D. (1982). 'Scientific Communities or Transepistemic Arenas of Research? A Critique of Quasi-Economic Models of Science', *Social Studies of Science*, **12** : 101–30.

171. Knorr, K.D., Krohn, R. and Whitley, R.D. (eds) (1980). *The Social Process of Scientific Investigation. Sociology of the Sciences Yearbook*, **4**. Dordrecht, Boston, Mass. & London : Reidel.

172. Knorr, Karin, Strasser, Hermann and Zilian, Hans Georg (eds) (1977). *Determinants and Controls of Scientific Development*. Dordrecht, Boston, Mass. & London : Reidel.

173. Kopp, Carolyn (1979). 'The Origins of the American Scientific Debate over Fallout Hazards', *Social Studies of Science*, **9** : 403–22.

174. Kornhauser, William (1962). *Scientists in Industry: Conflict and Accommodation*. Berkeley, Calif.: University of California Press; London : Cambridge University Press.

*175. Kreilkamp, Karl (1971). '*Hindsight* and the Real World of Science Policy', *Science Studies*, **1** : 43–66.

*176. Krohn, Roger G. (1971). *The Social Shaping of Science : Institutions, Ideology, and Careers in Science*. Westport, Conn. & London : Greenwood Publ.

*177. Krohn, Wolfgang, Layton, Edwin and Weingart, Peter (eds) (1978). *The Dynamics of Science and Technology. Sociology of the Sciences Yearbook*, **2**. Dordrecht, Boston, Mass. & London : Reidel.

*178. Kuhn, T.S. (1961). 'Sadi Carnot and the Cargard Engine', *Isis*, **52** : 367–74.

*179. Kuhn, Thomas S. (1962). *The Structure of Scientific Revolutions*. Chicago : The University of Chicago Press. Second edition, enlarged (1970).

*180. Kuhn, Thomas S. (1970). 'Postscript – 1969', *in* Kuhn (1962), Second edition (1970) : 174–210.

*181. Kuhn, Thomas S. (1974). 'Second Thoughts on Paradigms', *in* F. Suppe (ed.), *The Structure of Scientific Theories* : 459–82. Urbana, Ill.: University of Illinois Press. Reprinted *in* Kuhn (1977) : 293–319.

*182. Kuhn, Thomas S. (1977). *The Essential Tension : Selected Studies in Scientific Tradition and Change*. Chicago & London : The University of Chicago Press. Contains item 181.

183. Lakoff, Sanford A. (ed.) (1966). *Knowledge and Power : Essays on Science and Government*. New York : The Free Press; London : Collier-Macmillan.

184. Landes, David S. (1969). *The Unbound Prometheus : Technological Change and Industrial Development in Western Europe from 1750 to the Present*. Cambridge : Cambridge University Press.

185. Langrish, J., Gibbons, M., Evans, W.G. and Jevons, F.R. (1972). *Wealth from Knowledge*. London : Macmillan.

186. Lapp, Ralph E. (1965). *The New Priesthood : The Scientific Elite and the Uses of Power*. New York : Harper and Row.

*187. Latour, Bruno and Woolgar, Steve (1979). *Laboratory Life : The Social Construction of Scientific Facts*. Beverly Hills, Calif. & London : SAGE Publications.

*188. Laudan, Larry (1977). *Progress and its Problems : Towards a Theory of Scientific Growth*. London & Boston, Mass.: Routledge and Kegan Paul.

*189. Lavine, Thelma Z. (1942). 'Sociological Analysis of Cognitive Norms', *Journal of Philosophy*, **39** : 342–56.

*190. Law, John (1973). 'The Development of Specialties in Science : The Case of X-ray Protein Crystallography', *Science Studies*, **3** : 275–303.

*191. Law, John (1980). 'Fragmentation and Investment in Sedimentology', *Social Studies of Science*, **10** : 1–22.

192. Layton, David (1973). *Science for the People : The Origins of the School Science Curriculum in England*. London : George Allen and Unwin.

*193. Layton, E. (1977). 'Conditions of Technological Development', *in* Spiegel-Rösing & Price (1977) : 197–222.

194. Lear, John (1978). *Recombinant DNA : The Untold Story*. New York : Crown Publishers.

195. Lécuyer, Bernard (1978). 'Bilan et Perspectives de la Sociologie de la Science', *Archives Européennes de Sociologie*, **19**, 2 (September) : 257–336.

196. Lemaine, Gérard, McLeod, Roy, Mulkay, Michael and Weingart, Peter (eds) (1976). *Perspectives on the Emergence of Scientific Disciplines*. The Hague & Paris : Mouton; Chicago : Aldine. Contains item 372.

197. Lenoir, Timothy (1979). 'Quantitative Foundations for the Sociology of Science : On Linking Blockmodeling with Co-Citation Analysis', *Social Studies of Science*, **9** : 455–80.

198. Levy, Elizabeth (1973). *The People Lobby : The SST Story*. New York : Delacorte Press.

199. *Lex et Scientia* (1974). Special Issue on Technology Assessment. *Lex et Scientia*, **10**, 4.

200. Lilienfeld, Robert (1977). *The Rise of Systems Theory : An Ideological Analysis*. New York : John Wiley & Sons.

*201. Lindblom, C. (1968). *The Decision Making Process*. Englewood Cliffs, NJ : Prentice-Hall.

202. Long, Franklin A. and Oleson, Alexandra (eds) (1980). *Appropriate Technology and Social Values – A Critical Appraisal*. Cambridge, Mass.: Ballinger; London : Harper and Row.

203. Lubrano, Linda L. (1976). *Soviet Sociology of Science*. Columbus, Ohio : American Association for the Advancement of Slavic Studies.

204. Lubrano, Linda L. and Solomon, Susan Gross (eds) (1980). *The Social Context of Soviet Science*. Boulder, Col.: Westview Press; Folkestone, Kent : Dawson.

205. McCracken, Daniel D. (1971). *Public Policy and the Expert : Ethical Problems of the Witness*. New York : The Council of Religion and International Affairs.

*206. McDermott, John (1969). 'Technology : The Opiate of the Intellectuals', *New York Review of Books* (31 July) : 25–35. Reprinted *in* Bereano (1976) : 78–99; and *in* Teich (1977) : 180–207.

*207. MacKenzie, Donald A. (1981). *Statistics in Britain, 1865–1930 : The Social Construction of Scientific Knowledge*. Edinburgh : Edinburgh University Press.

*208. MacKenzie, Donald and Barnes, Barry (1979). 'Scientific Judgment : The Biometry – Mendelism Controversy', *in* Barnes & Shapin (1979) : 191–210. German version *in* Stehr & König (1975) : 165–96.

209. MacLeod, Roy and MacLeod, Kay (1979). 'The Contradictions of Professionalism : Scientists, Trade Unionism and the First World War', *Social*

Studies of Science, **9** : 1–32.

210. MacLeod, Roy and Moseley, Russell (1979). 'Fathers and Daughters : Reflections on Women, Science and Victorian Cambridge', *History of Education,* **8** : 321–33.

211. MacRae, Duncan, Jr (1976). 'Technical Communities and Political Choice', *Minerva,* **14** : 169–90.

212. Mahoney, Michael J. (1979). 'Psychology of the Scientist : An Evaluative Review', *Social Studies of Science,* **9** : 349–75.

213. Maier, Charles S. (1970). 'Between Taylorism and Technocracy : European Ideologies and the Vision of Industrial Productivity in the 1920s', *Contemporary History,* **5** : 27–61.

*214. Martin, Brian (1979). *The Bias of Science.* Canberra : ACT Society for Social Responsibility in Science.

215. Martin, Daniel (1980). *Three Mile Island : Prologue or Epilogue?* Cambridge, Mass.: Ballinger/Harper and Row.

216. Marvin, Ursula B. (1974). *Continental Drift : The Evolution of a Concept.* Washington, DC : Smithsonian Institution Press. Second printing, revised.

217. Mathias, Peter (1972). 'Who Unbound Prometheus? Science and Technical Change, 1600–1800', *in* P. Mathias (ed.), *Science and Society, 1600–1900* : 54–80. Cambridge : Cambridge University Press.

218. Mauskopf, Seymour H. (ed.) (1979). *The Reception of Unconventional Science.* Boulder, Col.: Westview Press, for the AAAS.

*219. Mazur, Allan (1977). 'Science Courts', *Minerva,* **15** : 1–14.

*220. Mazur, Allan (1981). *The Dynamics of Technical Controversy.* Washington, DC : Communications Press. Chapter 2 contains 'Disputes Between Experts', *Minerva,* **11** (1973) : 243–62.

221. Meadows, A.J. (1974). *Communication in Science.* London & Boston, Mass.: Butterworths.

222. Medvedev, Zhores A. (1969). *The Rise and Fall of T.D. Lysenko.* Translated by I. Michael Lerner. New York & London : Columbia University Press.

223. Mendelsohn, Everett, Weingart, Peter and Whitley, Richard (eds) (1977). *The Social Production of Scientific Knowledge. Sociology of the Sciences Yearbook,* **1**. Dordrecht, Boston, Mass. & London : Reidel.

224. Mendelsohn, Everett & Elkana, Yehuda (eds) (1981). *Sciences and Cultures : Anthropological and Historical Studies of the Sciences. Sociology of the Sciences Yearbook,* **5**. Dordrecht, Boston, Mass. & London : Reidel.

*225. Merton, Robert K. (1957). *Social Theory and Social Structure.* New York : The Free Press; London : Collier-Macmillan. Enlarged edition (1968).

*226. Merton, Robert K. (1973). *The Sociology of Science : Theoretical and Empirical Investigations.* Edited and introduced by Norman W. Storer. Chicago & London : The University of Chicago Press.

227. Merton, Robert and Gaston, Jerry (eds) (1977). *The Sociology of Science in Europe.* Carbondale, Ill.: Southern Illinois University Press; London & Amsterdam : Feffer and Simons.

228. Miller, Alan V. (1979). *The Genetic Imperative : Fact and Fantasy in Sociobiology : A Bibliography.* Toronto : Pink Triangle Press. Canadian Gay Archives Publication No. 2.

*229. Miller, Jonathan (1972). 'The Dog Beneath the Skin', *The Listener* (London) (20 July) : 74–76.

*230. Mitroff, Ian (1984a). 'Norms and Counter-Norms in a Select Group of the Apollo Moon Scientists', *American Sociological Review,* **39** : 579–95.

231. Mitroff, Ian I. (1974b). *The Subjective Side of Science : A Philosophical Inquiry into*

the Psychology of the Apollo Moon Scientists. Amsterdam : Elsevier.

232.   Molotch, Harvey and Lester, Marilyn (1975) 'Accidental News : The Great Oil Spill as Local Occurrence and National Event', *American Journal of Sociology*, **81** : 235–60.

*233.   Moore, John A. (1974). 'Creationism in California', *Daedalus*, **103**, 3 (Summer) : 173–89.

234.   Morison, Elting E. (1966). *Men, Machines and Modern Times*. Cambridge, Mass. & London : The MIT Press.

235.   Morley, David (1978). *The Sensitive Scientist : Report of a British Association Study Group*. London : SCM Press.

236.   Mulkay, M.J. (1972). *The Social Process of Innovation : A Study in the Sociology of Science*. London : Macmillan.

*237.   Mulkay, Michael (1974). 'Conceptual Displacement and Migration in Science : A Prefatory Paper', *Science Studies*, **4** : 205–34.

*238.   Mulkay, Michael (1976a). 'Norms and Ideology in Science', *Social Science Information*, **15** : 637–56. Reprinted *in* Gebhardt (1981).

*239.   Mulkay, Michael (1976b). 'The Mediating Role of the Scientific Elite', *Social Studies of Science*, **6** : 445–70.

*240.   Mulkay, M.J. (1977). 'Sociology of the Scientific Research Community', *in* Spiegel-Rösing & Price (1977) : 93–148.

*241.   Mulkay, Michael (1979a). *Science and the Sociology of Knowledge*. London : George Allen and Unwin.

*242.   Mulkay, Michael (1979b). 'Knowledge and Utility : Implications for the Sociology of Knowledge', *Social Studies of Science*, **9** : 63–80. German translation *in* Stehr & Meja (1981) : 52–72.

243.   Mulkay, Michael (1980). 'Interpretation and the Use of Rules : The Case of the Norms of Science', *in* Gieryn (1980) : 111–25.

*244.   Mulkay, Michael, Gilbert, G. Nigel and Woolgar, Steve (1975). 'Problem Areas and Research Networks in Science', *Sociology*, **9** : 187–203.

245.   Mulkay, Michael and Milić, Vojin (1980). 'The Sociology of Science in East and West', *Current Sociology*, **28**, 3 (Winter) : 1–342.

*246.   Mulkay, M.J. and Turner, B.S. (1971). 'Over-Production of Personnel and Innovation in Three Social Settings', *Sociology*, **5** : 47–61.

*247.   Mullins, Nicholas C. (1968). 'The Distribution of Social and Cultural Properties in Informal Communication Networks among Biological Scientists', *American Sociological Review*, **33** : 786–97.

*248.   Mullins, Nicholas C. (1972). 'The Development of a Scientific Specialty: The Phage Group and the Origins of Molecular Biology', *Minerva*, **10** : 51–82.

249.   Mullins, Nicholas C. (1973). *Science: Some Sociological Perspectives*. Indianapolis & New York: Bobbs-Merrill.

250.   Najarian, Thomas (1978). 'The Controversy over the Health Effects of Radiation', *Technology Review*, **81**, 2 (November) : 74–82.

251.   Nash, Roderick (ed.) (1976). *The American Environment : Readings in the History of Conservation*. Reading, Mass. & London: Addison-Wesley. Second edition.

252.   Nelkin, Dorothy (1972). *The University and Military Research: Moral Politics at MIT*. Ithaca, NY & London: Cornell University Press.

253.   Nelkin, Dorothy (1973). *Methadone Maintenance: A Technological Fix*. New York: George Braziller.

254.   Nelkin, Dorothy (1974). 'The Role of Experts in a Nuclear Siting Controversy', *Bulletin of the Atomic Scientists*, **30**, 9 (November) : 29–36.

*255.   Nelkin, Dorothy (1975). 'The Political Impact of Technical Expertise',

Social Studies of Science, **5** : 35–54. Reprinted *in* Boyle et al. (1977) : 189–205.

256. Nelkin, Dorothy (1977a). *Technological Decisions and Democracy: European Experiments in Public Participation*. Beverly Hills, Calif. & London: SAGE Publications.

257. Nelkin, Dorothy (1977b). *Science Textbook Controversies and the Politics of Equal Time*. Cambridge, Mass. & London: The MIT Press.

*258. Nelkin, Dorothy (1977c). 'Scientists and Professional Responsibility: The Experience of American Ecologists', *Social Studies of Science*, **7** : 75–95.

*259. Nelkin, Dorothy (1977d). 'Thoughts on the Proposed Science Court', *Harvard Newsletter on Science, Technology and Human Values*, No. 18 (January) : 20–31.

260. Nelkin, Dorothy (1978). 'Threats and Promises: Negotiating the Control of Research', *Daedalus*, **107**, 2 (Spring): 191–209.

*261. Nelkin, Dorothy (ed.) (1979). *Controversy: Politics of Technical Decisions*. Beverly Hills, Calif. & London: SAGE Publications.

262. Nelkin, Dorothy and Fallows, Susan (1978). 'The Evolution of the Nuclear Debate: The Role of Public Participation', *Annual Review of Energy*, **3** : 275–312.

*263. Nelkin, Dorothy and Pollak, Michael (1979). 'Public Participation in Technological Decisions: Reality or Grand Illusion?', *Technology Review*, **81**, 8 (August/September) : 55–64.

264. Nelkin, Dorothy and Pollak, Michael (1980). *The Atom Besieged: Extraparliamentary Dissent in France and Germany*. Cambridge, Mass. & London: The MIT Press.

265. Nelson, William R. (ed.) (1968). *The Politics of Science: Readings in Science, Technology, and Government*. New York & London: Oxford University Press.

266. Neustadt, Richard E. and Fineberg, Harvey V. (1978). *The Swine Flu Affair: Decision-Making on a Slippery Disease*. Washington, DC: US Department of Health, Education and Welfare.

267. Nichols, K. Guild (1979). *Technology on Trial: Public Participation in Decision-Making related to Science and Technology*. Paris: OECD.

268. Nicholson, William J. (ed.) (1981). *Management of Assessed Risk for Carcinogens. Annals of the New York Academy of Sciences*, **363** (30 April): 1–301.

*269. Noble, David F. (1977). *American by Design: Science, Technology, and the Rise of Corporate Capitalism*. New York: Alfred A. Knopf.

*270. Norman, Colin (1979). *Knowledge and Power: The Global Research and Development Budget*. Washington, DC: Worldwatch Institute, Worldwatch Paper No. 31. Abridged to 'Global Research: Who Spends What', *New Scientist* (26 July) : 279–81.

271. Nowotny, H. and Hirsch, H. (1980). 'The Consequences of Dissent: Sociological Reflections on the Controversy of the Low Dose Effects', *Research Policy*, **9** : 278–94.

272. Ophuls, William (1977). *Ecology and the Politics of Scarcity: Prologue to a Political Theory of The Steady State*. San Francisco, Calif.: W.H. Freeman.

273. Otway, H.J. (1977). 'Risk Assessment and the Social Response to Nuclear Power', *Journal of the British Nuclear Energy Society*, **16** : 327–33.

274. Passmore, John (1978). *Science and its Critics*. Brunswick, NJ : Rutgers University Press; London: Duckworth.

*275. Pavitt, Keith and Worboys, Michael (1977). *Science, Technology and the Modern Industrial State*. London & Boston, Mass.: Butterworths/SISCON.

276. Pearce, David, Edwards, Lynne and Beuret, Geoff (1969). *Decision Making for Energy Futures: A Case Study of the Windscale Inquiry*. London: Macmillan.

277. Penick, James L., Jr, Pursell, Carroll W., Jr, Sherwood, Morgan B. and Swain, Donald C. (eds) (1972). *The Politics of American Science: 1939 to the Present*. Cambridge, Mass. & London : The MIT Press. Revised edition.

278. Petersen, James C. and Markle, Gerald E. (1979). 'Politics and Science in the Laetrile Controversy', *Social Studies of Science*, **9** : 139–66.

*279. Pfetsch, Frank R. (1979). 'The "Finalization" Debate in Germany: Some Comments and Explanations', *Social Studies in Science*, **9** : 115–24.

280. Pilat, J.F. (1980). *Ecological Politics: The Rise of the Green Movement*. Beverly Hills, Calif. & London: SAGE Publications.

281. Polanyi, Michael (1958). *Personal Knowledge: Towards a Post-Critical Philosophy*. London: Routledge and Kegan Paul; Chicago: The University of Chicago Press. Reprinted New York: Harper Torchbooks (1964).

282. Polanyi, Michael (1967). *The Tacit Dimension*. London: Routledge and Kegan Paul.

*283. Price, Derek J. de Solla (1963). *Little Science, Big Science*. New York & London: Columbia University Press.

284. Price, Derek J. de Solla (1965). 'Is Technology Historically Independent of Science? A Study in Statistical Historiography', *Technology and Culture*, **6** : 553–68.

285. Primack, Joel and von Hippel, Frank (1974). *Advice and Dissent: Scientists in the Political Arena*. New York: Basic Books.

*286. Provine, William B. (1973). 'Geneticists and the Biology of Race Crossing', *Science*, **182** (23 November) : 790–96.

287. Raffel, Stanley (1979). *Matters of Fact: A Sociological Inquiry*. London & Boston, Mass.: Routledge and Kegan Paul.

*288. Ravetz, Jerome R. (1971). *Scientific Knowledge and its Social Problems*. Oxford: Clarendon Press.

289. Reid, Robert W. (1969). *Tongues of Conscience: War and the Scientists' Dilemma*. London: Constable; New York: Walker.

290. Reingold, Nathan and Molella, Arthur (eds) (1976). 'The Interaction of Science and Technology in the Industrial Age', *Technology and Culture*, **17**, 4 (October) : 621–745.

291. Reiser, Stanley Joel (1978). *Medicine and the Reign of Technology*. Cambridge & New York: Cambridge University Press.

292. Reynolds, W.C. (ed.) (1976). *The California Nuclear Initiative: Analysis and Discussion of Issues*. Stanford, Calif.: Stanford University Institute for Energy Studies.

*293. Robbins, David and Johnston, Ron (1976). 'The Role of Cognitive and Occupational Differentiation in Scientific Controversies', *Social Studies of Science*, **6** : 349–68.

294. Rogers, Everett M. (1962). *Diffusion of Innovations*. New York: The Free Press; London: Collier-Macmillan.

295. Rose, Hilary and Rose, Steven (1969). *Science and Society*. London: Allen Lane The Penguin Press. Reprinted Harmondsworth, Middx.: Pelican Books (1970).

296. Rose, Hilary and Rose, Steven (eds) (1976a). *The Radicalisation of Science: Ideology of/in the Natural Sciences*. London: Macmillan.

297. Rose, Hilary and Rose, Steven (eds) (1976b). *The Political Economy of Science: Ideology of/in the Natural Sciences*. London: Macmillan.

298. Rosenberg, Nathan (1976). *Perspectives on Technology*. London & New York: Cambridge University Press.

299. Rosenbloom, Joshua (1981). 'The Politics of the American SST Programme:

Origin, Opposition, and Termination', *Social Studies of Science,* **11** : 403–23.

300. Roy, Rustum (1981). 'The Technology of Nuclear-Waste Management', *Technology Review,* **83**, 5 (April) : 39–50.

301. Russell, C.A. and Goodman, D.C. (1972). *Science and the Rise of Technology since 1800.* Bristol: John Wright & Sons, with The Open University Press.

302. Sandbach, Francis (1980). *Environment, Ideology and Policy.* Oxford: Basil Blackwell; Montclair, NJ: Allanheld.

303. Sandman, Peter M. and Paden, Mary (1979). 'At Three Mile Island', *Columbia Journalism Review,* **18** (June/July) : 43–58.

*304. Sapolsky, Harvey M. (1968). 'Science, Voters, and the Fluoridation Controversy', *Science,* **162** (25 October) : 427–33.

*305. Sapolsky, Harvey (1972). *The Polaris System Development.* Cambridge, Mass.: Harvard University Press.

*306. Schäfer, W. (1979). 'Finalisation in Perspective: Toward a Revolution in the Social Paradigm of Science', *Social Science Information,* **18** : 915–43.

307. Schoeck, Helmut & Wiggins, James W. (eds) (1960). *Scientism and Values.* Princeton, NJ & London: D. Van Nostrand.

*308. Schon, Donald A. (1967). *Technology and Change: The New Heraclitus.* Oxford & New York: Pergamon Press.

309. Schroeder-Gudehus, Brigitte (1973). 'Challenge to Transnational Loyalties: International Scientific Organisation after the First World War', *Science Studies,* **3** : 93–118.

310. Schwartz, Leonard E. (1969). 'Artificial Weather Modification: A Case Study in Science Policy', *Technology and Society,* Bath University Press, **5** : 44–48.

311. Schwing, Richard C. and Albers, Walter A., Jr (eds) (1980). *Societal Risk Assessment: How Safe is Safe Enough?* New York: Plenum.

312. *SCLR* (1978). Symposium on 'Biotechnology and the Law: Recombinant DNA and the Control of Scientific Research', *Southern California Law Review,* **51**, 6 (September): 969–1554.

313. Shapin, Steven A. (1972). 'The Pottery Philosophical Society, 1819–1835: An Examination of the Cultural uses of Provincial Science', *Science Studies,* **2** : 311–36.

*314. Shapin, Steven (1979a). 'The Politics of Observation: Cerebral Anatomy and Social Interests in the Edinburgh Phrenology Disputes', *in* Wallis (1979): 139–78.

*315. Shapin, Steven (1979b). 'Homo Phrenologicus: Anthropological Perspectives on an Historical Problem', *in* Barnes & Shapin (1979): 41–71.

*316. Shapin, Steven (1980). 'Social Uses of Science', *in* G.S. Rousseau and Roy Porter (eds), *The Ferment of Knowledge: Studies in the Historiography of Eighteenth-century Science* : 93–139. Cambridge & New York: Cambridge University Press.

*317. Shapin, Steven (1982). 'History of Science and its Sociological Reconstructions', *History of Science,* **20** : in press.

318. Shapin, Steven and Barnes, Barry (1976). 'Head and Hand: Rhetorical Resources in British Pedagogical Writing', *Oxford Review of Education,* **2** : 231–54.

319. Shapin, Steven and Barnes, Barry (1977). 'Science, Nature and Control: Interpreting Mechanics' Institutes', *Social Studies of Science,* **7** : 31–74.

320. Shapo, Marshall S. (1979). *Experimenting with the Consumer: What the Hazards of Chemical Technology can Mean to You.* New York: The Free Press. Published in Britain as *A Nation of Guinea Pigs.* London: Collier-Macmillan.

*321. Shimshoni, D. (1970). 'The Mobile Scientist in the American Instrument Industry', *Minerva*, **8** : 60–89

*322. Sklair, Leslie (1973). *Organised Knowledge : A Sociological View of Science and Technology*. London : Hart-Davis, MacGibbon. Extracts from Chapter 7 reprinted *in* Boyle et al. (1977) : 172–85.

323. Skoie, Hans (ed.) (1979). *Scientific Expertise and the Public*. Oslo : Institute for Studies in Research and Higher Education.

324. Slack, Jonathan (1972). 'Class Warfare among the Molecules', *in* T. Pateman (ed.), *Countercourse* : 202–17. Harmondsworth, Middx.: Penguin Books.

325. *Social Studies of Science*, **6**, 3/4 (September 1976), 'Aspects of the Sociology of Science' (Special Issue) : 281–549. Contains items 133 (Part I), 239, 293, 371 & 375.

326. *Social Studies of Science*, **7**, 2 (May 1977), 'Citation Studies of Scientific Specialties' (Theme Issue) : 139–200, 223–40, 257–67.

327. *Social Studies of Science*, **8**, 1 (February 1978), 'Sociology of Mathematics' (Special Issue) : 2–142.

328. *Social Studies of Science*, **11**, 2 (May 1981), 'Soviet Science' (Theme Issue): 159–274.

329. Spiegel-Rösing, Ina and Price, Derek de Solla (eds) (1977). *Science, Technology and Society : A Cross-Disciplinary Perspective*. London & Beverly Hills, Calif.: SAGE Publications. Contains items 193 & 240.

330. Stander, S. and Clutterbuck, C. (1976). *Health Hazards in Industry*. SISCON Pamphlet. Manchester : Department of Liberal Studies in Science, Manchester University.

331. Stehr, Nico (1978). 'The Ethos of Science Revisited : Social and Cognitive Norms', *in* Gaston (1978) : 172–96.

332. Stehr, Nico and König, Rene (eds) (1975). *Wissenschaftssoziologie. Kölner Zeitschrift für Soziologie und Sozialpsychologie*, Sonderheft 18. Opladen : Westdeutscher Verlag. Contains German versions of items 79 (Chapter 10, abridged) & 208.

333. Stehr, Nico and Meja, Volker (eds) (1981). *Wissenssoziologie. Kölner Zeitschrift für Soziologie und Sozialpsychologie*, Sonderheft 22/1980. Opladen : Westdeutscher Verlag. Contains German translations of items 12, 24 & 238.

334. Steinhart, C. and Steinhart, J. (1972). *Blowout : The Santa Barbara Oil Spill*. Belmont, Calif.: Duxbury Press.

335. Storer, Norman (1966). *The Social System of Science*. New York : Holt, Rinehart and Winston.

336. Studer, Kenneth E. and Chubin, Daryl E. (1980). *The Cancer Mission : Social Contexts of Biomedical Research*. Beverly Hills, Calif. & London : SAGE Publications.

337. Sugden, T.M. and West, T.F. (eds) (1980). *Chlorofluorocarbons in the Environment : The Aerosol Controversy*. Chichester, Sussex : Horwood, for the Society of Chemical Industry.

338. Sullivan, Daniel (1975). 'Competition in Bio-Medical Science : Extent, Structure and Consequences', *Sociology of Education*, **48** : 223–41.

339. *Sunday Times* (London), Insight Team (1979). *Suffer the Children : The Story of Thalidomide*. London : André Deutsch.

*340. Surrey, John and Huggett, Charlotte (1976). 'Opposition to Nuclear Power : A Review of International Experience', *Energy Policy*, **4** : 286–307.

*341. Swatez, Gerald M. (1970). 'The Social Organisation of a University Laboratory', *Minerva*, **8** : 36–58.

342. Symes, John M.D. (1975). *Instruments, Progress and Policy in Science : The Cases of IR and NMR Spectroscopy in Chemistry.* Unpublished M Phil. Thesis, University of Sussex.

343. Symes, John (1975). 'Policy and Maturity in Science', *Social Science Information,* **15** : 337–47.

344. Teich, Albert H. (ed.) (1974). *Scientists and Public Affairs.* Cambridge, Mass. & London : The MIT Press.

345. Teich, Albert H. (ed.) (1977). *Technology and Man's Future.* New York : St. Martin's Press. Second edition. Contains items 43 & 206. Third, enlarged edition (1981) also contains item 368.

346. Thackray, Arnold (1974). 'Natural Knowledge in Cultural Context : The Manchester Model', *American Historical Review,* **79** : 672–709.

*347. Toulmin, Stephen (1972). *Human Understanding,* Volume 1. *The Collective Use and Evolution of Concepts.* Oxford : Clarendon Press ; Princeton, NJ : Princeton University Press.

*348. Travis, G.D.L. (1981). 'Replicating Replication? Aspects of the Social Construction of Learning in Planarian Worms', *Social Studies of Science,* **11** : 11–32.

349. Tribe, Laurence H. (1972). 'Policy Science : Analysis or Ideology?', *Philosophy and Public Affairs,* **2** : 66–110.

350. Turner, R. Steven (1971). 'The Growth of Professorial Research in Prussia, 1818 to 1848 – Causes and Context', *in* R. McCormmach (ed.), *Historical Studies in the Physical Sciences,* **3** : 137–82. Philadelphia : University of Pennsylvania Press.

351. van den Bosch, Robert (1978). *The Pesticide Conspiracy.* New York : Doubleday; Dorchester, Dorset : Prism Press (1980).

352. Vaughan, Paul (1970). *The Pill on Trial.* London : Weidenfeld and Nicolson.

353. von Hippel, Frank and Primack, Joel (1972). 'Public Interest Science', *Science,* **177** (29 September): 1166–71. Reprinted *in* Bereano (1976) : 507–18.

354. Waldbott, George L. (1978). *Fluoridation : The Great Dilemma.* Lawrence, Kansas ; Coronado Press.

*355. Wallis, Roy (1976). *The Road to Total Freedom : A Sociological Analysis of Scientology.* London : Heinemann.

356. Wallis, Roy (ed.) (1979). *On the Margins of Science : The Social Construction of Rejected Knowledge.* Keele, Staffs.: Sociological Review Monograph No. 27. Contains items 56, 314 & 378.

357. Watson, James D. (1968). *The Double Helix : A Personal Account of the Discovery of the Structure of DNA.* London : Weidenfeld and Nicolson; New York : Athenaeum.

*358. Weinberg, Alvin M. (1970). 'Scientific Teams and Scientific Laboratories', *Daedalus.* **99**, 4 (Fall): 1056–75.

359. Werskey, Gary (1978). *The Visible College.* London : Allen Lane.

360. Westrum, Ron (1977). 'Social Intelligence About Anomalies : The Case of UFOs', *Social Studies of Science,* **7** : 271–302.

361. Whiteside, Thomas (1979). *The Pendulum and the Toxic Cloud : The Course of Dioxin Contamination.* New Haven, Conn. and London : Yale University Press.

362. Whitley, Richard (ed.) (1974). *Social Processes of Scientific Development.* London & Boston, Mass.: Routledge and Kegan Paul.

363. Whittemore, Gilbert F., Jr (1975). 'World War I, Poison Gas Research, and the Ideals of American Chemists', *Social Studies of Science,* **5** : 135–63.

364. Wildavsky, Aaron and Tenenbaum, Ellen (1981). *The Politics of Mistrust :*

Estimating American Oil and Gas Resources. Beverly Hills, Calif. & London :
SAGE Publications.

365. Williams, Roger (1971). *Politics and Technology.* London : Macmillan.
366. Williams, Roger (1981). 'British Scientists and the Bomb : The Decisions of
1980', *Government and Opposition,* **17** : 267–92.
*367. Winner, Langdon (1977). *Autonomous Technology : Technics-out-of-Control as a
Theme in Political Thought.* Cambridge, Mass. & London : The MIT Press.
368. Winner, Langdon (1979). 'The Political Philosophy of Alternative
Technology', *Technology in Society,* **1** : 75–86. Reprinted *in* Teich (1981).
*369. Wittgenstein, Ludwig (1953). *Philosophical Investigations.* Edited by G.
Anscombe & R. Rhees, translated by G. Anscombe. Oxford : Basil
Blackwell.
370. Woolf, Patricia (1975). 'The Second Messenger : Informal Communication
in Cyclic AMP Research', *Minerva,* **13** : 349–73.
*371. Woolgar, S.W. (1976a). 'Writing an Intellectual History of Scientific
Development : The Use of Discovery Accounts', *Social Studies of Science,* **6** :
395–422.
*372. Woolgar, Steve (1976b). 'The Identification and Definition of Scientific
Collectivities', *in* Lemaine et al. (1976) : 223–45.
373. Wynne, Brian (1975a). *The Sociology of Science.* Three SISCON Pamphlets.
Manchester : Department of Liberal Studies in Science, Manchester
University.
*374. Wynne, Brian (1975b). 'The Rhetoric of Consensus Politics : A Critical
Review of Technology Assessment', *Research Policy,* **4** : 108–58.
*375. Wynne, Brian (1976). 'C.G. Barkla and the J Phenomenon : A Case Study in
the Treatment of Deviance in Physics', *Social Studies of Science,* **6** : 307–47.
*376. Wynne, Brian (1977). *C.G. Barkla and the J Phenomenon : A Case Study in the
Treatment of Deviance in Physics.* Unpublished M Phil. thesis, Edinburgh
University.
*377. Wynne, Brian (1978). 'The Politics of Nuclear Safety', *New Scientist* (26
January) : 208–11.
*378. Wynne, Brian (1979). 'Between Orthodoxy and Oblivion : The Normalis-
ation of Deviance in Science', *in* Wallis (1979) : 67–84.
*379. Wynne, Brian (1980). 'Technology, Risk and Participation : On the Social
Treatment of Uncertainty', *in* Conrad (1980) : 173–208.
*380. Wynne, Brian (1982). *Rationality or Ritual? Nuclear Decision-Making and the
Windscale Inquiry.* Chalfont St. Giles, Bucks.: The British Society for the
History of Science Monographs.
381. York, Herbert F. (1970). *Race to Oblivion : A Participant's View of the Arms Race.*
New York : Simon and Schuster.
382. Young, Michael F.D. (ed.) (1971). *Knowledge and Control : New Directions for
the Sociology of Education.* London : Collier-Macmillan.
383. Ziman, J.M. (1968). *Public Knowledge : An Essay Concerning the Social Dimension
of Science.* Cambridge : Cambridge University Press. Extract *in* Gebhardt
(1981).
384. Ziman, John (1976). *The Force of Knowledge : The Scientific Dimension of Society.*
London & New York : Cambridge University Press.
385. Zinberg, Dorothy S. (1976). 'Education through Science : The Early Stages
of Career Development in Chemistry', *Social Studies of Science,* **6** : 215–46.
386. Zinberg, Dorothy (1979). 'The Public and Nuclear Waste Management',
Bulletin of the Atomic Scientists, **35**, 1 (January) : 34–39.
387. Zuckerman, Harriet (1977a). 'Deviant Behavior and Social Control in

Science', *in* Edward Sagarin (ed.), *Deviance and Social Change :* 87–137. Beverly Hills, Calif. & London : SAGE Publications.

388. Zuckerman, Harriet (1977b). *Scientific Elite : Nobel Laureates in the United States.* New York : The Free Press; London : Collier-Macmillan.

389. Zuckerman, Lord (1980). *Science Advisers, Scientific Advisers and Nuclear Weapons.* London : The Menard Press. Extract, entitled 'The Deterrent Illusion', in *The Times* (London) (21 January 1980) : 10.

Addenda

390. Brannigan, Augustine (1981). *The Social Basis of Scientific Discoveries.* Cambridge and New York : Cambridge University Press.

391. Frankel, Henry (1981). 'The Paleobiogeographical Debate over the Problem of Disjunctively Distributed Life Forms', *Studies in History and Philosophy of Science,* **12** : 211–59.

392. Friedman, R.M. (1979). *Vilhelm Bjerknes and the Bergen School of Meteorology, 1918–1923 : A Study of the Economic and Military Foundations for the Transformation of Atmospheric Science.* Unpublished PhD dissertation, Johns Hopkins University. University Microfilms No. 79–06452. See also 'Nobel Physics Prize in Perspective', *Nature,* **292** (27 August 1981) : 793–98.

393. Graham, Loren R. (1972). *Science and Philosophy in the Soviet Union.* New York : Alfred A. Knopf.

394. Grover, Sonja C. (1981). *Toward a Psychology of the Scientist : Implications of Psychological Research for Contemporary Philosophy of Science.* Washington, DC : University Press of America.

395. Hallam, A. (1973). *A Revolution in the Earth Sciences : From Continental Drift to Plate Tectonics.* Oxford : Clarendon Press.

396. Jones, Robert A. and Kuklick, Henrika (eds) (1981). *Research in Sociology of Knowledge, Sciences and Art,* Vol. 3. London : JAI Press.

397. Lawless, Edward W. (1977). *Technology and Social Shock.* New Brunswick, NJ : Rutgers University Press.

398. Lynch, Michael (1982). *Art and Artifact in Laboratory Science : A Study of Shop Work and Shop Talk in a Research Laboratory.* London : Routledge and Kegan Paul.

399. Messeri, Peter (1981). *The Formation of Scientific Consensus : The Acceptance of Continental Drift.* Unpublished PhD dissertation, Columbia Univesity.

400. Moravcsik, Michael J. (1980). *How to Grow Science : An Insider Sizes up the Uses and Misuses of Modern Science.* New York : Universe Books.

401. Nickles, T. (ed.) (1980). *Scientific Discovery : Case Studies.* Dordrecht, Boston, Mass. & London : Reidel.

402. Nowotny, Helga (1981). 'Experts and Their Expertise : On the Changing Relationship Between Experts and Their Public', *Bulletin of Science, Technology and Society,* **1** : 235–41.

403. Petersen, J.C. and Markle, G.E. (eds) (1980). *Politics, Science, and Cancer : The Laetrile Phenomenon.* Boulder, Col. : Westview Press.

404. Rolph, Elizabeth S. (1979). *Nuclear Power and the Public Safety : A Study in Regulation.* Lexington, Mass. : D.C. Heath.

405. Schwartz, Robert (1981). 'Acupuncture and Expertise : A Challenge to Physican Control', *The Hastings Center Report,* **11**, 2 (April) : 5–7.

406. Franks, Felix (1981). *Polywater.* Cambridge, Mass. and London: The MIT Press.

Stop press

As we go to press, we hear that Bath University Press is going to publish a related Reader in 1982 : H.M. Collins (ed.), *Sociology of Scientific Knowledge : A Sourcebook*. It is hoped that it will include, among other items, numbers 22, 56, 238, 293, 314 and 371. See also:

Anderson, R.M., Perrucci, R., Schendel, D.E. and Trachtman, L.E. (1980). *Divided Loyalties : Whistle-Blowing at BART*. West Lafayette, Ind. : Purdue University.

Barbour, I.G. (1966). *Issues in Science and Religion*. London : SCM Press; Englewood Cliffs, NJ : Prentice-Hall.

Bynum, W.F., Browne, E.J. and Porter, R. (eds) (1981). *Dictionary of the History of Science*. London : Macmillan.

Collins, H.M. and Pinch, T.J. (1982). *Frames of Meaning : The Social Construction of Extraordinary Science*. London and Boston, Mass. : Routledge and Kegan Paul.

Graham, Loren R. (1981). *Between Science and Values*. New York : Columbia University Press.

Regens, James L. (1978). 'Energy Development, Environmental Protection, and Public Policy', *American Behavioral Scientist*, **22**, 2 (November/December) : 175–90.

Schnaiberg, Allan (1980). *The Environment : from Surplus to Scarcity*. Oxford & New York : Oxford University Press.

Stephens, M. (1981). *Three Mile Island*. New York : Random.

Story, R.D., with Greenwell, J.R. (1981). *UFOs and the Limits of Science*. New York : Morrow.

Woolf, P. (1981). 'Fraud in Science : How Much, How Serious?', *The Hastings Center Report*, **11**, 5 (October) : 9–14.

Name index

This index contains all the names referred to in the Editorial introductions, plus those in the main text of the Readings. It does not contain all the names in the notes and references to the Readings, nor those in the Bibliography, which is not indexed.

Surnames only used eponymously (e.g. Delaney Clause; Nobel Prize; Zweig rule) are not listed, nor are adjectival forms (e.g. Marxist; Durkheimian).

367

Thorpe, E. 308, 310, 332
Toulmin, S.E. 67
Toynbee, A.J. 170, 176
Train, R.E. 305, 311, 318, 320–24, 335
Travis, G.D.L. 246
Tswett, M. 33
Turner, B.S. 20
Tyndall, J. 225, 227

Vansina, J. 263
Vinci, L. da 158
Vonnegut, K. 96

Walker, A. 308, 310, 332
Wallis, R. 246
Watson, J.D. 209
Watt, J. 158
Weber, M. 31, 33, 34

Weinberg, A.M. 14
Weinberg, M.G. 235, 250
Wellington, Duke of 117
Wellstone, P.D. 244, 282, 289
Whately, R. 117
Whitley, R.D. 61, 62, 113, 144
Wilson, A.H. 178, 180–84
Winch, P. 62, 112, 114
Winner, L. 245
Wittgenstein, L. 62, 70, 112, 113, 123, 124
Wöhler, F. 203
Woolgar, S.W. 16, 19, 35, 72
Worboys, M. 14
Wynne, B.E. 67, 193–95, 212, 222, 225, 227, 228, 231, 246–48
Wynne-Edwards, V.C. 272, 275

Yetka, L.R. 284
Young, R.M. 212, 231, 334

Subject index